C. Fest · K.-J. Schmidt The Chemistry of Organophosphorus Pesticides

C. Fest K.-J. Schmidt

The Chemistry of Organophosphorus Pesticides

Reactivity · Synthesis · Mode of Action · Toxicology

Springer-Verlag New York Heidelberg Berlin 1973

Dr. Christa Fest

Dr. Karl-Julius Schmidt

Bayer AG, Wissenschaftliches Laboratorium für Pflanzenschutz
Wuppertal-Elberfeld

With 46 figures

ISBN 0-387-05858-3 Springer-Verlag New York Heidelberg Berlin
ISBN 3-540-05858-3 Springer-Verlag Berlin Heidelberg New York

 Library of Congress Catalog Card Number 72-80294. Printed in Germany. Monophoto typesetting and offset printing: Werk- und Feindruckerei Dr. A. Krebs, Hemsbach/Bergstr. and Bad Homburg v. d. H. Bookbinding: K. Triltsch, Graphischer Betrieb, Würzburg.

Dedicated to

Dr. G. Schrader

on his
seventieth birthday

Preface

Our intention has been to provide a short introduction to the chemistry and mode of action of insecticidal phosphoric acid compounds, with particular reference to the relationship between structure and activity. The yearly production of these pesticides is now approaching 100,000 tons and thus offers an important example of applied research. If, however, one examines the historical development of these compounds, it is apparent that this was preceded by a hundred years of pure chemistry of phosphorus. The utility of the phosphoric acid pesticides is undisputed today – and furthermore it can be expected that they will solve many of the world's nutritional problems, yet from this field of applied research many paths are now leading back into basic research in chemistry, biochemistry, biology and toxicology etc. This clearly illustrates the problem of attempting to define pure and applied research.

Originally, this book was conceived for students of chemistry who, on completion of their study, were uncertain about the place of applied research in industry but it was soon clear that such material, when supplemented with further data, would serve as an introduction to the field of pesticidal phosphoric acid compounds for many technicians, officials and scientists who, in various authorities in agriculture, in chemical and biological research, are concerned with the problems of crop protection and more recently with questions of pollution of the environment. We assume that anyone wishing to explore this whole field more deeply will refer to specialized literature, for example, SCHRADER's monograph or the Houben-Weyl handbook.

The existence of several names for the same product presents a special problem. Industry is interested in using the registered name, while in scientific literature the common name is preferred. We are of the opinion that anyone concerned with crop protection should have full command of both trade names and common names. We have, therefore, made no special effort to distinguish between either type of designation. We have, however, prepared two separate tables collating these names. It is sufficient if, when using a name, one is aware of its legal significance.

We are indebted to Professor R. WEGLER who prompted us to compile this volume on Organophosphorus Pesticides. (An abridged version appears in the book "Chemie der Pflanzenschutz- und Schädlingsbekämpfungsmittel", edited by R. WEGLER. Berlin-Heidelberg-New York: Springer 1970.)

We are indebted also to many colleagues for their co-operation and helpful criticism. We would like to thank Dr. G. SCHRADER (Farbenfabriken Bayer AG) for his advice and encouragement, as well as Professor G. UNTERSTENHÖFER (Farbenfabriken Bayer AG) and particularly Mr. J. DIXON who has translated this very difficult text into English. We are also grateful to Mr. J. EDWARDS for the translation of the first pages of the chapter "Biochemistry", and especially to Dr. H. MARTIN (London) for a final criticism and correction of the manuscript.

The Authors

Elberfeld, January 1973

Notes added in press:

1) The name "Farbenfabriken Bayer AG" should be read as "Bayer AG" throughout the text.
2) For legal reasons the trade name "®Valexon" (Bayer AG) should be read as "®Volaton".

Contents

First Aid see page 261

1. Introduction

Crop protection by means of chemicals has become increasingly an object for public discussion from about the time when RACHEL CARSON published her book "Silent Spring". According to the understanding of the basic concepts involved in this difficult scientific field, two lines of argument follow which tend to be ideological or of a more objective nature. One may regard chemical crop protection simply as a profit-induced poisoning of the environment or one can try to analyze the various factors that have compelled us to use chemical agents in the production of food.

The deciding factor is the development of the world population to be expected during the next decade (Fig. 1).

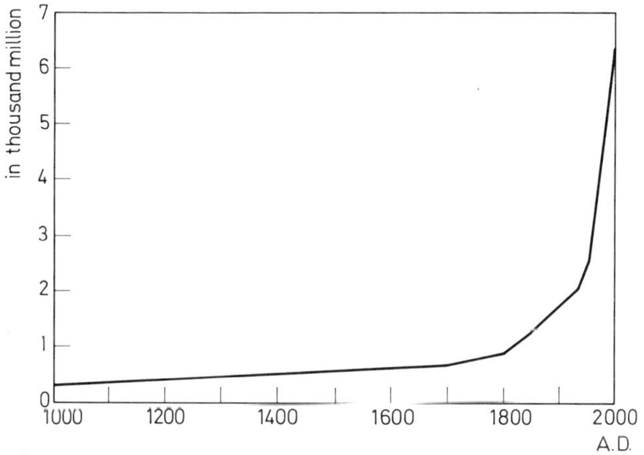

Fig. 1. Numerical development of the world population in the period from 1000 – 2000 A.D.

Whether these figures are absolutely correct is of secondary importance, what is quite certain is the expected development. The continuation of the curve must remain open, for the growth of population will stabilize under the influence of retarding factors. Otherwise the populated areas of Egypt would according to FUCKS [301] (see Table 1) possess a population density of more than 2000 people per square kilometer, which is equivalent to the density in the European capitals. Provision of food in the under-industrialized countries with a high birth rate will become the most urgent problem of the next decades.

If curves such as that in Fig. 1 are expressed in kilocalories instead of number of persons versus time and if a mean daily requirement per head of world population is taken as 2420 cal, then, in 1970 9 – 10 thousand million kcal must be produced, in the year 2000 about 16 thousand million kcal and by the year 2040 about 22 thousand million kcal. This does not, however, take into consideration the protein content of the food [301]. Food production must therefore be doubled in the next three decades. This objective is technically feasible if every agricultural possibility is exploited. CRAMER [218] cites the following measures:

1) Expansion of cultivated areas
2) Improvement of soil cultivation
3) Mineral fertilization
4) Propagation of improved varieties
5) Improved irrigation
6) Improved agricultural structure
7) Modern crop protection.

In addition there will in future be the introduction of nonclassical techniques which are independent of the soil, e.g. exploitation of the protein reserves of the sea, hydroponic culture etc.

Table 1. Population density of various countries between 1950 and 2040 according to FUCKS

Country	Inhabitants per square kilometer				
	1950	1963	1975	1995	2040
USA	19.9	20.2	25	30	35
Russia in Europe	27.0	30.4	35	40	45
Russia in Asia	2.6	3.0	3.5	4	4.5
USSR	8.6	10.0	11	13	15
China	56.45	74.9	100	160	270
Belgium	276.7	304.5	320	340	390
W.-Germany	189.5	223.5	240	260	290
United Kingdom	206.1	219.0	230	260	280
France	75.6	86.8	90	100	110
Italy	154.7	167.5	180	200	240
Netherlands	300.5	356.1	400	480	710
United Arab Republic	482.6	764	1050	1550	2350

Modern crop protection not only comprises the use of insecticides but also the introduction of fungicides and herbicides which in mono-cultivation have considerable influence in determining the type and techniques of agriculture.
The second fact to be considered is that, since biblical times, the history of agriculture has been a history of catastrophes. Here we follow CRAMER [218] who to date has published the most comprehensive and careful estimate of crop losses due to pests and pathogens. CRAMER cites numerous examples, some of which have almost become classical, as for example the "turnip winter" of 1917 in Germany. Due to a severe attack of *Phytophthora*, practically the entire potato crop was destroyed, so that turnips became the basic source of nutriment. This event demonstrates, as do many others, that such an occurrence can have extensive political consequences.

The famine in Ireland in the middle of the 19th century, in which more than a quarter of a million Irish perished, was the cause of one of the biggest migrations to the USA. Once more the culprit was *Phytophthora infestans*.

CRAMER's [218] description of the history of European viticulture is of great interest; with the aid of new sources of information he illustrates the extent of the catastrophe which occurred in the second half of the 19th century, predominantly in France. In particular there were three main events:

1) the arrival of powdery mildew, *Uncinula necator (Oidium tuckeri)*, around 1850,
2) introduction of grape phylloxera *(Phylloxera vitifoliae)* after 1860 and
3) downy mildew *(Plasmopara viticola)* after 1870.

According to CRAMER the economic and sociological consequences are still visible in France today. For many vineyard owners this was a reason for emigrating to the newly conquered Algeria and the formation of a colony there.

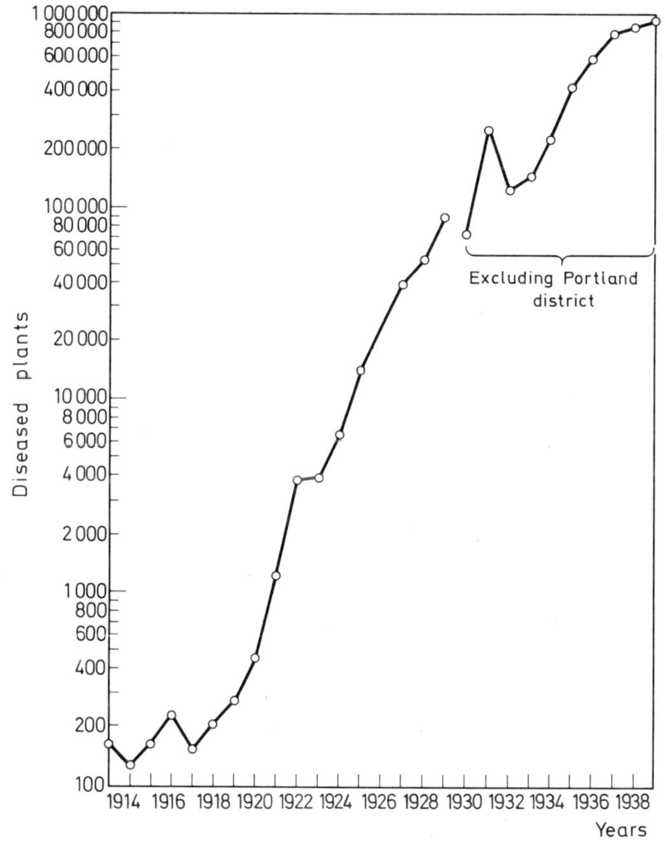

Fig. 2. Spread of Panama disease on Jamaica from 1913 to 1938

Fig. 2 gives an impression of the explosive course of Panama disease in the banana crops of Jamaica. Large areas in the Caribbean had to be completely abandoned or new crops introduced (PADWICK, cited by CRAMER).
The number of diseased trees is given on a logarithmic scale, excluding from 1929 Portland where banana cultivation was discontinued.
The original variety Gros Michel, lost due to *Fusarium oxysporum var. cubense*, could not be completely replaced by more resistant varieties: the variety Lacatan is resistant to Fusarium, but is susceptible to Sigatoka disease *(Mycosphaerella musicola)*. The variety Cavendish is becoming increasingly susceptible to attack by nematodes.
The conditions in citrus cultivation are very instructive. While, in North America, 94% crop protection is applied to areas requiring such treatment, and the losses are correspondingly low, in South America, citrus cultivation suffers badly due to plant diseases, especially the virus "Tristeza". In particular those trees are attacked which are grafted to the root stock of the sour orange. This stock was popular because it was regarded as resistant to *Phytophthora citrophthora*. After 1930 Tristeza destroyed 10 million trees in Argentina, between the years 1937 and 1958 seven million out of a total of 13 million trees were destroyed in Brazil. In European citrus cultivation the Mediterranean fruit fly *(Ceratitis capitata)* plays a considerable role; in African and Asian citrus cultivation it is the scale that reduces the harvest.
In the last century the Asian coffee harvest almost succumbed to the coffee rust *(Hemileia vastatrix)*. In Ceylon, for example, tea was introduced as a substitute crop. The English became a nation of tea drinkers and their habit of taking milk with tea is said to originate from this time.
In tobacco cultivation an example is to be found in Europe. In 1957 or 1958 the blue mould *(Peronospora tabacina)* found its way to the Continent. In the year 1959 in Holland there was a complete failure of the harvest in some areas. In 1960 in Germany the main areas of cultivation were attacked, many plants were destroyed and by 1961 the German cultivated area had diminished by a third.
Finally mention should be made of the boll weevil *(Anthonomus grandis)*. Its passage through the cotton belt of the USA at the end of the last century led to the ruin of whole communities. Trade, agriculture and economy collapsed. The terror of these years passed into folklore, and, for example, the "Ballad of the Boll Weevil" is said to have been the basis for the early blues.
From the examples just cited – and there are many more – it can readily be appreciated how chemical crop protection became established under the pressure of catastrophes, largely during the industrial revolution. Crop protection is a problem that can be discussed only from a worldwide point of view. Intensification and industrialization of agriculture also mean intensification of attack by pests and pathogens. Mono-cultivation of plants which are often bred for the highest yields, offers ideal environmental conditions which can lead to an explosive outbreak of pests.
From all types of cultivation cases are known where immediate measures were called for to protect the harvest. For example, at the beginning of the 60's an air-lift had to be set up between Cologne and Cairo because the entire Egyptian cotton harvest was in danger of destruction by a severe *Prodenia* infestation. The use

of ®Dipterex *(trichlorfon)* saved the largest part of the harvest. The economic significance of such happenings, especially for countries whose export depends upon a few crops, hardly needs to be emphasized.

Chemical crop protection offers therefore the possibility of preventing a catastrophe or of obtaining a rapid increase in harvest, in so far as pests or pathogens are the limiting factors. Here lies the extremely important tactical significance of crop protection.

Planned increase of the yield per unit area requires additional measures, as mentioned on p. 2. Varieties cultivated to give high yields are often particularly susceptible to pests and pathogens, so that the full yield can only be achieved with the aid of chemical crop protection. The achieved yield can only be assured by protection of both the plants and the harvest itself. Lasting success can only be obtained by purposeful combination of all agricultural methods, i.e. choice of variety, fertilization, use of insecticides, fungicides and herbicides. It is in this field that the real, the strategic significance of crop protection lies.

The following figures illustrate quite clearly the role of short-term and long-term application of pesticides. Fig. 3 shows the yield per hectare for potatoes in the USA and in India and Pakistan.

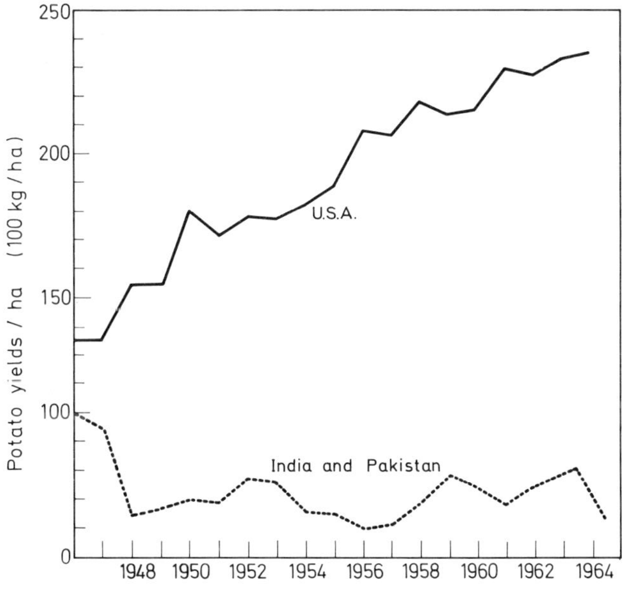

Fig. 3. Potato yields per hectare, according to CRAMER [218]

The large differences in yield are very noticeable and are certainly not attributable only to crop protection. Here other factors such as choice of variety, fertilization, cultivation methods etc. play a large role. On the other hand, the balance of the USA curve and the steady upward trend of production must be attributed to planned crop protection. The yields are predictable with great certainty. This is otherwise in India and Pakistan. The uneven course of the curve indicates a strong dependence upon climatic influences and attack by pests. The position of the lower curve in the diagram also gives an idea of the considerable reserves that can still be mobilized. Similar conditions are found in rice cultivation (Fig. 4). Once more the balance of the USA curve is evident, showing very stable yields. In contrast, the Indian rice production for 20 years has varied between 1000 and 1500 kg/ha with large deviations. Because of intensive cultivation measures, the Japanese curve lies considerably higher, but the extreme variation indicates the powerful influence of the yearly attack by the rice stem borer *(Chilo suppressalis)* and the rice blast (Brusone disease, *Piricularia oryzae*). In 1953 there were considerable losses due to *P. oryzae* (see arrow in Fig. 4). The beginning of planned crop protection in Japan can be accurately fixed at this point. Infestation by *P. oryzae* led to the importation of mercurial leaf fungicides (®Ceresan slaked lime) and the routine treatment of seed. In the same year organophosphate insecticides (®E 605) were flown in to control the rice stem borer. It is noteworthy that this is the first example of an air-lift in the history of crop protection. Since both pesticides were responsible for the steep rise in the rice production after 1953, monuments were erected to commemorate the event in Nangoku for Ceresan and in Zentsuji for *parathion*. For the first time in 1960 the Japanese rice yield was greater than that of the American and in 1970 the first restrictions on cultivation became effective.

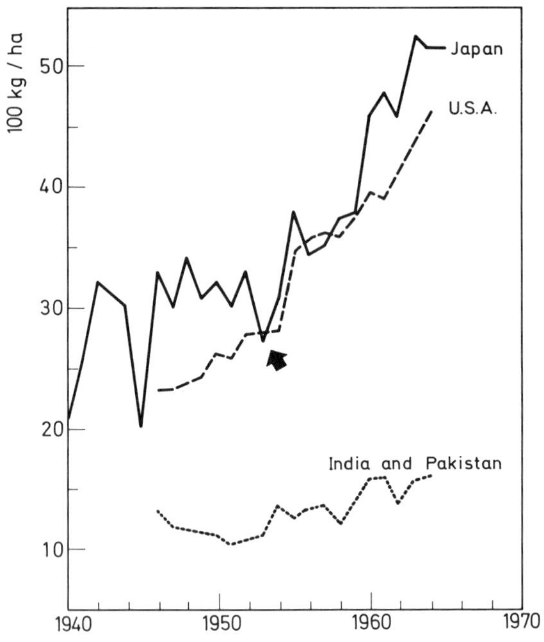

Fig. 4. Rice yields per hectare, according to CRAMER [218]

As the rice supply was assured, the toxicological requirements of the Japanese authorities were intensified. At the beginning of this year, the mercury-containing leaf fungicides were prohibited with the exception of products for treating seed. However, in the years before the law came into effect, industry was able to develop non-accumulating insecticides for rice cultivation. These were also phosphoric acid esters which will be discussed further on p. 144f.

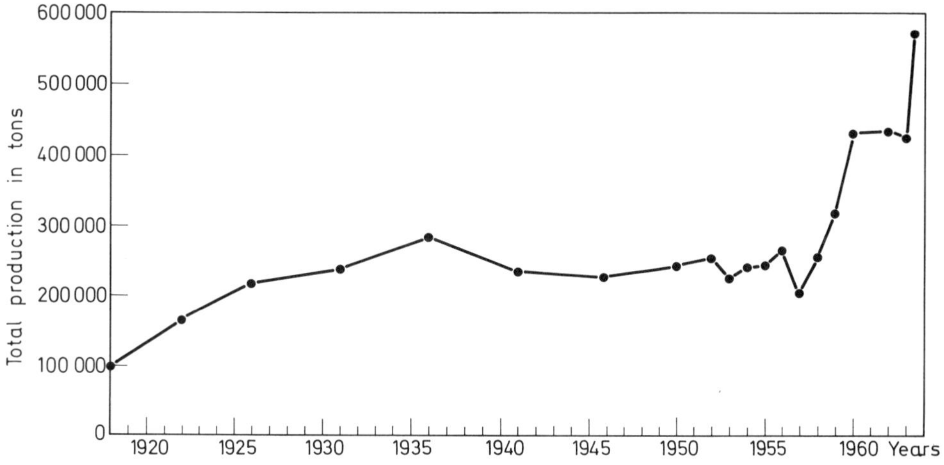

Fig. 5. Cocoa production in Ghana from 1918 to 1964

A similar situation existed in Ghana (Fig. 5) for the development of cocoa production. Until 1954 *Sahlbergella singularis* and *Distantiella theobroma (Heteropt., Myridae)* as well as *Phytophthora palmivora* and the "swollen shoot" virus were the limiting factors. The losses were estimated at 51% of the possible yield. In 1957 a crop protection program was started. In the first year the average yield for the country rose by 110%, in the second year by a further 50%. This increase can be attributed exclusively to an active crop protection; i.e. by employing better agricultural measures, the yield could still be considerably increased.
A last example is the use of hybrid seed corn in the USA. As can be seen from Fig. 6 the increase in the acreage planted to hybrid corn corresponds initially to the increase in yield with, however, wide variations: in 1947 there was extreme drought and a severe attack by grass-hoppers. By using insecticides such as DDT the yield was further increased, but the curves for area cultivated and yield have moved apart. It was not until 1955 that production began to follow the same pattern as the lower curve, which indicates the use of soil insecticides. Although there would appear to be a very good correlation, it is not entirely relevant, because the use of herbicides and fungicides can be expressed by very similar curves. The increase in yield per acre after 1956 cannot therefore be attributed solely to the use of soil insecticides, but once more to the purposeful combination of different measures. The yield per hectare in 1955 was only 2.55 tons and in 1962 over 4.0 tons.

A comparison with other maize-producing countries, e.g. the USSR, would be very interesting but there are no detailed figures available. The yield there was 1.3 tons/hectare in 1963/64 and today has increased to about 2.7 tons/hectare. The average African yield for 1962/63 was 1.1 tons/hectare, which, compared with the American figures, does not represent an optimum and there are, in Africa, still large reserves present that could readily be mobilized.

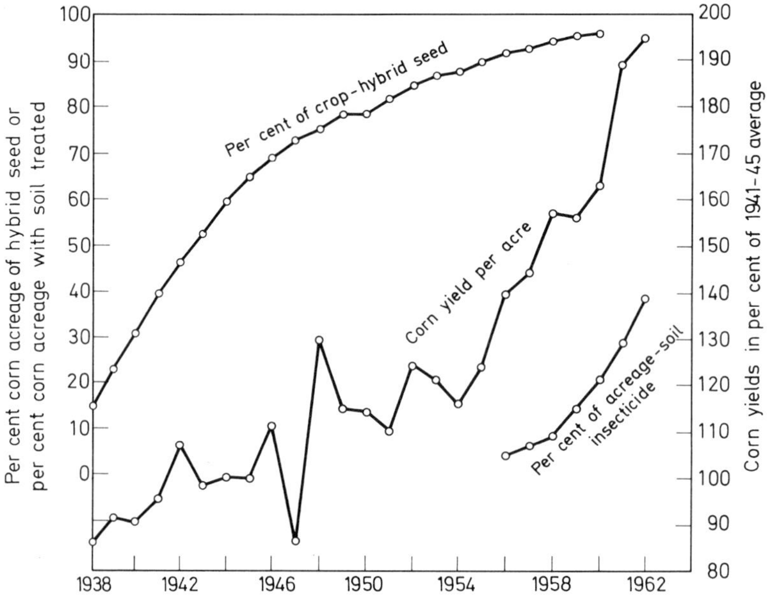

Fig. 6. A comparison of the increased use of hybrid seed and soil insecticides in relation to rising corn yields in the USA, presented as a percentage of 1941 – 1945 yields, according to CRAMER [218]

Figs. 7 and 8 represent the actual production and the losses arranged according to countries and crops.

From the two figures it can be seen that the real losses due to pests and pathogens vary between 30 and 35%. This figure, however, gives little information on the basic possibilities for increasing the yield of various crops. From an absolute point of view the greatest reserve lies almost certainly in grain crops, particularly in rice. The present conditions have been investigated more accurately by CRAMER [219]. Fig. 9 offers a comparison of various yields and their respective percentage of the world area cultivated with rice, and of the world rice harvest. It shows that the countries having low yields possess a high proportion of the world rice harvest, but that this result is obtainable only by cultivation of very large areas: on more than 40% of the total area cultivated with rice less than 1.4 tons/ha of rice is harvested, and on more than 90% of the total area the yield lies below 3 tons/ha. All these examples show quite clearly that, at the present time, crop protection is already a *conditio sine qua non*, although the population explosion is still only impending. Positively formulated, the problem consists in developing and making available optimal agents according to activity, toxicity, economy and ecology. In this respect the phosphoric acid esters present a very favourable class of compounds, for – as will be shown in this volume – by combining four different substituents on the central phosphorus atom, it is possible to fit the properties of an active agent to the special practical requirements.

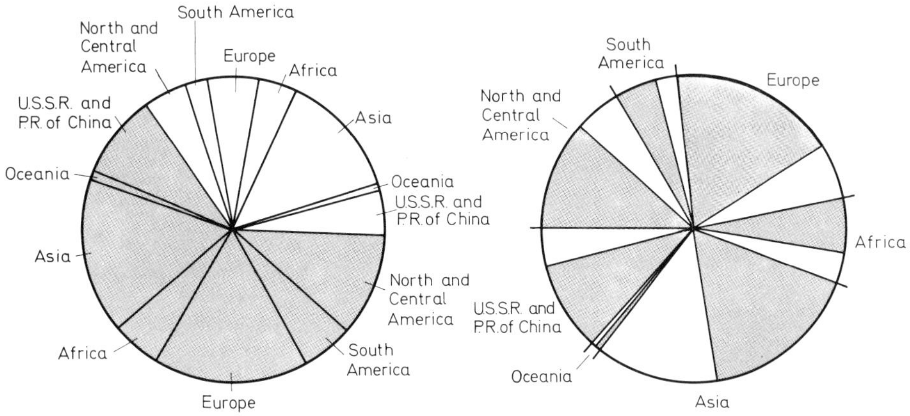

Fig. 7. Distribution of crop production (dark) and crop losses (white) in the different regions of the world, according to CRAMER [218]

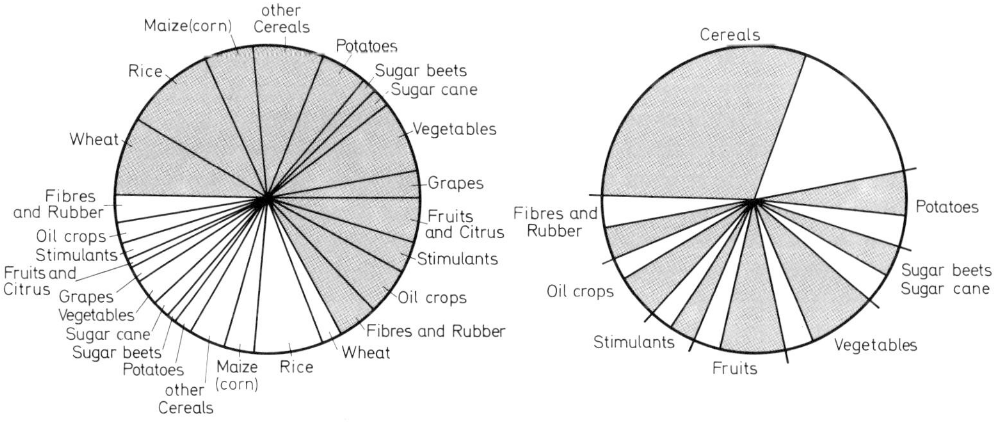

Fig. 8. Actual production (dark) and losses (white) for different crops or groups of crops, according to CRAMER [218]

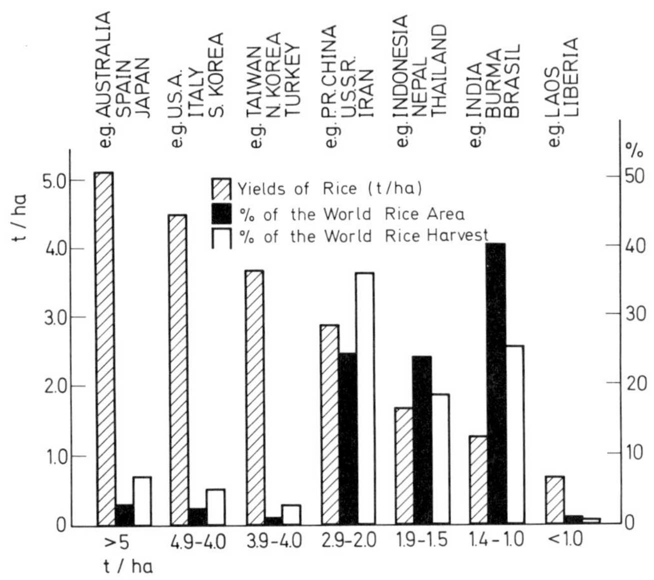

Fig. 9. Rice yields and rice areas of different groups of countries compared with the world rice area. The inverse ratio of yield to percentage of world rice area can be seen clearly [219]

The insecticidal action of the very earliest compounds such as *parathion* and *diazinon* have set a standard which, even 20–30 years after their discovery, is difficult to surpass. Toxicologically this class of compounds has the advantage, due to their ester nature, of being metabolized in the living organism to inorganic phosphates. The price for this favourable property is a certain acute toxicity of the phosphoric acid esters, for, because of their mechanism of action, the toxicity for mammals and that for insects cannot basically be separated. This acute toxicity is not, however, a great disadvantage in practice, since effective antidotes are now available. Also, the toxicologist prefers acute poisoning with its symptomatic treatment rather than chronic toxicity with ensuing irreversible damage. Meanwhile, relatively non-toxic compounds have been developed, and there is little doubt that within certain limits this trend will continue.

From an ecological point of view too the organophosphates represent a very interesting group. In contrast to certain chlorohydrocarbons, the problem does not consist in shortening persistence, but on the contrary of extending the duration of action, particularly in the soil. Harris [372, 375], who developed a laboratory method of investigating this problem, published the data reproduced here in Fig. 10 showing the variable behaviour of some insecticides in the soil. Between the conditions in sandy loam (Fig. 10) and muck soil there may be quantitative, but not really basic, differences.

The danger therefore of the accumulation of organophosphates in our environment up to a toxic level is very slight indeed. Meanwhile this class of compounds has found use into other fields, although the emphasis will certainly remain on their use as insecticides. Esters are known with a marked miotic action such as paraoxon or phosphonic esters of *p*-nitrophenol, used in human medicine for *glaucoma*; a series of less toxic compounds is also used in veterinary medicine. A broad development began in the field of fungicides. The first practically applicable herbicides and growth regulators are just becoming available and recently, in phosphonomycin, even an antibiotic has been found.

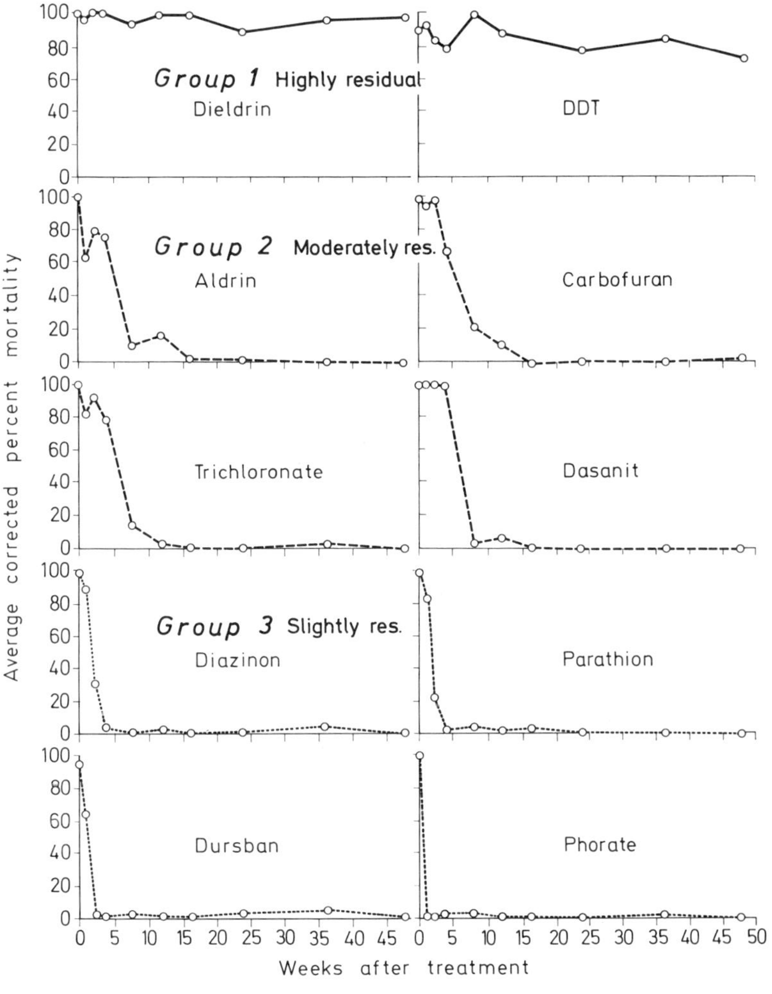

Fig. 10. Persistence of biological activity of some insecticides in sandy loam soil, according to HARRIS [372]

It would certainly not be an exaggeration to claim the organophosphates to be the most fruitful class of chemical compounds that man has found in his efforts to secure his physical existence.

1.1. Historical Development

Barely 150 years have passed since 1820 when I. L. LASSAIGNE [537] reacted alcohol with phosphoric acid in reaction analogous to that with sulfuric acid and therewith launched the chemistry of the organophosphorus compounds.

In 1847 a paper by P. E. THÉNARD [950] on phosphines appeared. At the same time M. CLOEZ [197] discovered a thiophosphoric acid ester; he had suspected the existence of sulfur derivatives of phosphorus in analogy to arsenic.

In 1854 PH. DE CLERMONT [195, c.f. 220], prompted by WURTZ, synthesized tetraethyl pyrophosphate (later TEPP) (I) by alkylating the silver salt of the pyrophosphoric acid with alkyl halides (Clermont method). MOSCHNIN [21, 665] is said to have previously prepared TEPP in WURTZ' laboratory. Clermont did not, however, recognize the important physiological activity of this compound.

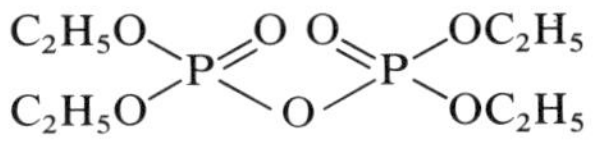

(I)

This ester, which may be regarded as a link between inorganic and organic chemistry, was often described in the following decades and yet almost 80 years were to pass before its insecticidal possibilities were discovered.

In 1872 A. W. HOFMANN [413] reported the oxidation of methyl and ethyl phosphine with nitric acid to give the corresponding phosphonic acids (II).

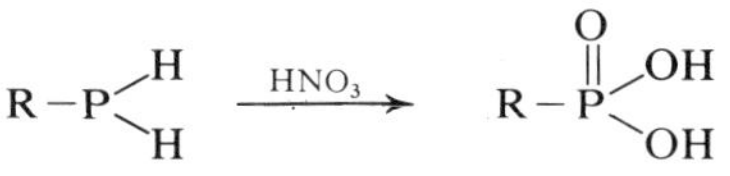

(1)

(II)

C. A. A. MICHAELIS in Germany and A. E. ARBUSOV in Russia can be named as the founders of classical phosphoric ester chemistry.

In 1897 MICHAELIS and TH. BECKER [641] reacted sodium dialkyl phosphite with ethyl iodide according to the following scheme:

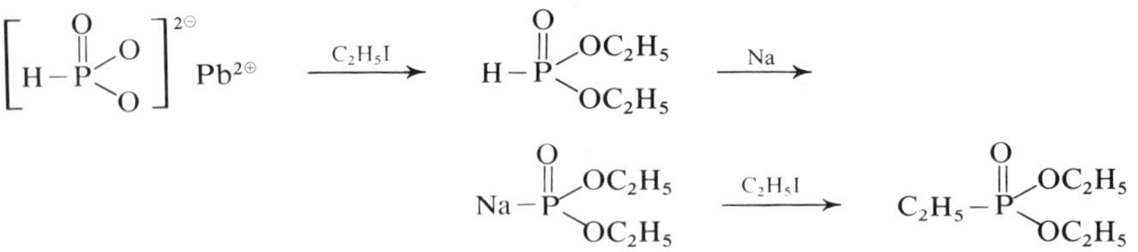

(2)

(III) (IV)

Scheme 1

(The reaction of sodium salts of dialkyl phosphites (III) with alkyl halides became known as the Michaelis-Becker reaction and will be described in greater detail on p. 67.) The compound so obtained (a phosphonic acid ester) (IV) was not identical with the phosphorous acid ester (V),

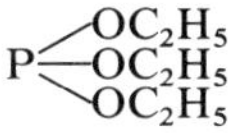

(V)

which they synthesized from phosphorus trichloride and sodium ethylate.
In 1898 MICHAELIS and R. KAEHNE [642] isolated a compound from trialkyl phosphite and methyl iodide, whose structure they did not recognize as being analogous to (IV) cf. Scheme I). The problem was taken up again seven years later by A. E. ARBUSOV [52].
In 1905 he repeated the reaction of phosphorus trichloride with sodium ethylate. He considered the boiling range (188°C – 192°C) to be too wide and fractionated his yield thirty times. ARBUSOV succeeded in reacting this highly purified triethyl phosphite (V) with ethyl iodide at room temperature by the following method:

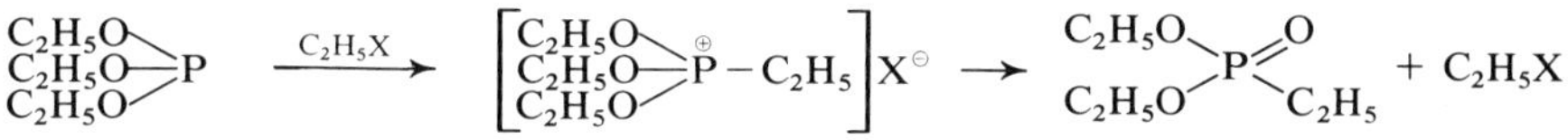

(V) (VI) (3)

Scheme 2

(The reaction of trialkyl phosphite with alkyl halides is called the Arbusov reaction and will be discussed more fully on p. 66.) The phosphonate synthesis analogous to Scheme 2 is also possible with the phosphoroamidites:

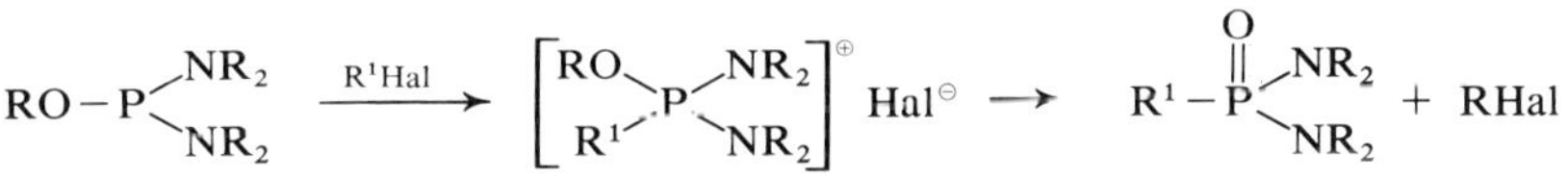

(4)

(VII)

Scheme 3

It is astonishing that without modern methods both MICHAELIS and ARBUSOV and their colleagues were able to elucidate these fundamental reactions of organophosphorus chemistry by simple deduction aided by exact laboratory techniques, and undoubtedly an intuition confirmed only by years of experiment.
At about the same time, i.e. in 1903, MICHAELIS [639] published the synthesis of phosphorus-nitrogen compounds from phosphorus trichloride, pentachloride, phosphorylchloride, thiophosphorylchloride and ammonia or amines. In the reaction of phosphorus trichloride with alkyl amines he obtained N-alkylaminodichlorophosphine (VIII), which he oxidized with chlorine to the tetrachlorides (IX).

(VIII) (IX) (X) (5)

Atmospheric moisture sufficed to hydrolyze these tetrachlorides to dialkylamino-phosphorodichloridates (X).
(Instead of trichloride, the pentachloride can be used.) Also in the reaction of phosphoryl chloride with aliphatic amines he found dialkylamino-phosphoro-dichloridate, in addition to the phosphorodiamidochloridate (XI), an important

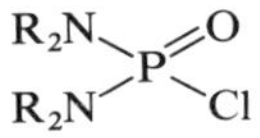

(XI)

starting material for phosphorylation reactions. In a comprehensive paper MICHAELIS [639] described the reaction of N,N-diethyl phosphoroamidodi-chloridate with potassium cyanide in absolute alcohol. As end product he gave a mixture consisting of O,O-diethyl N,N-diethyl phosphoroamidate (XII),

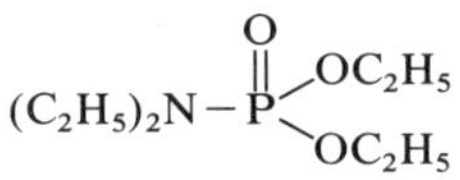

(XII)

and O-ethyl N,N-diethyl phosphoroamidocyanidate (XIII).

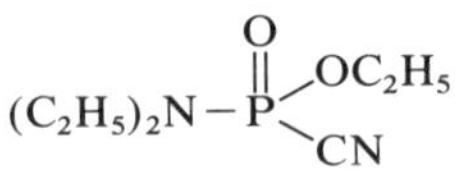

(XIII)

Surprisingly, he did not report the high toxicity of this compound (see p. 78). One can only assume that he had obtained a mixture of the diethyl ester with the N,N-diethyl phosphoroamidodicyanidate.
In 1895 H. N. STOKES [927] published a paper on the formation of phosphorus-nitrogen rings from phosphorus pentachloride and ammonium chloride, a reaction discovered by J. v. LIEBIG and A. WÖHLER in 1832:

$$PCl_5 \xrightarrow{NH_4Cl} [PNCl_2]_n \quad n = 3-8 \qquad (6)$$

(XIV)

In 1914 an article appeared by D. BALAREFF [63] on the structure of pyrophosphoric acid for which both symmetrical and asymmetrical forms were proposed.

In 1917 the same author [64, 65] published several papers concerning the action of phosphoryl chloride on methyl and ethyl alcohol but the final structure of pyrophosphoric acid was not clarified until 1930 by P. NYLÉN [685] who established the symmetric form. He synthesized the tetraalkyl pyrophosphate and mixed pyroesters.

In 1932, W. LANGE and GERDA V. KRUEGER [534] prepared the esters of monofluorophosphoric acid from its silver salts with alkyl iodide. They were the first to draw attention to the highly toxic properties of these compounds, including respiratory distress, clouding of the consciousness, temporary blindness and photophobia.

In 1941 during the Second World War, B. C. SAUNDERS and his group [611, 612, 615] worked on esters and ester amides of phosphoric acid fluoride.

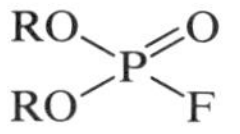

(XV)

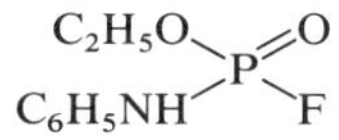

(XVI)

R = i-Propyl, sec. Butyl

They discovered the miotic action and very high inhalation toxicity of these substances. Their work, which remained secret until after the war and was only known to the Ministry of Supply, seemed to have placed emphasis on the pharmacological properties of the alkyl fluorophosphates.

Quite independently G. SCHRADER had long been working on acid fluorides in the search for compounds with acaricidal and aphicidal activity. In this field he was first successful with the methane sulfonyl fluoride [232, 841, 870] which is still used today in special cases as a fumigant.

$$CH_3SO_2F$$

(XVII)

By changing from sulfuric acid to phosphoric acid he was led to what became later his main work. As starting material he used N,N-dimethyl phosphoroamidodichloridate (XVIII) [231, 324, 640, 643] which is easily convertible to the difluoride.

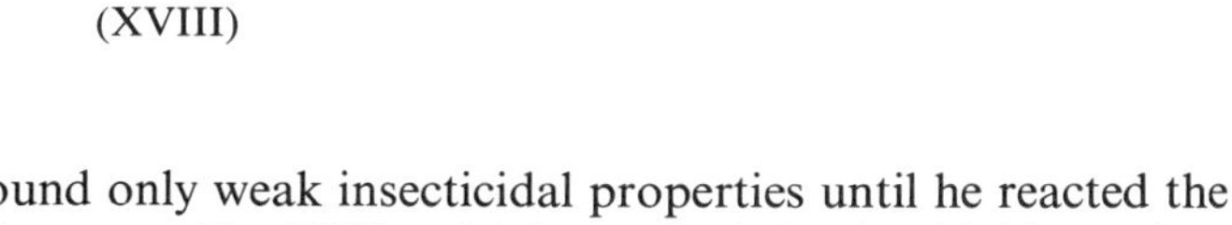

(7)

(XVIII)

At first he found only weak insecticidal properties until he reacted the dichloride with potassium cyanide [828], which resulted in the highly toxic and miotic Tabun, [875, 878] (XIX).

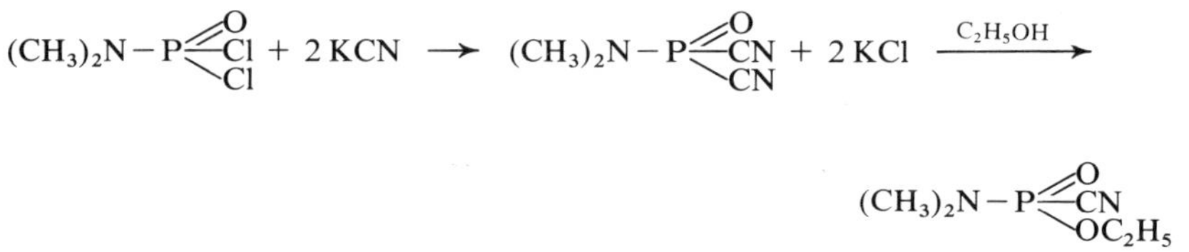

(8)

(XIX)

Then he replaced the dialkylamino group by an alkyl group [211] and so arrived, in 1937, at the physiologically extremely potent compound Sarin (XX).
Soman (XXI), however, did not originate from SCHRADER, but was synthesized in 1944 in Heidelberg by commission of the "Heereswaffenamt" [775]. On account of their high mammalian toxicity, neither Sarin nor Soman were used as insecticides [555].

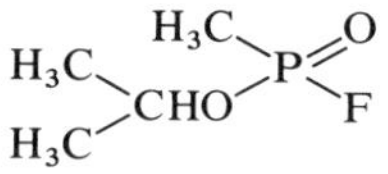

(XX) Sarin

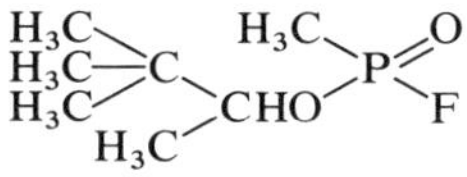

(XXI) Soman

These phosphonic esters were related to the compounds investigated by SAUNDERS et al. [794]. Diethyl phosphorofluoridate (XXII) and bis-dimethylamidophosphorofluoridate (XXIII) are highly toxic contact insecticides, but seldom used.

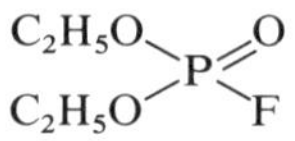

(XXII)

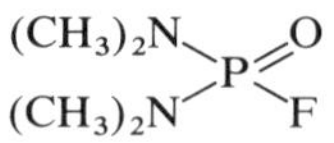

(XXIII)

In 1941 dimethyl phosphoroamidodichloridate served SCHRADER as the key substance for the synthesis of octamethyl pyrophosphoroamidate (XXIV) [825] and other pyrophosphoric acid derivatives. In honour of SCHRADER, scientists of Pest Control Ltd., in 1950 gave the name *schradan* [726] to OMPA or ® Pestox III.

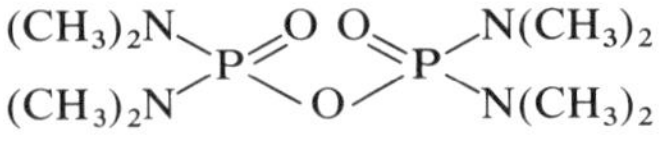

(XXIV)

On reflection, the main significance of OMPA has proved to be its potent systemic properties which were recognized by H. KÜKENTHAL in 1941 [881]. Later OMPA was superseded as a systemic insecticide by the ®Systox group (*sys*temically

toxic). SCHRADER was able, on the other hand, to synthesize tetraethyl pyrophosphate TEPP [827] by the reaction of phosphoryl chloride with triethyl phosphate, a reaction which has become known as the "Schrader process".
In 1941 E. GROSS [356], E. D. ADRIAN, W. FELDBERG, B. A. KILBY [7], and W. WIRTH [1030] discovered the cholinesterase inhibiting action of the organophosphates. This was occasioned by an attempt to clarify the miotic action of DFP (XXV).

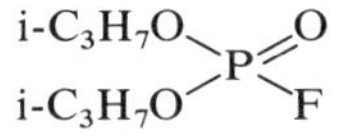

(XXV)

In 1944 SCHRADER synthesized the diethylthionophosphoric acid ester of p-nitrophenol (XXVI) (E 605) [879]

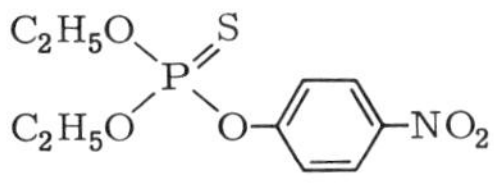

(XXVI)

Its insecticidal activity was recognized at the beginning of 1945, but in view of events in Berlin at this time, a patent application could not be filed. By the time it was possible to apply for a patent in 1948, American firms had already adopted E 605 under such names as ®Thiophos, (ACC), ®Niran (Monsanto Chem. Corp.), etc.
Great interest in the E 605 series was stimulated by the outstanding insecticidal properties and very broad spectrum of activity of these compounds which is reflected in the production figures of these phosphoric acid esters (Fig. 12, p. 50).
In 1952–54 W. PERKOW [721, 724] described the common reaction of α-halogen carbonyl compounds with triethyl phosphite and was the first to formulate correctly the end products as dialkylvinyl phosphates (XXVII). In contrast A. E. ARBUSOV and A. RAZUMOV [55] in 1934 had incorrectly attributed the phosphonate structure (XXVIII) to these products. The reaction given in the following equation has, therefore, become known as the Perkow reaction:

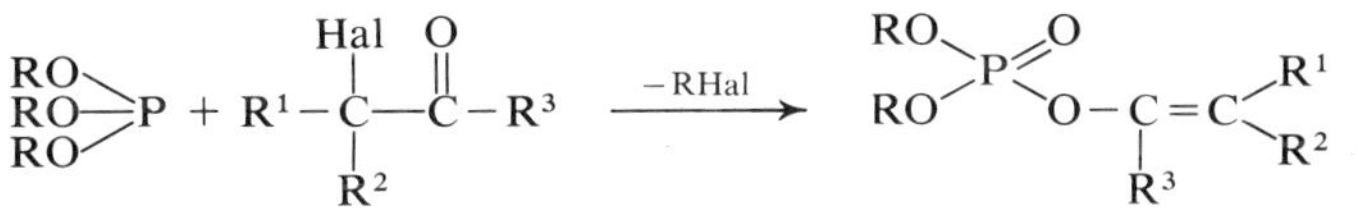

(9)

(XXVII)

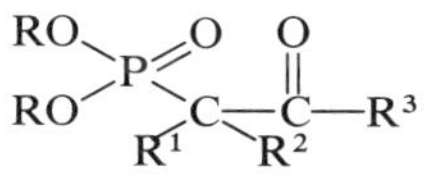

(XXVIII)

Historically, the enol phosphates were discovered by several teams of research workers at almost the same time. In the first papers, the structure was presumed to be an α-ketophosphonate, probably under the influence of the results published by Arbusov et al. As far as we have been able to establish from the literature, the first papers giving the correct structure appeared in 1952. It is therefore necessary to consider filing dates or receiving dates in order to establish priorities, because single compounds like DDVP or its diethyl congener are also involved in addition to the Perkow reaction. Below a short historical table of the basic papers is presented:

1949	(field 3. 9. 49)	Ladd and Harvey (U. S. Rubber) claimed in A. P. 2.631.162 the reaction of trialkyl phosphites with chloral, (no $(CH_3O)_3P$, structure is described as ketophosphonates) [529].
1951		Arbusov and Alimov described the reaction of triethyl phosphite with chloral as a Michaelis-Arbusov reaction. (Isvest. Akad. Nauk. USSR 1951, p. 530) [54].
1951	(filed 16. 1. 51) (pat. 24. 5. 56)	Perkow and Koddebusch claimed diethyl dichlorovinyl phosphate as final product of the reaction of triethyl phosphite with chloral (D.B.P. 944.430) [722].
1951	(filed 25. 9. 51) (pat. 16. 12. 55)	In the Swiss Pat. 310.395 (Ciba A.G.) the reaction $(CH_3O)_3P + CCl_3CHO \rightarrow (CH_3O)_2PO-CCl_2CHO$ was described. Although the structure was incorrect, at least the insecticidal properties of the product were mentioned [30].
1952	(filed 29. 2. 52) (pat. 3. 8. 54)	Whetstone and Harman (Shell Develop. Corp.) described the general reaction of phosphites and phosphonites with halogenated aldehydes (A. P. 2.765.331) [1013]:

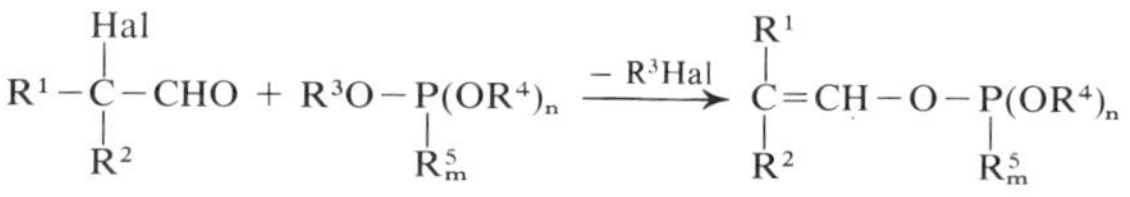

1952	(filed 29. 2. 52) (pat. 3. 8. 54)	Stiles (Shell Develop. Corp.) described in A. P. 2.685.552 the synthesis of Phosdrin and its insecticidal properties [926].
1952	(filed 29. 2. 52) (pat. 1. 5. 56)	Morris and van Winkle claimed compounds of the formula

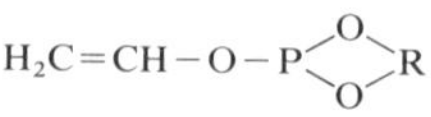

		but for the synthesis they referred to A P. 2.765.331 [663].
1952	(rec. 14. 6. 52)	Perkow, Ullerich and Meyer described the diethyl congener of DDVP (Naturw. *39*, 353 (1952)) [724].
1952	(U.S. Priority: 29. 10. 52)	The Food Machinery and Chem. Corp. applied for a patent (E. P. 783.697, filed 28. 10. 53) concerning the reaction of triethyl phosphite with numerous halogen carbonyl compounds. DDVP is also given as an example. The biocidal properties of the products are well established [34].
1952	(U.S. Priority: 29. 10. 52)	The same company claimed enol phosphates "useful as monomers and chemical intermediates" derived from monochloroacetaldehyde (E. P. 784.985) [35].
1952	(U.S. Priority: 29. 10. 52)	In E. P. 784.986 the addition of halogen to the double bond of enol phosphates was described, yielding for instance Dibrom [36].
1952	(rec. 26. 11. 52)	Corey, Dorman, Hall, Glover and Whetstone (Science *118*, 28 (1953)) published the synthesis of O,O-diethyl O-(2-chlorovinyl) phosphate and O,O-dimethyl O-(2-carbomethoxyvinyl) phosphate [214].

1953	(filed 5.2.53) (pat. 2. 11. 55)	Perkow and Koddebusch claimed dichlorovinyl phosphates, e.g. from triethyl phosphite and chloral, in E. P. 739.726 [723].
1954	(Germ. Priority: 25. 5. 54)	Lorenz (Farbenfabriken Bayer A.G.) filed application for a patent in respect of rearrangement of trichlorfon to dichlorvos (A. P. 2.865.943) [559].
1954		In Chem. Ber. *87*, 755 (1954) an article appeared concerning the Perkow reaction [721].
1955		Allen and Johnson (Food Mach.) published the synthesis described in the E. P. 783.697 in 1955 (J. Am. Chem. Soc. *77*, 2871 (1952)) [15].
1955		Kharasch and Bengelsdorf (J. Org. Chem. *20*, 1356 (1955)) published the thesis of Bengelsdorf involving O,O-diethyl O-(2-dichlorovinyl) phosphate [479].
1955		The rearrangement of trichlorfon to dichlorvos published in A. P. 2.865.943 (equiv. to D.B.P. 1.003.720) was confirmed ny Lorenz, Henglein and Schrader (J. Am. Chem. Soc. *77*, 2554 (1955)) [564].
1955		Barthel, Alexander, Giang and Hall published a second paper dealing with the alkaline rearrangement of trichlorfon to dichlorvos (J. Am. Chem. Soc. *77*, 2424 (1955)) [71].
1955		Mattson, Spillane and Pearce (J. Agr. Food Chem. *3*, 319 (1955)) reported the occurrence of DDVP as a by-product of Dipterex (trichlorfon); they had already done so earlier in a paper presented at the 126th Meeting of the American Chemical Society in 1954 [609].
1956		Rosin and Haus (Montrose Chem. Comp.) claimed a new process for the synthesis of trichlorfon (see equ.) and described the alkaline rearrangement of trichlorfon to dichlorvos (A. P. 2.899.456) [780]:

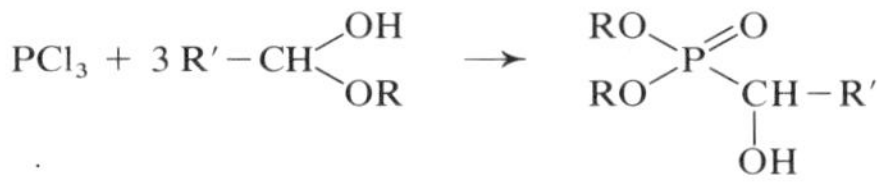

This type of reaction provides the insecticidal enol phosphates. Commercial compounds such as ®Phosdrin, *phosphamidon*, ®Birlane and many others have been introduced by various firms.

Several groups of workers had been occupied for many decades with the chemistry of the organophosphorus compounds. In the early thirties, on the other hand, the first objective of crop protection was to provide intensifive agriculture with readily-producible material which would serve as substitutes for the natural insecticides, such as nicotine, rotenone, pyrethrum etc., which were rapidly becoming scarce.

Only after these objectives had been reconciled with Schrader's rule co-ordinating structure and activity was it possible for the stormy development of the organophosphorus insecticides to commence.

The present-day importance of the phosphoric acid esters in the field of chemical crop protection is, to a large extent, attributable to the systematic work of Schrader and his attempts to find less toxic compounds with otherwise unchanged biological action.

2. General Section

2.1. Background

Phosphorus plays a central role in the living organism; it is sufficient to mention photosynthesis, metabolism, saccharide synthesis, nucleic acid helices, involvement in coenzyme systems, etc.

If the reduction and oxidation reactions between carbon and oxygen taking place in the organism are regarded in simplified terms as being responsible for gain and expenditure of energy then, disregarding their structural function, the phosphorus-oxygen compounds serve predominantly for the transport and storage of energy.

Two factors are decisive: firstly the condensed phosphates, anhydrides or esters are thermodynamically stable under such redox conditions, and secondly compounds of this type hydrolyze in most of the biochemical reactions taking place in aqueous solution or at aqueous interfaces. Depending upon their structure and the type of enzyme involved, these compounds hydrolyze over many kinetic stages and under very mild conditions. Expressed in more general terms, this means that phosphorus compounds can exert a phosphorylating action on nucleophilic molecules. The best known example of this is the interplay between adenosine triphosphate (ATP) and adenosine diphosphate (ADP). As will be shown in Chapter IV, the mechanism of action of the phosphoric acid ester insecticides also follows such hydrolysis and esterification equilibria.

a) Electronic Structure, Types of Compounds

On the basis of its position in the periodic system, phosphorus in its fundamental state possesses the external configuration $3s^2\,3p^3$. In contrast to the elements of the second period, $3d$ orbitals may participate during the formation of phosphorus compounds. However, the transfer from $3s^2\,3p^3$ to $3s^2\,3p^2\,3d$ requires a relatively high promotion energy of about 200 kcal/mol (~9 eV) (Fig. 11) [431, 430].

In accordance with the structure $3s^2\,3p^3$ of the neutral phosphorus atom, primarily compounds with the co-ordination number 3 are to the expected. Derivatives of type PX_3 form a distorted tetrahedron with participation of the s electrons and not a trigonal pyramid corresponding to the pure p^3 bond.

$\sphericalangle$ X–P–X for pure p^3 bonding 90°
for sp^3 hybridization 109° 28′
for most PX_3 types 100 ± 1°
for the nitrogen compound 106–108°

From an estimation of the spatial conditions, it becomes evident that trivalent phosphorus is coordinatively unsaturated and has the tendency to reach co-ordination number 5, by forming a new bond with the *s* electrons.

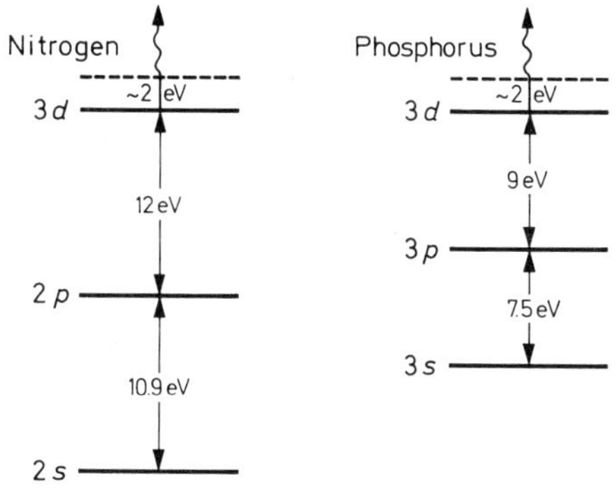

Fig. 11. Atomic energy levels for nitrogen and phosphorus (from HUDSON, R. F.: Structure and mechanism in organo-phosphorus chemistry. London-New York: Academic Press 1965).

This means that in the case of PX_3 compounds such as PCl_3, $RO-PCl_2$, $R-PCl_2$, $(RO)_3P$ or R_3P, very reactive nucleophiles are involved which are often used for the synthesis of derivatives of higher coordination numbers.

Examples:

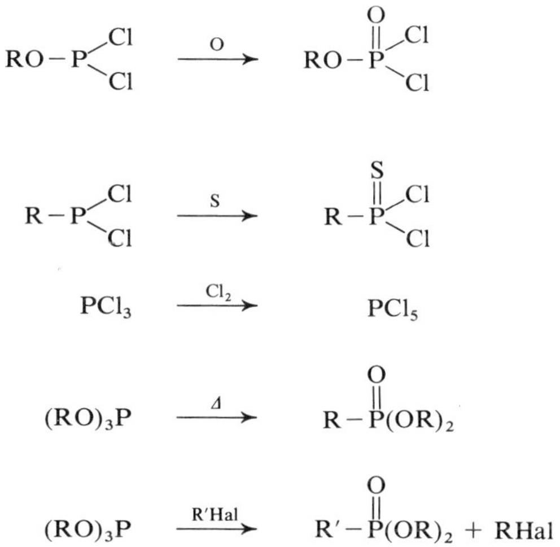

As these few examples show, most of the resulting products are derivatives of covalent tetravalent phosphorus which possess a tetrahedral structure corresponding to sp^3 hybridization. Numerically these exceed the compounds with other co-ordination numbers by several powers of ten.

Derivatives of phosphorus with co-ordination numbers 5 and 6, indicate an effort to return to structures with a lower co-ordination:

Examples:

$$PBr_{5(cryst.)} \longrightarrow [PBr_4]^+ \, Br^-$$

$$PBr_{5(gas)} \longrightarrow [PBr_3 + Br_2]$$

$$2\,PCl_{5(cryst.)} \longrightarrow [PCl_4]^+ \, [PCl_6]^-$$

Only a few compounds do, in fast, possess the co-ordination number 5; these include, for example, PCl_5 (gas) and pentaphenyl phosphorus with the structure of a trigonal bi-pyramid which is now considered verified.
RAMIREZ [763] succeeded in synthesizing a phosphoric acid ester with the co-ordination number 5. From PCl_5 and phenol and using γ-collidine as base, he obtained the pentaphenol ester (I) for which a long search had been made.

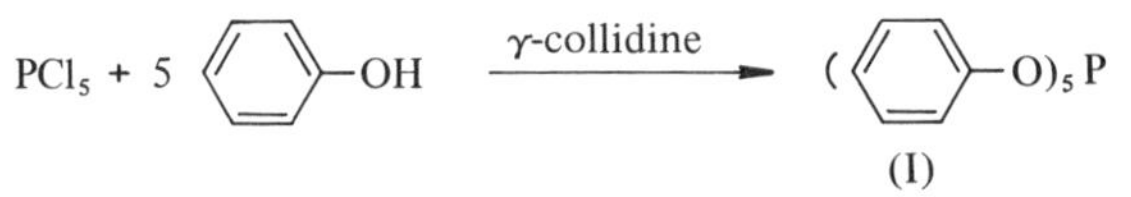

(I) (1)

Other types of hypervalent phosphorus [762], also found by RAMIREZ, are the cyclic addition products (II) from 1,2-diketones and tri-alkyl phosphites which are considerably more stable than the acyclic pentaphenol esters. In all cases their structure is that of a trigonal bi-pyramid, which in the cyclic derivatives is somewhat distorted. Eq. (2) illustrates the simplest example:

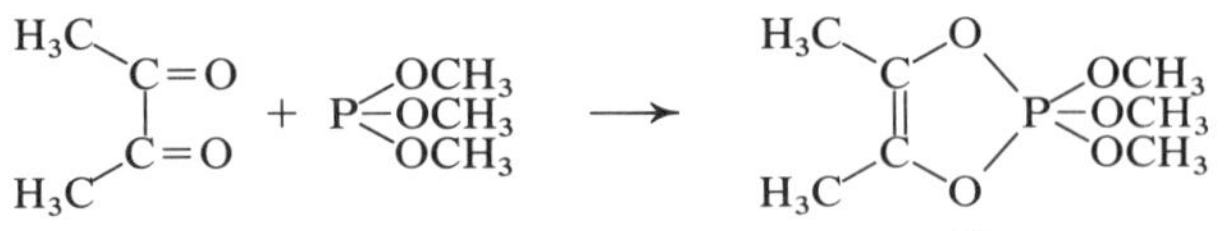

(II) (2)

More important in this connection are the structures of the transition or intermediate states in nucleophilic replacement reactions on the electrophilic phosphorus which will be discussed again later. They are consistent with a sp^3d hybridization [see formula (II a), p. 30].
The co-ordination number 6 is only possible with phosphorus anions, for example, PF_6^-, a substance relatively resistant to hydrolysis, or PCl_6^-. Such anions possess an octahedral structure corresponding to sp^3d^2 hybridization [712, 919].

b) Bond Properties

In order to satisfy valency considerations, the phosphorus-oxygen bond in derivatives of the type X_3PO was, like the carbonyl compounds, originally formulated as a $P{=}O$ double bond. When LEWIS put forward his octet rule, it

was believed that a semi-polar $P \rightarrow O$ bond should be assumed [708]. This was stated in the American literature and in more recent German handbooks, although as early as 1950 there was new thought on this problem. PAULING [712] had pointed out that in MO_4^{n-} ions, $3d$ orbitals of the central atom can form π bonds with $2p$ orbitals of oxygen. VAN WAZER applied this concept to phosphorus compounds of similar structure.
Accordingly the $3s$ and $3p$ orbitals of the central phosphorus atom hybridize to $4\,sp^3$ orbitals directed towards the oxygen atoms, which with suitable $2p$ orbitals of the oxygen overlap to four σ bonds. A π-bonding system is superimposed on this σ-bonding skeleton, to which the phosphorus contributes an electron forming a $d_\pi - p_\pi$ bond to oxygen. An oxygen atom bound only to phosphorus participates which, with three π electrons and an oxygen atom exerts two σ bonds, takes part in the bonding system with 2π electrons.
In a series of tetrahedrally constructed compounds including phosphorus, CRUICKSHANK (cf. [203]) considered the geometrical properties of two of the five $3d$ orbitals of phosphorus for the qualitative assessment of the bond distances and structural proportions. COLLIN [203] went a step further and took into account not only the two strongly bonding orbitals but all five $3d$ orbitals of phosphorus. Following a Hückel-MO method he used self-consistent field-modified Hückel parameters to estimate the charge distribution and π-electron energies. With these parameters it is also possible to assess chemical reactivity, thermodynamic parameters, and conformation.
From UV spectra it is possible to prove the lack of any analog between phosphoryl and carbonyl compounds, since $n-\pi^*$ transitions cannot be found [897]. Reliable information on the presence of $p_\pi - d_\pi$ bonding is not revealed by IR spectroscopy.
Further information can be obtained from spectroscopic investigations, especially from the valence force constants (mdyn/Å). GOUBEAU reported the results on various phosphorus-sulfur bonds [347]. The bonding order 2.0 (double bond) is to be expected when the sum of PAULING's electronegativities of two bonding partners is five or more and the difference between their electronegativities does not exceed 1.5. For the element combination phosphorus-sulfur the respective values 4.6 and 0.4 are found, i.e. the element pair is a borderline case of the double-bonding rule. Therefore all factors must be taken into consideration which might influence the force constants:

1) Hybridization
2) Inductive effects
3) Ion charges
4) π bonds

In all phosphorus compounds hybridization increases the force constants with increasing s character of the bond by a maximum of 40%, in the case of $P-S$ compounds by a maximum of 25%.
Like the $P-O$ force constants, the $P-S$ force constants are also dependent upon the electronegativity of the remaining bonding partners. By increasing electronegativity (for example replacement of $-SCH_3$ by $-OCH_3$, of $-SCH_3$ by $-Cl$) the force constants are increased by 0 to 30%. In the transition series $PS_4^{3-} \rightarrow PO_4^{3-}$

π-bonding effects are superimposed on the effect exerted by the ion charge so that it is not possible to generalize.
In order to investigate the influence of π bonding, compounds of the type $S-PX_3$ are particularly suitable. In the molecule $S-P(SCH_3)_3$ the force constant $k_{(P-S)}$ is some 63% greater than $k_{(P-SCH)}$, i.e., as in OPX_3, there must be a substantial π component in the S–P bond. Using SIEBERT's method [903] it is possible to determine from both force constants the bond order for $P-SCH_3$ as 1.16 (approaching the single bond) and for S–P as 1.57, a value considerably greater than that of the single bond.
The S–P force constant is strongly dependent upon the electronegativity of the neighbouring atom. In SPF_3 $k_{(S-P)}$ is 5.21 mdyn/Å, in $SP(CH_3)_3$ it is 3.33 mdyn/Å from which, according to SIEBERT, bond orders of 1.89 and 1.33 respectively are derived. For the analogues O–P the respective values 2.44 and 1.88 are found. Bond partners with *d* electrons, like the O–P bond, exert a noticeably enhanced effect.
If these results are regarded in the context of how a single oxygen or sulfur atom is bound to the phosphorus, then they also support the formulation of an X=P double bond and not that of a semi-polar bond. In more specialized investigations the bond orders of all substituents on the phosphorus must, however, be taken into consideration.
The most correct way of depicting these compounds would be to write four σ bonds and to show the associated π component separately. It is not possible to estimate the absolute extent of this π component, on the other hand changes within homologous series are covered. However, the actual state prevailing is certainly better given by the formulation(III) than by the structure (IV) [99].

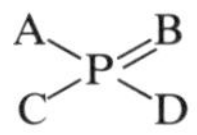

(III)

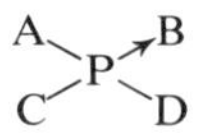

(IV)

Derivatives of phosphorus with the co-ordination numbers 4 and 5 can be considered from a more general point of view. With these compounds the lowest stable valency of phosphorus is exceeded, which as donor atom utilizes more bonding electron pairs than are required by the Lewis-Langmuir theory to reach a stable configuration. In the case of compounds such as $POCl_3$, $PSCl_3$, $P(OR)_5$, PCl_5 etc., we are concerned with molecules that may be regarded as addition products of a stable molecule with two univalent or bivalent ligands. Therefore the term hypervalent molecule is used; these molecules possess different types of chemical bonds and are able to form geometric isomers on a single atom, because one bonding type can change by oscillation or rotation into another bonding type. A detailed review on the theory and practical aspects of hypervalent molecules of the 5.–8. main group of the Periodic Table has been published by MUSHER [671].
With regard to the "elasticity" of the mode of reaction of phosphoric acid esters, the result of COLLIN is of importance in that, by minor energy additions (10 kcal/mol), changes in conformation can be achieved which for their part induce changes

of the $2p-3d$ interaction and thus the charge distribution. However, this means considerable variance in chemical reactivity. Such changes in conformation and their effects on $2p-3d$ π bonds might be evoked in biochemical systems by hydrogen bridges or electrostatic forces, and thus play a significant role in the catalytic activity of enzymes or at biochemical interfaces.

2.2. Reactivity

a) Hydrolysis, Alcoholysis

The hydrolysis of organophosphorus compounds follows several patterns, depending upon the type of ester (see Table 2), the solvent, the pH range or upon catalytically active additives. A qualitative knowledge of the various grades of reactivity in relationship to structure facilitates the synthesis of products with special properties for practical application, e.g. for use in rather strongly alkaline dip, in acid soils, to regulate the rate of degradation in plants or animals (residue tolerances, waiting periods, toxicological effects, etc.).

In the following, the hydrolysis of the ester types summarized in Table 2 will be discussed, then the reaction with other nucleophiles as hydroxyl ion, some nucleophilic replacement reactions with ester halides, and finally intramolecular C–O cleavage reactions.

Emphasis will not be placed on the kinetic aspects of these reactions but on the pragmatic comparison of the reactivities.

Table 2. Organophosphate types used in crop protection

Type	Structure		Type	Structure	
A	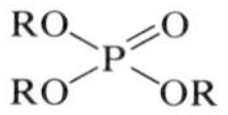	Three alkoxy groups	H	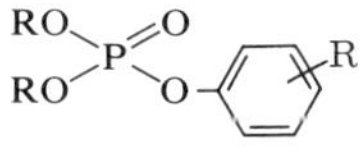	Phenol esters
B	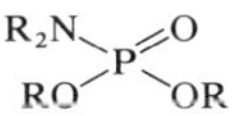	Amide esters	I	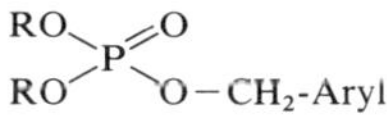	Benzyl esters
C	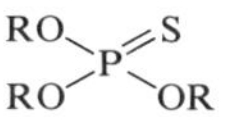	Thionoesters	K	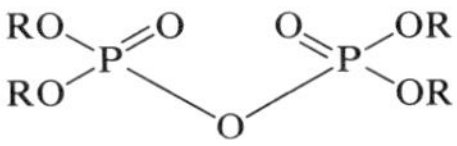	Pyro esters
D	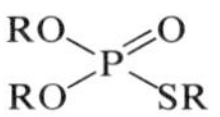	Thiolesters	L	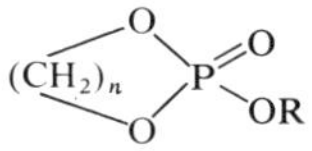	Cyclic esters
E	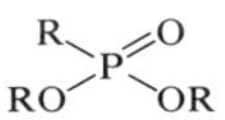	Phosphonic esters	M	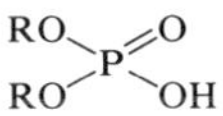	Dialkyl ester acid
F	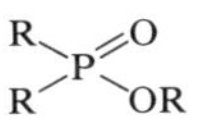	Phosphinic esters	N	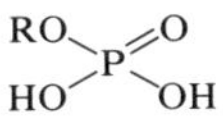	Monoalkyl ester acid
G	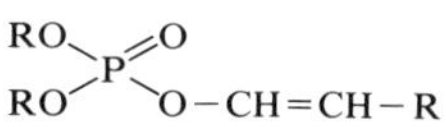	Enol esters			

The hydrolysis of the trialkyl phosphates of type A in *aqueous alkaline medium* is a first-order reaction in respect of both hydroxyl ion and ester, whereby only the P–O bond is cleaved [104]. In the rate-determining step, the hydroxyl ion attacks the phosphorus; an exchange of oxygen between the phosphoryl group and the hydroxyl ion can be excluded [67]. According to LARSSON [535] the process is a one-step reaction, where the degree to which a group is leaving rests solely upon the approach of the hydroxyl ion to the phosphorus atom. The alternative two-step process, in which the intermediate decomposes more rapidly than equilibrium is reached with the solvent, is not distinguishable experimentally from a one-step reaction.
Therefore, the alkaline hydrolysis of trialkyl phosphate can be given by Eq. (1):

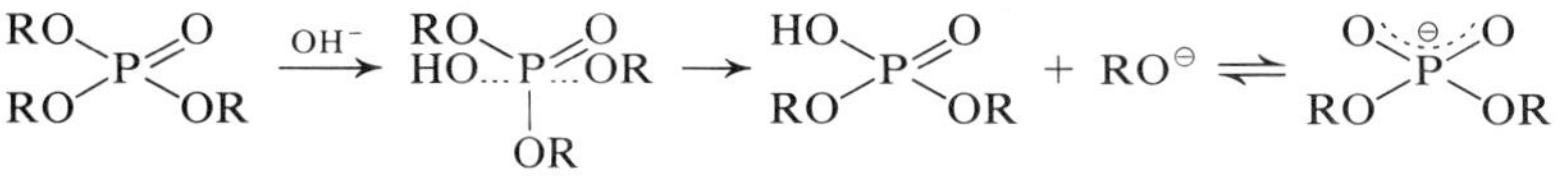

(I) (1)

This reaction is consistent with the definition of an SN_2 process but too close a comparison with carbon chemistry should be avoided. The intermediate or transition state corresponding to Formula (I) is favoured by the fact that phosphorus can form structures of co-ordination number 5, with the configuration $3s^2\,3p^3\,3d^1$ corresponding to a trigonal bi-pyramid.
This pattern involves non-equivalent bonds: with the same bond length, the axial bonds must be stronger [255, 361]. If on the other hand, one assumes complete sp^3d hybridization, i.e. bonds of the same energy, then the axial bond distances are longer than those of the equatorial bonds. This means that, from an energy aspect, the SN_2 transition state at the phosphorus favours substituting and leaving groups in the axial position. With these prerequisites, Walden inversion should occur during hydrolysis, which in fact HUDSON and GREEN [353, 432] were able to demonstrate.
In *neutral or acid aqueous solution* the conditions are not so clear. In principle it appears that protonation at the ester oxygen first takes place. The hydroxyl ion then attacks the carbon atom in an SN_2 reaction for, with hydrolysis in water containing ^{18}O, the activity is found almost exclusively in the methanol [67]. The leaving group would accordingly be the dialkyl phosphate anion:

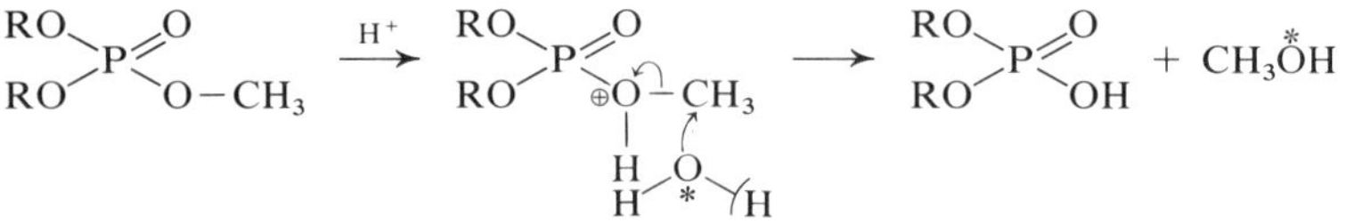

(2)

Since many nucleophils compete with water in the C–O cleavage, the hydrolysis of trialkyl to dialkyl phosphates is accelerated on addition of buffer substances or acids, whose anions are strongly nucleophilic toward carbon. According to PEARSON's concept of hard and soft acids and bases C–O cleavage is to be expected when soft bases attack the soft acid R^+ (cf. p. 33 f.). Hard bases react pref-

erentially with the hard acid R_2PO^+ and result in P–O cleavage as was amply illustrated by TEICHMANN and HILGETAG [947].
In *strong acid medium* the direct attack of the hydroxyl ion on the phosphorus is superimposed on SN_2 reaction at the carbon atom.

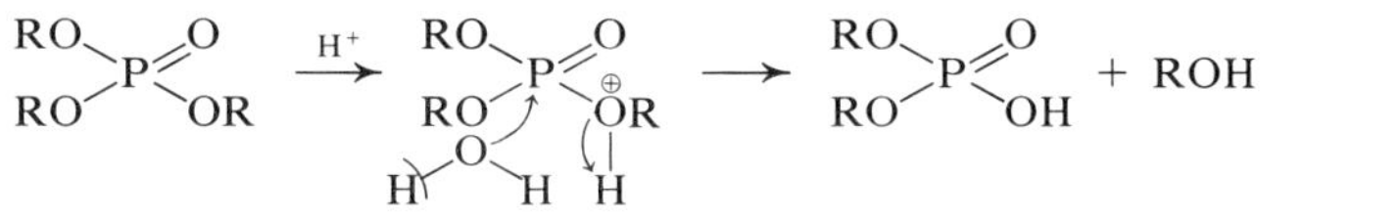
(3)

On considering other ester types and the details expounded on page 24 the rule generally holds good that substituents on phosphorus which readily participate in π bonds and thus reduce the positive charge on the phosphorus must render attack on nucleophilic groups more difficult. A typical example is the ester amides of type B, in which the nitrogen can accept the positive charge of the phosphorus. Amidophosphates are therefore relatively difficult to hydrolyze in alkaline solution. It is different in acid medium; protonation on the nitrogen facilitates attack on the phosphorus with the formation of the corresponding ammonium salts.

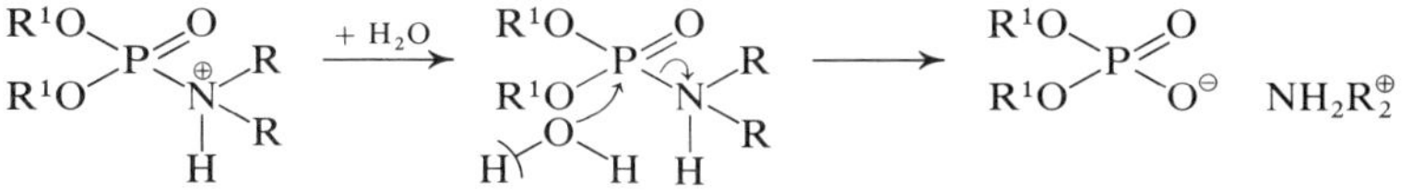
(4)

The most probable mechanism is that proposed by GARRISON and BOOZER [325] on the basis of kinetic investigations of the following series of reactions (Eq. (5) and (6)):

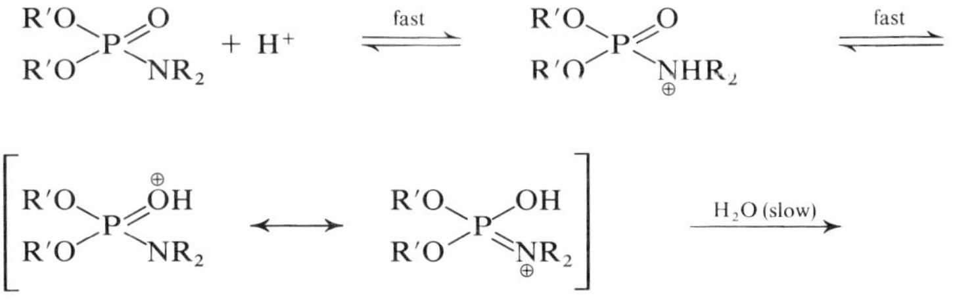
(5)

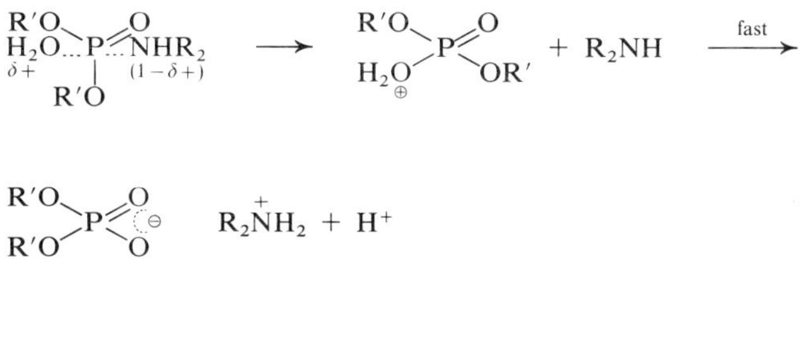
(6)

Taking into consideration the activation energies and entropies, the authors postulate direct substitution of the phosphorus by a water molecule. This takes place in a single step that determines the overall rate in which the P–N bond is cleaved at the same moment at which the attack of the water molecule occurs (Eq.(6)).
Proton equilibrium precedes this step (Eq. (5)). The influence of the N-alkyl substituents are considered in the light of linear free-energy relationships.
A further example is provided by the thiono-esters of type C. Since the double-bound sulfur is less electronegative and more readily polarized than double-bound oxygen, the effective charge on the phosphorus atom is reduced. Therefore the rate of alkaline hydrolysis of the thiono-esters must be decreased in comparison to the P=O derivatives, which is a well-known fact in preparative chemistry.
Compounds of type D contain sulfur in the thiol form. Here the same argument is valid regarding the lowered electronegativity of the ester sulfur in comparison to the ester oxygen. The ability to participate in π bonding is diminished, the positive character of the phosphorus atom is maintained, favouring a basic attack. Within homologous series the thiol compounds do, in fact, hydrolyze faster than the corresponding oxygen ester. Another fact to be taken into consideration is that the mercaptide ion is more stable as the leaving group than the alkoxide ion.
The alkaline cleavage of malathion to dithiophosphate and fumaric acid ester is a special case. This intramolecular C–S cleavage most likely proceeds by way of an elimination catalyzed by the carbethoxy group, for a P–S cleavage ought normally to occur [359].

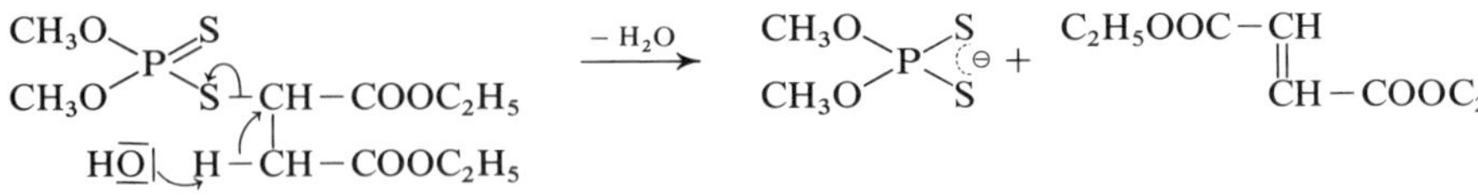

(7)

P–S cleavage, on the other hand, takes place by acid hydrolysis and can be exploited preparatively to obtain γ-mercapto-succinic acid, which is otherwise difficult to synthesize [215].

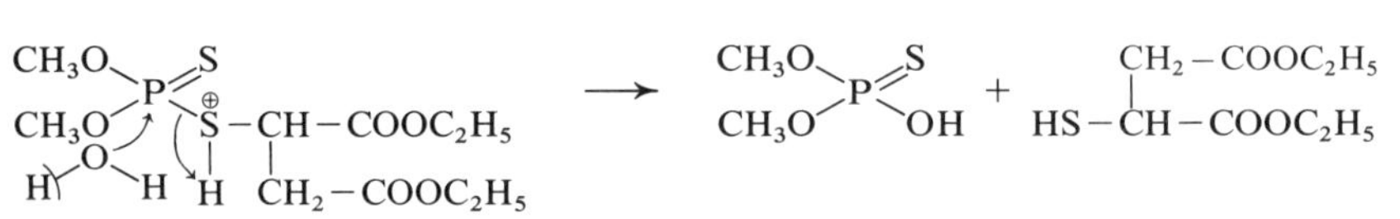

(8)

Other substituents with reduced ability to participate in π bonds are alkyl groups, as in the phosphonic acid esters (Type E, one P–C bond) and phosphinic acid esters (Type F, two P–C bonds). The π character of the remaining P–O bonds ought to be increasingly strengthened since the P–C bond is practically non-polar and hydrolytically inert (disregarding exceptions such as the trichloromethyl group).
The consequence is an increased positive charge on phosphorus in comparison to the trialkyl phosphates. The "hardness" of the acyl group increases in the order phosphoryl – phosphonyl – phosphinyl.

Alkaline hydrolysis ought to be favoured, while in contrast acid hydrolysis must be rendered more difficult. In fact, experimentally, a decreasing alkaline and an increasing acid stability is found in the order:

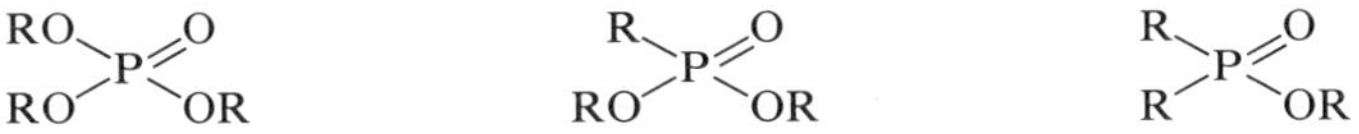

Phosphates Phosphonates Phosphinates

From the point of view of practical crop protection, the next three types belong to the most important series of compounds. The enol phosphates of type G, like the aryl phosphates (Type H), differ from the examples so far discussed in that inductive effects are superimposed on the mesomeric effects of the double bond systems and on any of its possible substituents.

Generally alkaline hydrolysis of the enol phosphates [549] proceeds exclusively with loss of the enol.

In comparison to the trialkyl phosphates, acid hydrolysis is favoured (half-lives of the order of 1 : 10) [550] and consists largely in a cleavage of the P–O bond in contrast to the predominantly C–O cleavage in the trialkyl phosphates:

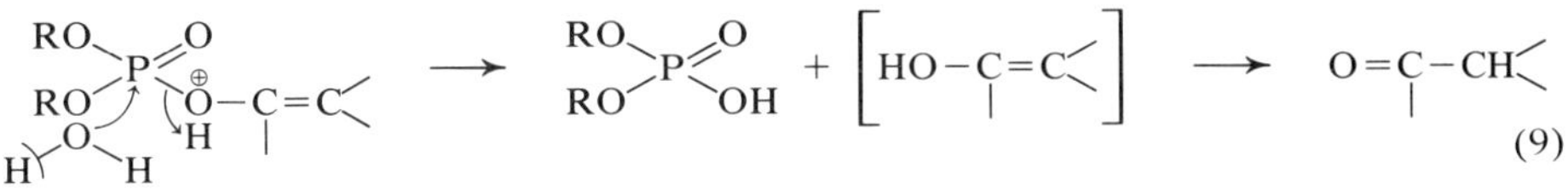

(9)

According to Bunton and Robinson [147] the acidic hydrolysis of diethyl α-phenylvinyl phosphate follows an A–SE 2 mechanism, i.e. the proton transfer is rate limiting:

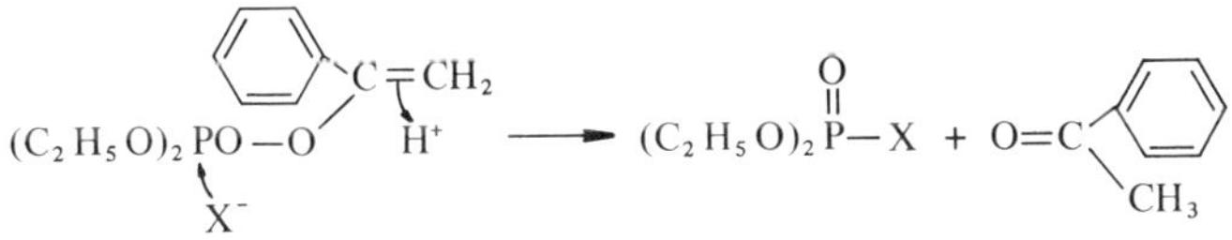

(10)

On the basis of then kinetic data, it appears questionable whether a water molecule is involved as a nucleophile in the transition state.

Aryl esters of the type H hydrolyze under P–O cleavage. Here also, mesomeric and inductive effects overlap. If the benzene ring is substituted with electron-withdrawing groups, then the rate of hydrolysis of the corresponding phosphoric acid esters follows a linear free-energy relationship of the Hammett type [308]. Substituted phenols can be replaced by enolisable heterocyclic oxo-compounds. With appropriate structures one can expect an increase in rate of acid hydrolysis, which, analogous to the enol phosphates, is probably accompanied by P–O cleavage:

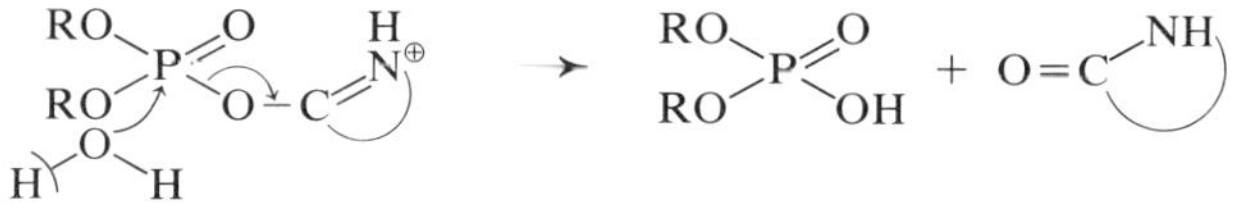
(11)

Both ester types (G and H) fall within the term "Acyl" of SCHRADER [863], i.e. the P–O bond possesses somewhat of an anhydride character. The significance of this property for biological action is discussed in greater detail on p. 40. Hydrolysis of benzyl phosphates of type I is accompanied by C–O cleavage which is consistent with the chemical behaviour of the benzyl carbon for the benzyl cation is a typical soft acid.

Since most compounds comparable with this group are in practice the esters of N-methylol heterocyclics, benzazimide has been chosen as example:

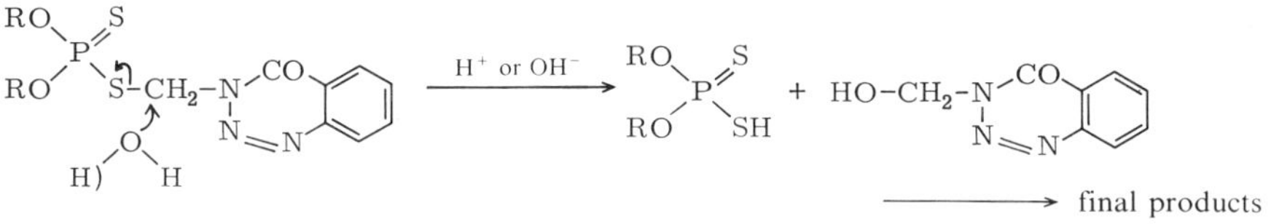
(12)

Whereas P-ester cleavage should result in an O,O-dialkyl monothiophosphoric acid, in the case of ®Gusathion, however, dimethyl dithiophosphoric acid is obtained. Its intensely yellow copper salt serves for the analytical determination of the active substance [406].

Other classic representatives of SCHRADER'S "Acyl" rule are the pyrophosphates of type K ("Acyl" = phosphoryl): during hydrolysis they react like normal acid anhydrides with P–O cleavage.

However, the cyclic phosphates of type L are somewhat exceptional, being characterized by a surprisingly high rate of hydrolysis if the phosphorus is contained in a five-ring system [203]:

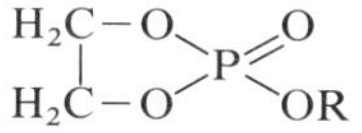
(II)

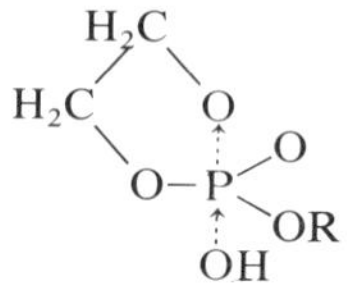
(IIa)

Since the hybridization on the phosphorus cannot differ in principle from that in trimethyl phosphate, the alkaline hydrolysis of (II), which is higher by a factor of 10^8, should be attributed to the ring strain and also to the fact that the intermediate state (IIa) is sterically favoured. Cleavage takes place at the P–O and not at the C–O bond.

This also applies to analogously constructed aromatic esters, e.g. (III):

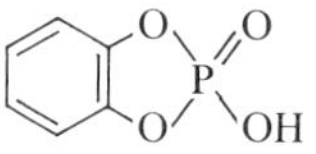

(III)

According to KAISER and KUDO [468] the alkaline hydrolysis of these compounds is faster than that of the corresponding diphenyl ester by a factor of 6×10^6. KUGEL and HALMANN [522] found, however, that with the cyclic esters of glycerine (IV),

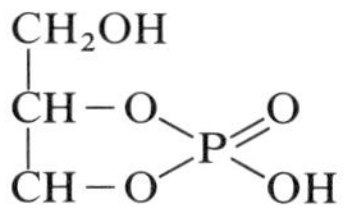

(IV)

the hydrolytic activity between pH 3 and 12 is only slight and comparable with that of the dimethyl phosphate.
Participation of free primary alcohol groups in the hydrolysis may be responsible for this result.
WESTHEIMER reviewed the experimental results in connection with the problem of cyclic phosphate esters. BOYD [121] contributed molecular orbital calculations on cyclic and acyclic phosphate esters, which support the mechanisms formulated by WESTHEIMER (c.f. [1011]) to explain the reactivity of cyclic esters towards hydrolysis and oxygen exchange. In this respect, however, ring strain still seems to be of importance. It leads to a lowered occupation of the P 3*d* orbitals, which effects deshielding of the phosphorus nucleus. This in its turn accounts for the high rate of nucleophilic attack in hydrolysis and oxygen exchange. The following mechanism involves the lowest activation energies [121]:

1) The nucleophile approaches the phosphate ester on the reverse side of one of the P–O ring bonds.
2) This bond lengthens to an apical bond and the other P–OC bonds become basal bonds of a trigonal bipyramidal intermediate.
3) The intermediate undergoes pseudorotation more easily when the phosphoryl oxygen serves as pivot (i.e. remains in a basal position).
4) The apical oxygens prefer to be protonated.
5) The hydroxyl group formed in step 4 moves away from phosphorus.
6) The local geometry at phosphorus relaxes towards a tetrahedron.

The exocyclic demethylation of (V) is a side-reaction of the ring-opening and can be explained by pseudorotation on the central phosphorus atom. The rate of ring-opening increases linearly with the acidity, whereas the rate for exocyclic cleavage may reach a plateau with increasing acidity if pseudorotation becomes rate-limiting.

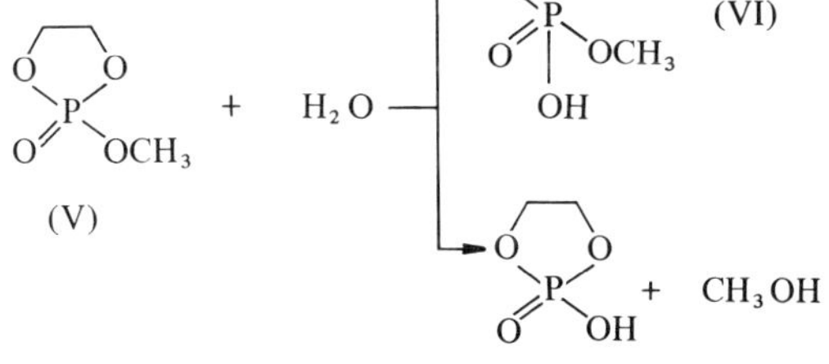

(VII)

(13)

The product ratio of (VI): (VII) therefore should be shifted to (VII) if acidity increases [493].

For practical application in crop protection such high rates of hydrolysis are disadvantageous because an insecticidal ester on its passage through the organism would be rapidly detoxified by enzymatic hydrolysis reactions before it was able to reach its target.

The question now arises of the further course of hydrolysis after the first stages so far discussed, i.e. the behaviour of ester acids of the type M. Apart from a few exceptions, these compounds are hydrolytically extremely stable. In many cases their salts can be isolated by evaporation of aqueous alkaline solutions without notable degradation taking place.

Amongst the exceptions are compounds which are able to form five-membered rings with sterically fitting neighbouring groups. A very important example biochemically is the rapid depolymerization of ribonucleic acids in alkaline medium. Cox and Ramsay [216] have reviewed this problem:

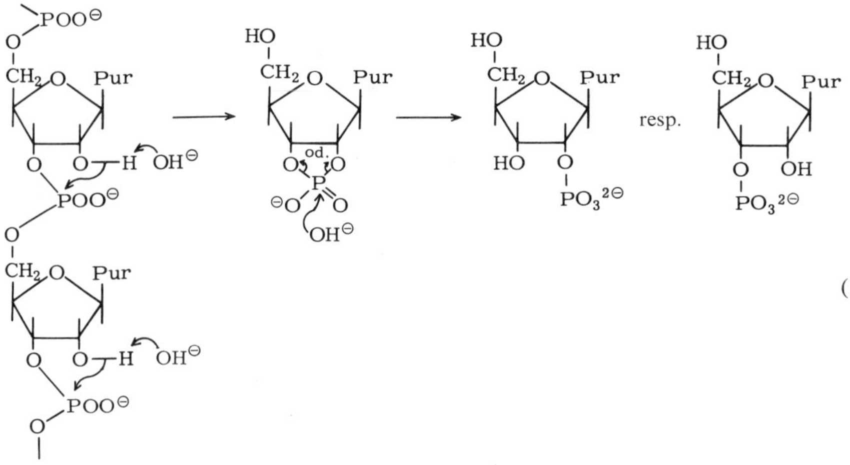

(14)

The hydrolytic degradation of monoester acids of Type N, discussed recently by Kirby and Varvoglis [487] as well as by Bunton [146], then proceeds very rapidly. The conditions are not easily surveyed because, according to the pH conditions, either the neutral molecule, mono- and di-anion may react, or via preceding proton equilibria, the conjugated acids may be involved. These hydrolyses are certainly more interesting from a kinetical aspect than from the practical application point of view chosen here.

Furthermore, for all ester types so far discussed, a general rule applies to the relationship between structure and hydrolytic reactivity: with increasing size of the substituents, irrespective of whether they are bound to the phosphorus directly or via oxygen, sulfur and nitrogen atoms, respectively, the reaction rate is reduced by steric hindrance.

Nucleophiles other than the Hydroxyl Ion

The attack of an alkoxide ion on the phosphorus leads to a trans-esterification which, in principle, almost certainly proceeds in the same manner as the hydrolysis.

Trans-esterifications are particularly easy to achieve when the esters contain strongly acidic groups. For example, with alcohol in an alkaline medium, a dialkyl *p*-nitrophenyl ester is readily converted to the trialkyl ester with formation of the alkaline salt of *p*-nitrophenol. A technical process for the manufacture of ®Systox is based on this reaction (see p. 121). It is also the fundamental reaction in enzyme inhibition by organophosphates and the real sense of SCHRADER'S "Acyl" rule, as will be shown on page 40:

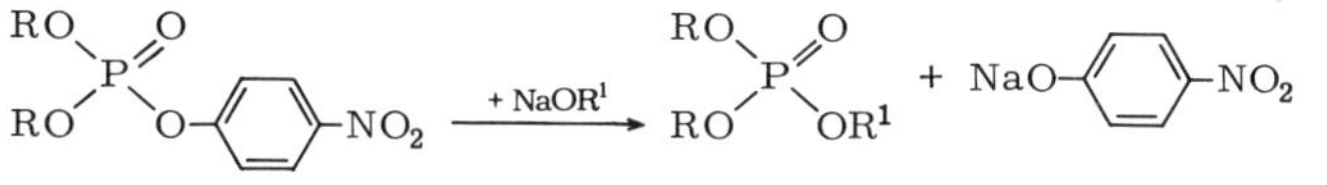
(15)

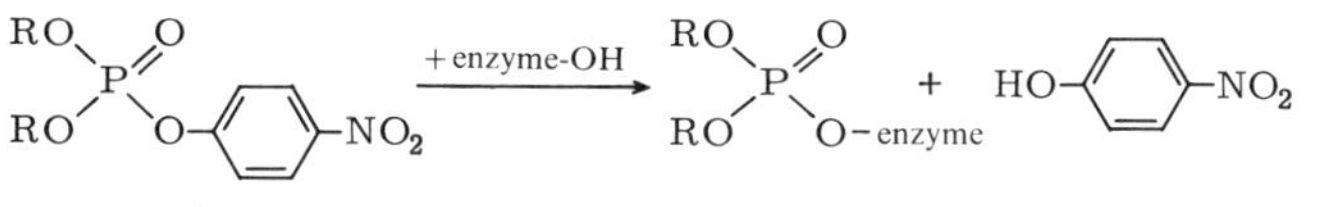
(16)

If the nucleophilicity of the proffered ion suffices for an attack on the ester carbon, then the alkylating properties of the phosphoric acid esters come to the fore. A good example of both reactions is the solvolysis of tetrabenzyl pyrophosphate in propanol [254]. The ratio of the rate constants of Eqs. (17) and (18) is 3 : 1:

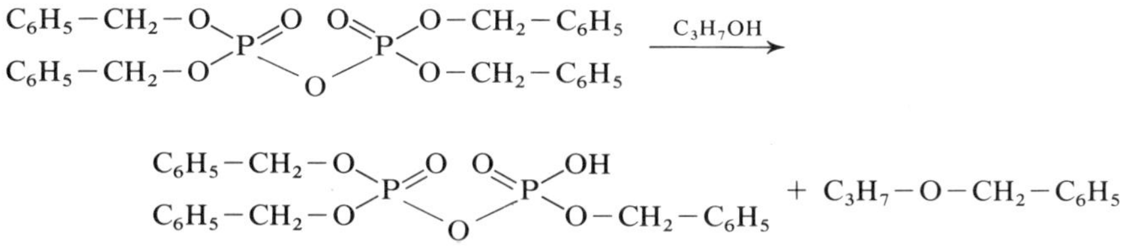
(17)

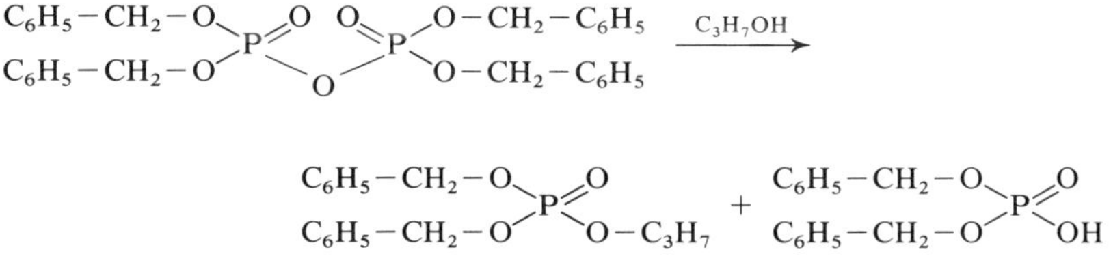
(18)

Mercaptans react mostly via C–O cleavage to give the thioesters [652, 782]:

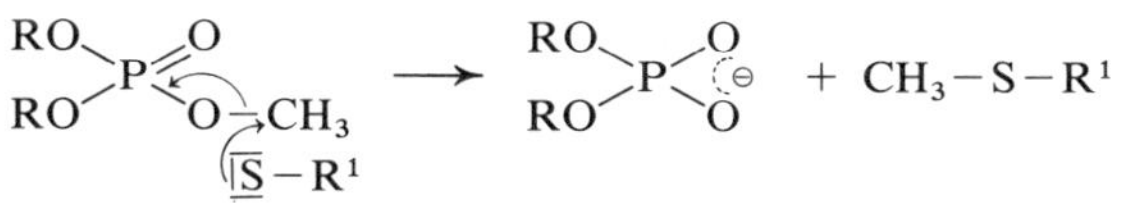
(19)

In the same manner tri-esters are dealkylated by amines [179]:

$$(RO)_2P{\displaystyle\genfrac{}{}{0pt}{}{\nearrow O}{\searrow O-R'}} + R_2NH \longrightarrow (RO)_2P{\displaystyle\genfrac{}{}{0pt}{}{\nearrow O}{\searrow O^{\ominus}}} \quad R_2R'H\overset{\oplus}{N} \qquad (20)$$

In some cases ammonia itself also attacks via P—O cleavage, resulting in the amide [490]:

$$(RO)_2P{\displaystyle\genfrac{}{}{0pt}{}{\nearrow O}{\searrow OR'}} + NH_3 \longrightarrow (RO)_2P{\displaystyle\genfrac{}{}{0pt}{}{\nearrow O}{\searrow NH_2}} + R'OH \qquad (21)$$

Iodide reacts with triesters always via C—O cleavage and, in a similar manner to the amines, may be of interest for the synthesis of dealkyl esters.
Iodide, mercaptide and many amines are typical soft bases which attack R^+ (soft acids) and therefore bring about C—O cleavage.
In aqueous solution the case quoted on page 26 applies: The nucleophilic molecules compete with the hydroxyl ion and thus catalyze the hydrolysis reaction. Heterocyclic amines are particularly effective, especially imidazole, but also pyridine, histidine and their derivatives. Whether a nucleophilic or a general base catalysis is involved, has not yet been clarified. It is possible with many hydrolases, including cholinesterase, that the imidazolering of histidine is responsible for the hydrolytic activity of these enzymes (see p. 170f.).
Using esters of N-acetyl serinamide as a model, MILSTIEN and FIFE [656] attempted to clarify this question and found that imidazole in its basic form acts as a classical general base in the imidazole-catalyzed hydrolysis of this ester series. If a heteroatom possessing at least one free-electron pair is situated in the alpha position to the nucleophilic atom, then the molecule shows considerably higher reactivity than might be expected from its basicity. Important examples of this α-effect [259] are hydroxylamine and oxim anions: these are compounds which play a considerable therapeutic role in the reactivation of blocked cholinesterases, as will be shown on page 254. EDWARDS and PEARSON [259] attributed the increased reactivity of such nucleophiles to the stabilization of the active complexes by the electron pair on the α-atom. They assume that, in the course of the displacement reaction, an electron pair attempts to pass from the nucleophilic to the electrophilic atom, causing positive polarization of the nucleophilic atom during the transition state. An electron pair available on the α-atom would then stabilize the positive charge on the nucleophilic atom.

b) Alkylating Properties

The Eqs. (19) and (20) illustrate the ability of (thio) phosphoric acid esters to alkylate suitable partners. HILGETAG and TEICHMANN [408] have made a careful and comprehensive review of the alkylating properties of (thio) phosphoric acid esters. As in the case of other alkylating agents, the lower alkyl (1—4C), benzyl, and allylic groups are preferentially transferred. Alone for the reasons discussed in connection with hydrolysis on page 28, P=S esters are somewhat weaker alkylating agents than the P=O compounds. This gradation in alkylating action is weakened or may even be reversed by inductive effects of the substituents R^1 and R^2. If they are of a strong electron withdrawing character, then the P=O polarity is reduced. For esters of the type:

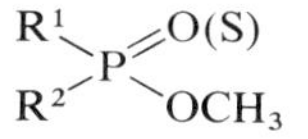

HILGETAG and TEICHMANN give the following order for decreasing capacity to alkylate:

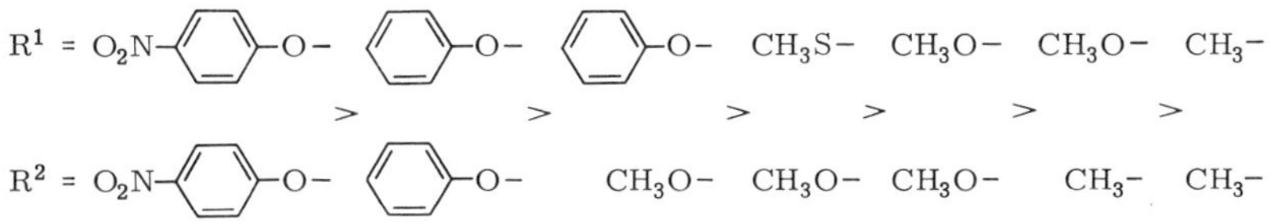

The alkylating potential runs parallel to the acidity of the corresponding acid (see p. 62). Also this potential can be greatly increased by raising the polarity of the C—O bond using LEWIS adducts. For example, when ether is reacted with trialkyl phosphorothionate and antimony(V) chloride, it is converted quantitatively into triethyloxonium hexachloroantimonate [407]:

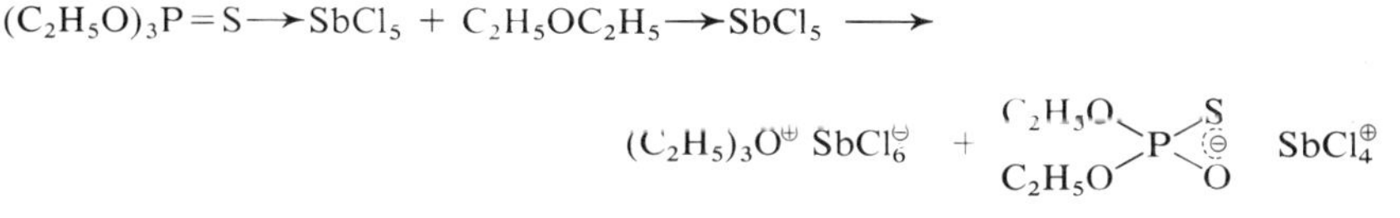

(22)

HILGETAG and TEICHMANN suggest the following distinction between the various alkylation types [408]:

1. Re-Alkylation

This occurs when nucleophilic substances are used, which in the alkylated form themselves become alkylating agents, for example J^- or CH_3SCH_3:

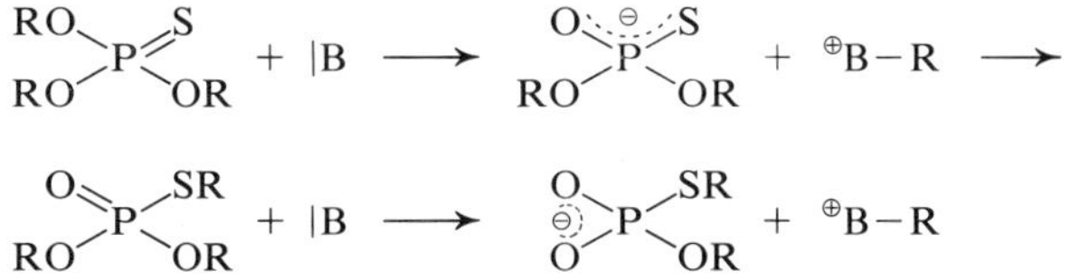

(23)

2. Alkylation of Diester Anions by triesters

This is likely when the alkyl group taken over by the nucleophile is so strongly bound that re-alkylation is not possible. The resulting diester anion then serves as substitute nucleophile, and is converted into the thiol ester:

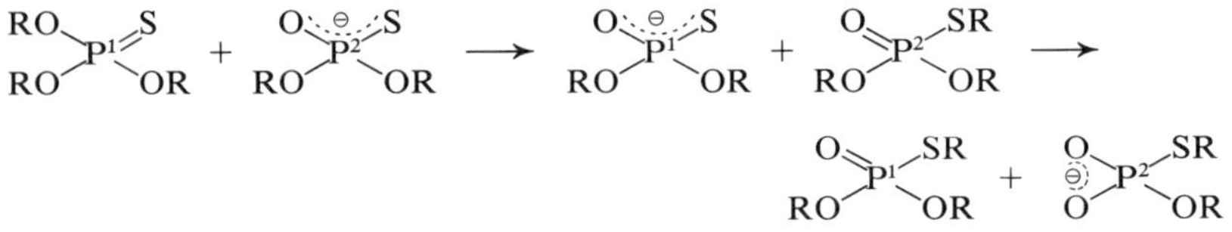

(24)

Accordingly the dithiophosphates yield two further isomeric anions:

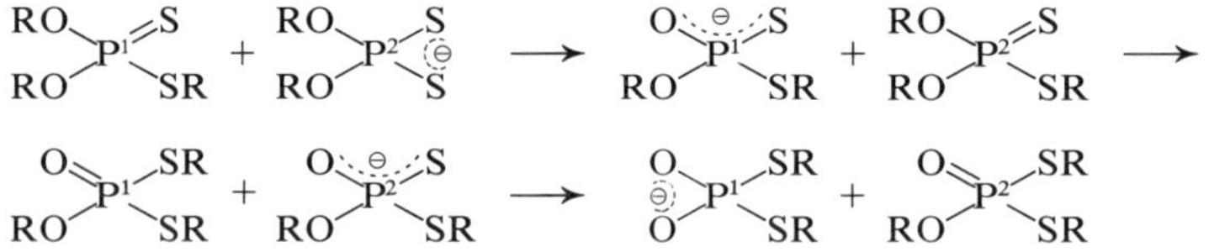

(25)

3. Self-Alkylation (Isomerization)

Even without nucleophilic partners, thiono-phosphoric acid esters themselves can participate in alkylation reactions. In order to obtain intermolecular C–O cleavage, vigorous reaction conditions are required, but such conditions (prolonged reaction time, high temperatures) are readily achievable in technical synthesis. Here, the nucleophilic center is the thiono-group:

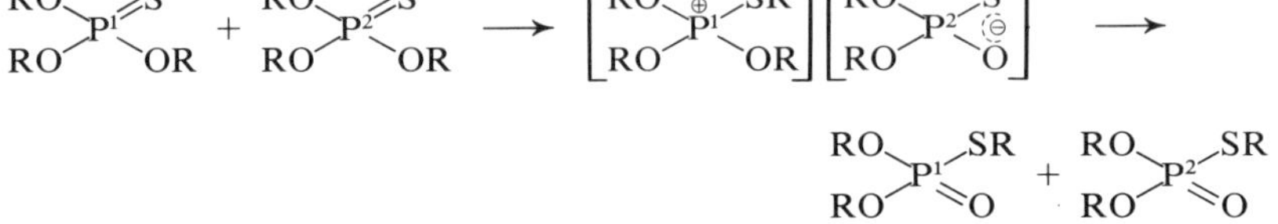

(26)

Once the reaction has started, the resulting thiol esters are the more powerful alkylating agents and accelerate the reaction.

This mechanism may be regarded as a special case of the Pistschimuka reaction [735] in which the thionophosphates are isomerized by alkyl iodide to the thiol ester:

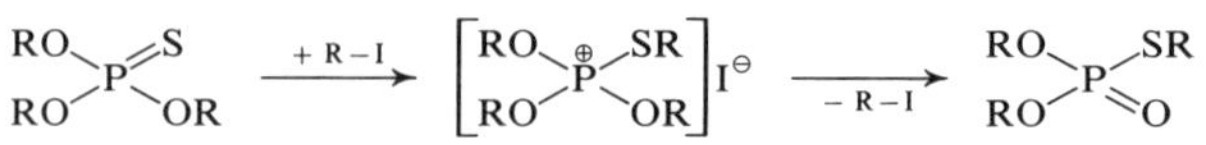

(27)

On the other hand there is an obvious relationship to the Arbusov reaction:

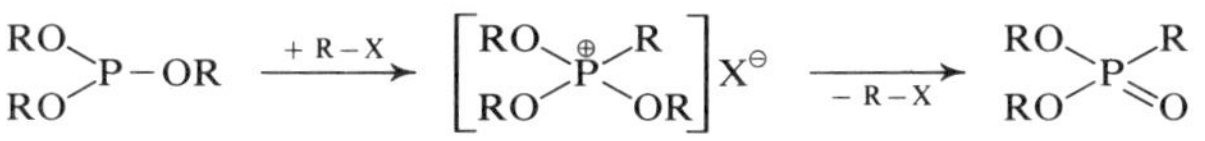
(28)

By addition-complexes with LEWIS acids, the self-alkylation of thiono-esters is possible at room temperature. Important examples of an intramolecular C–O cleavage are the thiono-thiol rearrangements in the *demeton* series [310, 400] ("Systox rearrangement") and with the phosphoryl cholines [157, 318, 942]:

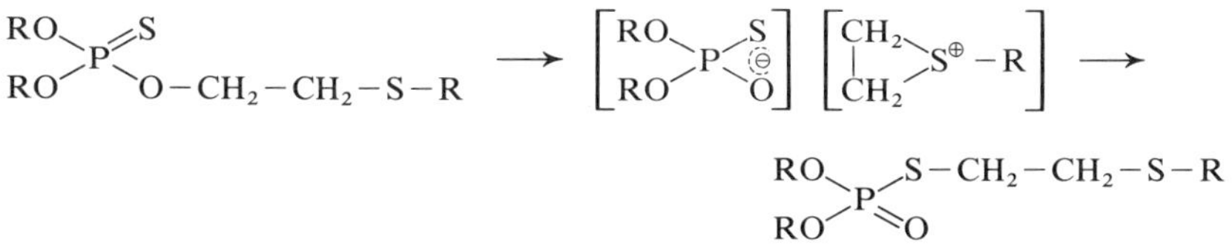
(29)

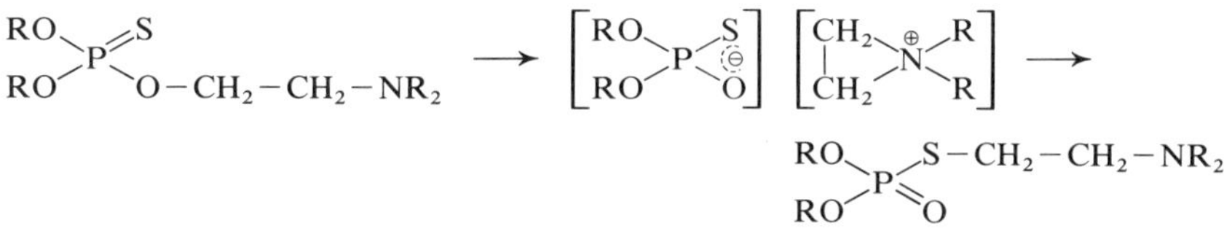
(30)

The Eqs. (29) and (30) are examples which demonstrate the rule that soft acids combine preferentially with soft bases.
The end products of both reactions are such thermodynamically stable combinations. TEICHMANN and HILGETAG investigated PEARSON's concept in relation to the nucleophilic reactivity of the thiophosphoryl group, also for the de-alkylation of trialkyl phosphates in the PISTSCHIMUKA and MICHALSKI reactions [735, 647, 648, 650].
HEATH [386] found that the thiol esters resulting from these rearrangements are able, as strong alkylating agents, to further alkylate the nitrogen or sulfur atom in the side chain:

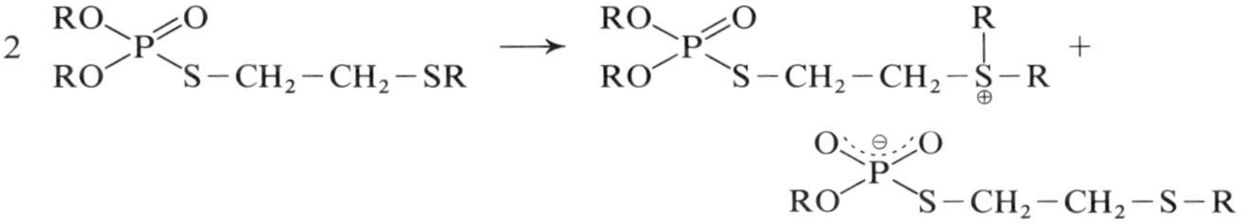
(31)

The result of all these alkylation reactions is a conversion of thiono-esters into thiol esters. This is of practical importance because their susceptibility to hydrolysis is thereby increased and in physiologically active series is accompanied by an increased toxicity to mammals and insects. Consequently, for the technical synthesis and formulation of the active substance, water and nucleophilic molecules must be avoided, especially on long storage and, for the same reason, mild conditions during manufacture should be chosen.

c) Phosphorylating Properties

1. Ester Halides

The reactivity of the ester halides follows, *mutatis mutandis*, the rules for ester hydrolysis. In phosphoric and thiophosphoric acid chlorides the chlorine atoms are not, to any notable extent, capable of participating in a π bond, because their spatial arrangement does not favour overlapping with the $3d$ orbitals, for the bond distances are greater than with oxygen atoms. The positive charge on the phosphorus is thereby stabilized and nucleophilic attack facilitated. With increasing substitution by oxygen atoms, this effect is increasingly diminished, so that the chlorine atoms can be successively replaced by hydroxy compounds. In many cases the third chlorine atom is bound so tightly that it is relatively stable against aqueous alkali. In view of what has been said so far, this lowered activity is to be expected above all with the thiono compounds as with amides.
The first chlorine atom in phosphoryl chloride can be exchanged with alcohol or even with ammonium chloride, whereas for the second chlorine atom, alcoholate or two molecules of amine are required.
In the case of the phosphonic and phosphinic acid ester chlorides a series is obtained, analogous to that for alkaline ester hydrolysis, with increasing capacity to acylate nucleophilic substances.

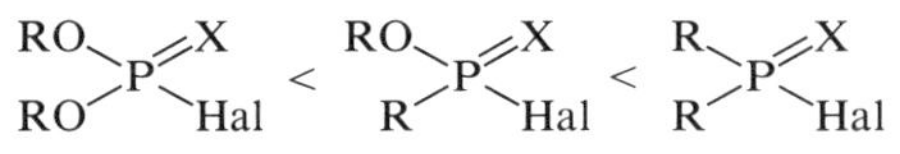

According to MILLER [653] the deciding factor for the exchange of a halogen atom in the ester chlorides is not, as in carbon chemistry, the nucleophilicity, but rather the basicity of the attacking molecule. Between oxygen and sulfur anions, no other difference in activity exists towards O,O-diphenyl phosphorochloridate than that attributable to the basicity measured as acidity of the conjugated acids. (Here C–O cleavage can practically be excluded.)
DRAGO, MODE, KAY and LYDY [250] investigated the rate of exchange of chloride in compounds of the type (I):

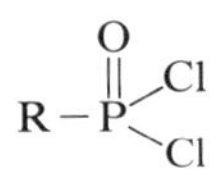

(I)

In bimolecular exchange reactions they found the following order for the rate-diminishing activity of the radical R:

$$C_6H_5 < CH_3 < OCH_3 \sim Cl < OC_6H_5$$

In general, ester fluorides are more difficult to attack hydrolytically than the analogous chlorides. This is readily explained by the ability of the fluorine atom via π bonds to reduce the positive charge on the phosphorus atom. However, interesting effects are obtained with the amido-fluorides. N-monosubstituted amides (II) hydrolyze a thousand times faster than the corresponding disubstituted amides (III) [385]:

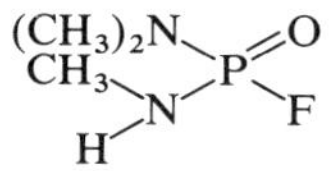
(II)

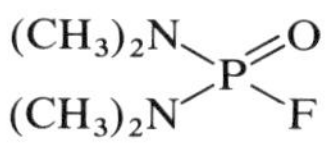
(III)

WESTHEIMER [1010] therefore postulated the withdrawal of a proton in (II) to give the reactive intermediate (IV) which rapidly reacts further to the final products.

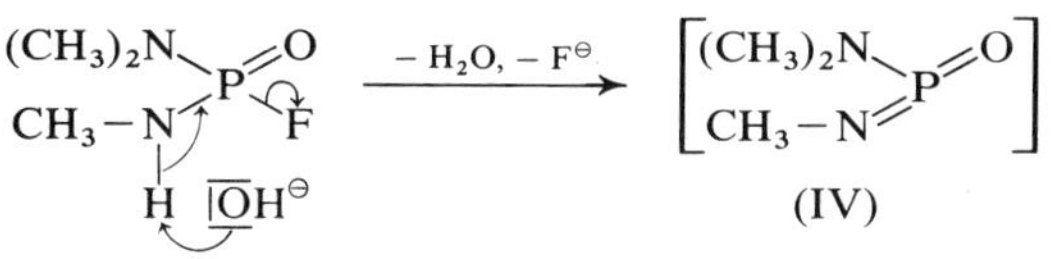

(1)

During acidic hydrolysis, cleavage of the fluoride ion is favoured in comparison to chloride, because as with the amides, protonation on fluorine facilitates P–F cleavage:

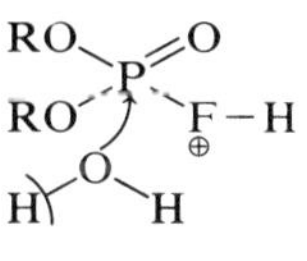

As is the case with esters, the nucleophilic replacement of halogen can also be catalyzed by many bases such as pyridine. A well-known effect is the catalytic action of substituted carboxylic acid amides in phosphorylation reactions in which Vilsmeier-analogue complexes may be involved (Method of CRAMER and WINTER) [222]:

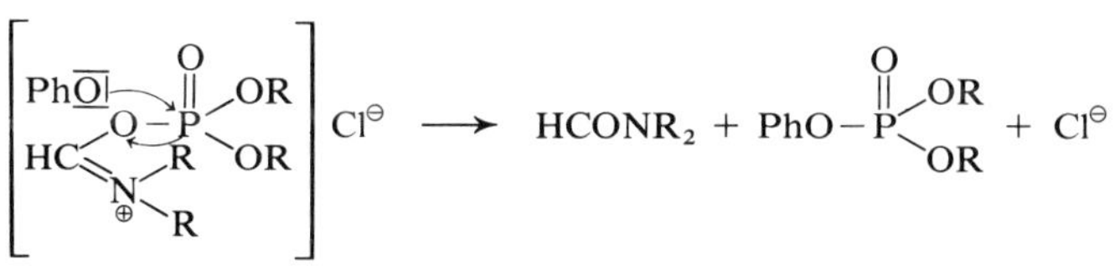

(2)

For a detailed discussion of the hydrolysis of phosphoric acid ester derivatives, reference must be made to the literature, in particular to the reviews by LOSHADKIN and SMIRNOV [578], by COX, Jr. and RAMSAY [216], as well as a paper by HUDSON [429] which views phosphorylation in the light of a simplified interpretation of nucleophilic reactivity. HUDSON links the rate constant with the solvation energy and the electron affinity of the nucleophiles, he also relates the rate constant to the energy of the bond formed between the nucleophilic and electrophilic center.
SCHRADER [863] provides a wealth of experimental data on special trade products and their analogues.

2. *Triesters*

SCHRADER'S empirical rule [863], according to which biologically active phosphoric acid esters must possess an "acyl" moiety, was the first indication that the chemical mechanism of insecticidal action might depend upon the phosphorylation of biologically important targets.
In 1937/8 [877] SCHRADER gave a specific formulation (V, VI), and in 1950 [837] a somewhat more generalized form (VII):

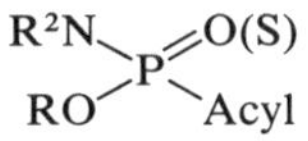

(V)

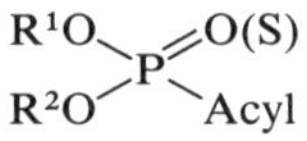

(VI)

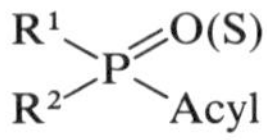

(VII)

In 1963 [863] he described this formula (VII) as follows:
"It is likely that a biologically active phosphoric acid ester will be obtained when the following prerequisites are satisfied: Either sulfur or oxygen must be directly bound to the pentavalent phosphorus, R^1 and R^2 may be alkoxy groups, alkyl groups or amines, while the "acyl" may be represented by the anions of organic or inorganic acids such as fluorine, cyanate, and thiocyanate, or of other acidic compounds (enolates, mercaptides etc.)".
If "acyl" stands for a phosphoryloxy group, then this formula readily leads to the pyrophosphoric acid esters. SCHRADER'S "acyl", however, has a different meaning from the present-day one:

e.g. 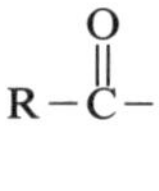

but comprises more generally the anions of a variety of H-acidic compounds.
Although a more schematic way of depicting the compound is to be preferred, e.g. the formula (VIII) used by LOHS [554], the concept "Acyl" is so well-known that its use can be considered established.
"Acyl" includes many groups: for example the fluoride ion, a second phosphoryloxy radical, aryloxy groups or heterocyclic oxo-compounds. The "acyl" condition, i.e. the simple model of Y as leaving groups,

 (3)

(VIII) Y = "Acyl"

does not suffice in all cases to explain insecticidal action. Examples of this type are DFP (*D*i-isopropyl *fluorop*hosphates) and *paraoxon*, or the insecticidal pyroesters TEPP and *schradan*. Thirty years were to pass before a scheme was suggested on the phosphorylating (not biocidal) properties of a esterphosphate molecule, supplementing SCHRADER's simple working hypothesis, which has proved most fruitful in the synthesis of new compounds. CLARK, HUTCHINSON, KIRBY and WARREN [190] described phosphorylating compounds as P−XYZ systems in which the electrons of the P−X bond can be accepted by Z. X, Y and Z are usually H, C, N, O, S or halogen. The phosphorylating potential is also enhanced when the P−X bond is naturally weak, i.e. when there is no $p_\pi - d_\pi$ overlapping of the lone electron pairs (N, O, S) with the phosphorus atom. (Contribution of the X atom to a π bond – see p. 24). The Z group should therefore be as electronegative as possible or become electronegative through the influence of electrophilic agents (e.g. protons) or oxydizing substances. A third possibility of reducing basicity of the X atom consists in the introduction of a sp^2-hybridized Y atom

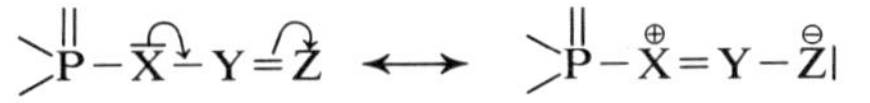 (4)

Examples of insecticides which phosphorylate by this system are the enol esters, providing R^2 and R^3 are suitable for the withdrawal of electrons from the Y=Z double bond. An important member of this group is DDVP:

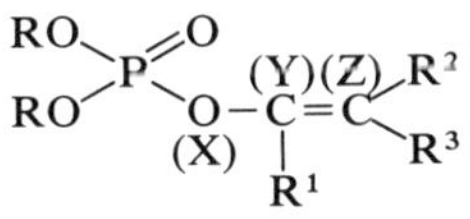

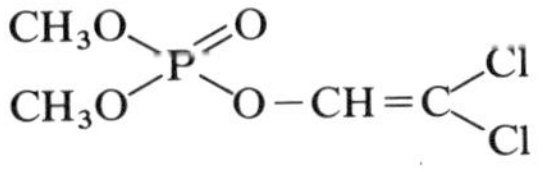 DDVP

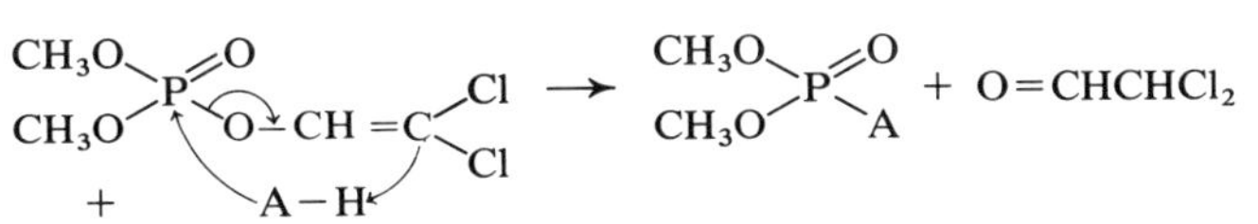 (5)

If one examines the phosphorylation of acetylcholinesterases in which the imidazole ring of histidine is involved (see p. 170f.), then the phosphorylating property of N-phosphoryl imidazole in acidic medium proves of interest:

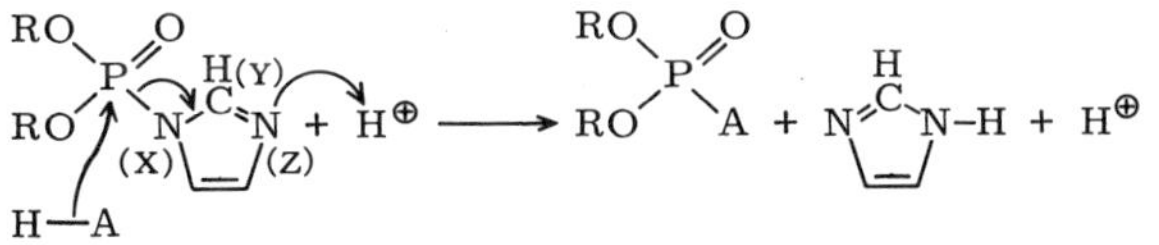 (6)

Single bonds between Y and Z may lead to a fragmentation of the molecule:

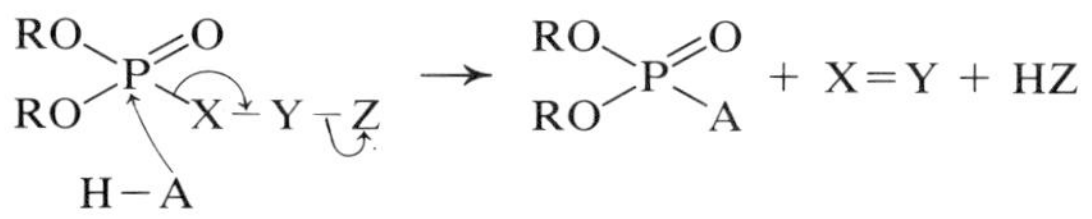

(7)

Examples of phosphorylating agents fitting into this scheme are the *azinphos* analogues:

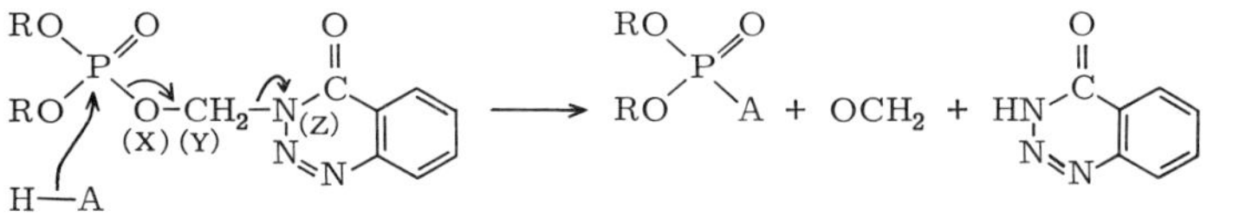

(8)

If the X, Y, Z grouping leaves as a resonance-stabilized anion, phosphorylation of H–A is described by Eq. (9).

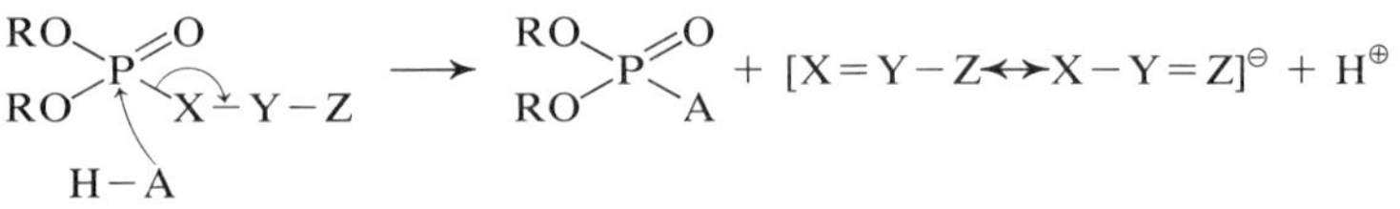

(9)

The most important examples are provided by the parathion group and other phenol esters:

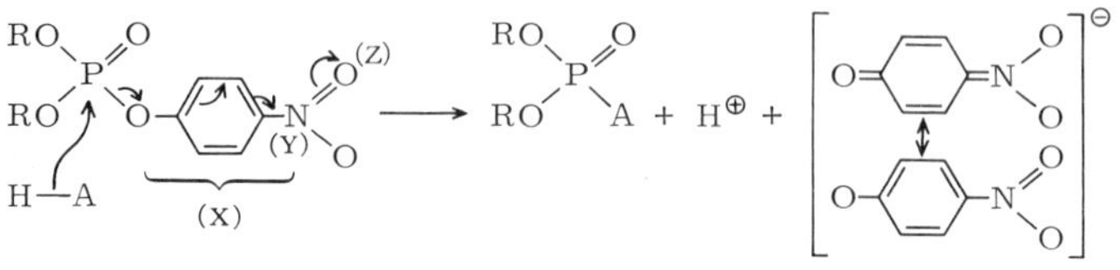

(10)

According to the P–XYZ scheme, the ®Dipterex-DDVP rearrangement (see p. 72) may be formulated as an internal phosphorylation, the second step represents a phosphorylation of the enzyme site by the DDVP scheme (Eq. (5)):

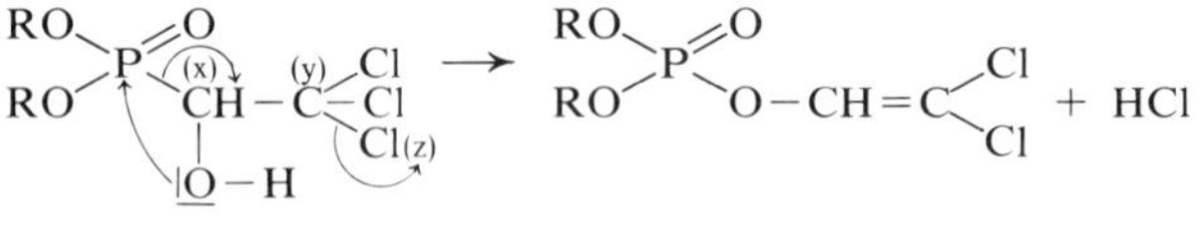

(11)

There is little doubt that the Schrader rule in the form of the (chemical) modification of Clark *et al.* is extremely useful when applied to the synthesis of new compounds and offers more possibilities than the suggestions regarding Hammett

σ values (see p. 187). Here pKa values in a clearly defined range are required for the starting compounds, a rule which experience has often proven to be invalid: for instance, if one leaves the class of phenol phosphates since here protonation, for example, is not taken into account as a possible step of "lethal synthesis" at the site of action (see p. 230).
On the other hand, CLARK's scheme provides very definite evidence that protonation of phosphorylating substances shortly before or at the site of action must play a substantial role in the development of the biological activity of a phosphorylating system, e.g. in the case of ®Bayrusil [817] and related compounds [814].

2.3. Nomenclature

A simple, logical and generally accepted nomenclature of the organophosphorus compounds does not exist. Tradition amongst individual teams plays a great role, as is apparent from a comparison of the Russian (ARBUSOV) and German Schools (MICHAELIS).
The type-I compound resulting from the Arbusov reaction

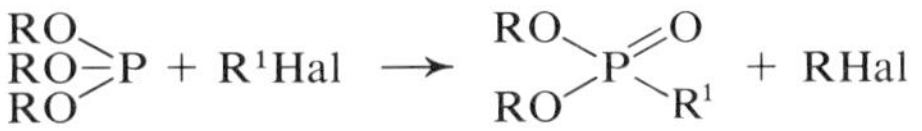

(1)

(I)

is designated phosphinic acid ester by some Russian authors. In Germany, on the other hand, the compound is described as phosphonic acid ester, and phosphinic acid derivatives are understood to be compounds with two P–C bonds:

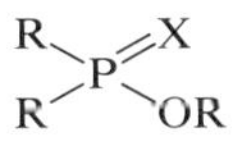

(II)

Such differences are particularly disturbing when making a study of the earlier literature. As a consequence many attempts have been made in recent years to devise a uniform nomenclature, with the result that the literature became even more difficult to survey, for no system managed without at least some arbitrary axioms.
The most consequential attempt was made in Scandinavia [536] where all organic derivatives of trivalent phosphorus are designated phosphines, and all derivatives of pentavalent phosphorus which contain oxygen as phosphinoxides. This attempt was stimulated by the gross inconsistencies between the American nomenclature based on Chemical Abstracts, KOSOLAPOFF [513] and VAN WAZER [1003] on the one hand and the British system on the other, suggested by the British Chemical Society [39]. For example, while a compound of the type

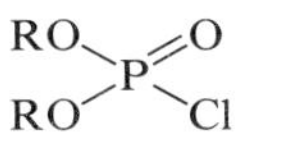

(III)

is regarded in the German and American literature as an acid chloride and is considered therefore as a phosphate, according to the British designation the chlorine atom is regarded as an alkyl substituent, as the name phosphonate for compound III illustrates.
In the American nomenclature, the concept phosphinic acid is applied to compounds of type (IV) which, according to the British designation, are called phosphonous acid.

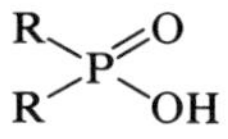
(IV) USA: phosphinic acid
Brit.: phosphonous acid

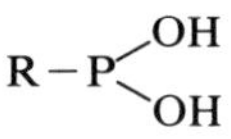
(V) USA: phosphonous acid
Brit.: phosphinic acid

With compounds of type V the situation is exactly the reverse.
The German nomenclature, laid down in Beilstein and later in Houben-Weyl [421], is based, like the American system, on the acids. An inconsistency between structure and German nomenclature is found in the compounds VI and VII.

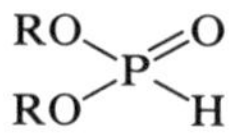
(VI)

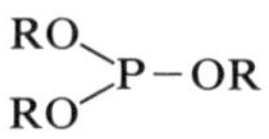
(VII)

While with the trialkyl ester (VII) it is quite clear that a derivative of phosphorous acid is involved which becomes manifest in the German designation "Trialkylphosphit", the name "Dialkylphosphit" for (VI) is derived from the long-used formula (VIII):

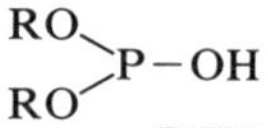
(VIII)

In numerous investigations into the structure of these dialkyl esters, the form (VI) was confirmed, so that compound VI may be classified as the basic member of the phosphonic acid dialkyl esters. The American nomenclature (C. A.) also follows this principle.
Since the number of types of compounds which are important in crop protection is relatively limited, only a few rules suffice. Therefore the halides, cyanides as pseudo-halides, amides, esters or thiol esters are classified according to the corresponding acids which they yield on hydrolysis.
In the American literature, acids with two P–C bonds receive the ending -inate in the pentavalent form, and the ending -inite in the trivalent form.
Acids with one P–C bond are correspondingly named phosphonates and phosphonites respectively.
Where English is used, the nomenclature proposed by IUPAC is now largely established. Its advantage consists in the intrinsic logic of the system.

The scientific names are formed in two steps:

1) By the prefix O-alkyl for each alkylated ester oxygen atom, e.g. O-methyl O-ethyl O-propyl phosphate.
2) If other atoms or functional groups than alkoxy are involved, the names are formed by insertion of the corresponding terms into the key-word, e.g. phosphoro-ate:

Key-word	Phosphoro | ate	Phosphoro | ate
Functional groups other than alkoxy	thio amid(e) fluorid(e) chlorid(e) etc.	thio amido chloro fluoro

Two or more groups of the same type are designated by the prefixes di, tri, and so on:

Phosphoro | thioate — di, tri

Phosphoro | ate — diamido, dithio

In a similar manner the names for free acids are based, for instance, on the key-word phosphoro-ic acid:

Phosphoro | ic acid — thio, amid(e), etc.

Some other key-words are for example:

$(RO)_2PO-R$	phosphon(o)-ate	$R-PO{<}^{OH}_{OR}$	phosphono-ic acid
$RO-PO(R)_2$	phosphin(o)-ate	$(R)_2PO-OH$	phosphino-ic acid
$(RO)_3P$	phosph(oro)-ite	$(RO)_2P-OH$	phosphoro-ous acid
$R-P(OR)_2$	phosphon(o)-ite	$R-P{<}^{OH}_{OR}$	phosphono-ous acid

Therefore, the main problem of nomenclature rather consists of naming the more complex substituents at the ester oxygen (i.e. the phosphorylated compounds) (cf. p. 48 f.).

A principal disadvantage of the IUPAC nomenclature, however, is more of a philological nature, for in languages other than English it is often impossible to form words with analogous syllables. We fear, therefore, that the IUPAC system may not bring about the international unity hoped for in the phosphorus nomenclature.

In the following tables a comparison of the American, British and Scandinavian nomenclature is given.

Table 3. Examples of the nomenclature of phosphoric acid derivatives*

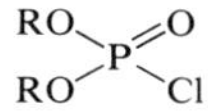	Scand.	Dialkoxy-chlor-phosphinoxid
	Amer.	Dialkyl-phosphorochloridate
	Brit.	Dialkyl chlorophosphonate
	Beilstein	Phosphorsäure-dialkylester-chlorid
	Houben-Weyl	Phosphorsäure-dialkylester-chlorid
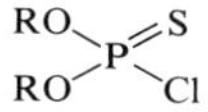	Scand.	Dialkoxy-chlor-phosphinsulfid
	Amer.	Dialkyl phosphorochloridothioate
	Brit.	Dialkyl chlorothiophosphonate
	Beilstein	Thiophosphorsäuredialkylester-chlorid
	Houben-Weyl	Thiophosphorsäure-O,O-dialkylester-chlorid
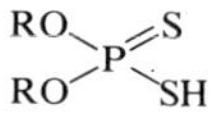	Scand.	Dialkoxy-mercapto-phosphinsulfid
	Amer.	O,O-Dialkyl hydrogen phosphorodithioate
	IUPAC	O,O-Dialkyl phosphoro dithioic acid
	Brit.	–
	Beilstein	Dithiophosphorsäure-O,O-dialkylester
	Houben-Weyl	Dithiophosphorsäure-O,O-dialkylester
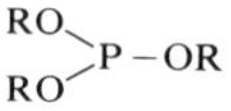	Scand.	Trialkoxy-phosphin and Trialkylphosphit
	Amer.	Trialkyl phosphite
	Brit.	Trialkyl phosphite
	Beilstein	Trialkylphosphit
	Houben-Weyl	Phosphorigsäure-trialkylester and Trialkyl-phosphit
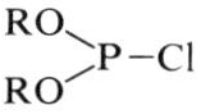	Scand.	Dialkoxy-chloro-phosphin
	Amer.	Dialkyl phosphorochloridite
	Brit.	Dialkoxychlorophosphine
	Beilstein	Dialkylphosphorigsäure-chlorid
	Houben-Weyl	Phosphorigsäure-dialkylester-chlorid
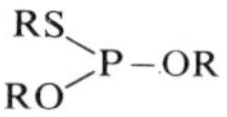	Scand.	Dialkoxy-alkylmercapto-phosphin
	Amer.	O,O,S-Trialkyl phosphorothioite
	Brit.	Trialkylthiophosphite
	Beilstein	Thiophosphorigsäure-O,O,S-trialkylester
	Houben-Weyl	Thiophosphorigsäure-O,O,S-trialkylester
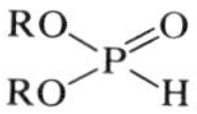	Scand.	Dialkoxy-phosphinoxid
	Amer.	Dialkyl phosphonate
	Brit.	Dialkyl phosphonate
	Beilstein	Dialkylphosphit
	Houben-Weyl	Phosphorigsäure-dialkylester
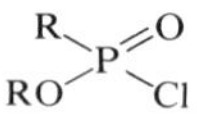	Scand.	Alkoxy-alkyl-chloro-phosphinoxid
	Amer.	Alkyl alkylphosphonochloridate
	Brit.	Alkyl alkoxychlorophosphine oxide
	Beilstein	Alkylphosphonsäure-alkylester-chlorid
	Houben-Weyl	Alkanphosphonsäure-alkylester-chlorid
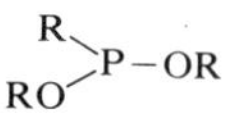	Scand.	Dialkoxy-alkyl-phosphin
	Amer.	Dialkyl alkylphosphonite
	Brit.	Dialkyl alkylphosphinate
	Beilstein	Alkyl phosphinigsäure-dialkylester
	Houben-Weyl	Alkanphosphonigsäure-dialkylester

Table 3 (Continued)

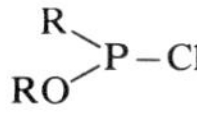	Scand.	Alkoxy-alkyl-chlor-phosphin
	Amer.	Alkyl alkylphosphonochloridite
	Brit.	–
	Beilstein	Alkylphosphinigsäure-alkylester-chlorid
	Houben-Weyl	Alkanphosphonigsäure-alkylester-chlorid
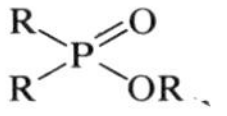	Scand.	Alkoxy-dialkyl-phosphinoxid
	Amer.	Alkyl dialkylphosphinate
	Brit.	Alkyl dialkylphosphonite
	Beilstein	Dialkylphosphinigsäure-alkylester
	Houben-Weyl	Dialkyl-phosphinsäure-alkylester
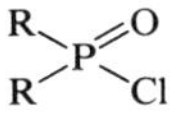	Scand.	Dialkyl-chlor-phosphinoxid
	Amer.	Dialkyl phosphinochloridate
	Brit.	Dialkylchlorophosphine oxide
	Beilstein	Dialkylphosphinigsäure-chlorid
	Houben-Weyl	Dialkylphosphinsäurechlorid
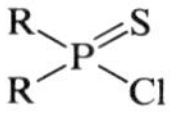	Scand.	Dialkyl-chlor-phosphinsulfid
	Amer.	Dialkyl phosphinochloridothioate
	Brit.	Dialkylchlorophosphine sulfid
	Beilstein	Dialkylthiophosphinigsäurechlorid
	Houben-Weyl	Dialkylthiophosphinsäurechlorid
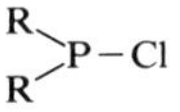	Scand.	Dialkyl-chlor-phosphin
	Amer.	Dialkylphosphinochloridite
	Brit.	–
	Beilstein	Dialkylphosphin-chlorid
	Houben-Weyl	Dialkylphosphinigsäurechlorid

* In all cases R may also be Aryl.

Several possibilities are open for the nomenclature of dimeric phosphoric acid derivatives. The formation of the name with "anhydride" is generally applicable. In the special case of a combination of acids of the same type, the designation "pyro" may be applied. In order to show that both acids are bound together via a sulfur atom, the term "anhydride" may be replaced by the term "anhydro-sulfide".

Examples:

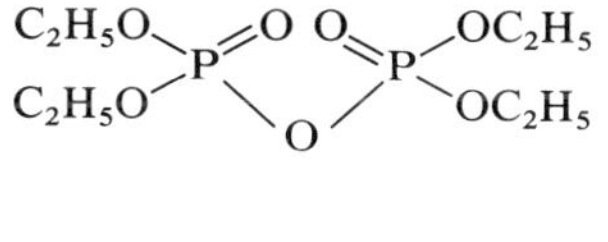

(IX)

Possible: (O,O-Diethyl phosphoric acid) (O,O-diethyl phosphoric acid) anhydride
Usual: Tetraethyl pyrophosphate

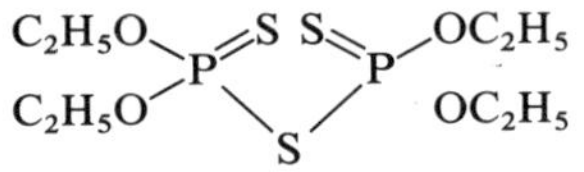

(X)

Possible: (O,O-Diethyl phosphorothionic acid) (O,O-diethyl phosphorothionic acid) anhydrosulfide
Possible: O,O,O′,O′-Tetraethyl pyrophosphorotrithioate

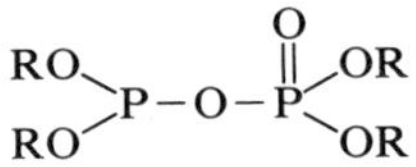

(XI)

Possible: (O,O-Dialkyl phosphorous acid) (O,O-dialkyl phosphoric acid) anhydride

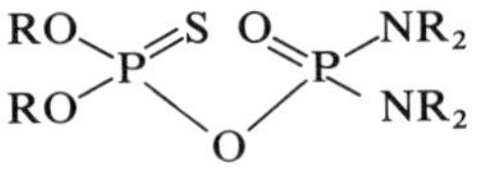

(XII)

Possible: (O,O-Dialkyl phosphorothioic acid) (N,N,N′,N′-tetraalkyl phosphoroamidic acid) anhydride

As the examples show, a sulfur atom may be bound to the phosphorus by either a single or double bond. Although the designations, e.g. "O,O,S-trialkyl phosphorothioate, O,O,O-trialkyl phosphorothioate and O,O,S-trialkyl phosphorodithioate, appear quite clear, it has become common practice to designate P=S compounds as "thiono" and P−SR compounds as "thiolo" derivatives.
In many cases if phosphorus is considered as a central atom the use of the nomenclature already discussed leads to very unwieldy names as the following examples illustrate (XIII, XIV, XV). Such names are better derived by regarding the organic molecule as substituted with a "phosphoryl, phosphonyl or phosphinyl" group:

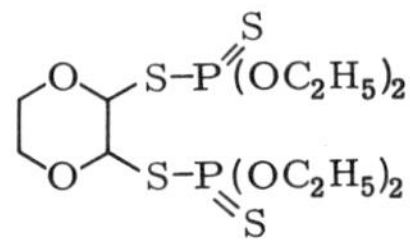

(XIII) *dioxathion*

JSC: 1,4-dioxan-2,3-ylidene bis(O,O-diethyl phosphorothiolothionate)
Manufacturer: 2,3-*p*-dioxanedithiol-S,S-bis(O,O-diethyl phosphorodithionate)
"Phosphoryl": 2,3-bis-(O,O-diethylthionophosphorylthio)-1,4-dioxane

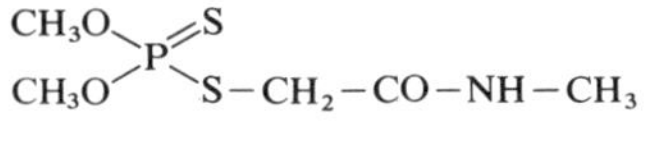

dimethoate

JSC: O,O-dimethyl S-(methylcarbamoylmethyl) phosphorothiolothionate
Manufacturer: O,O-dimethyl S-(N-methyl-carbamoylmethyl) phosphorodithioate
"Phosphoryl": (O,O-dimethyl thionophosphorylthio) N-methyl acetamide

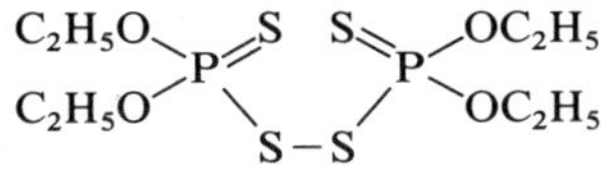

(XV)

Bis-(O,O diethyl-thiono-phosphoryl) disulfide

As a consequence of this situation we have, on the one hand, VAN WAZER's philosophy [1002]:

> "– what's in a name? That which we call a rose by any other name would smell as sweet".... (Romeo and Juliet II,2)

and on the other hand, the recommendation that structural formulae always be used when discussing problems.

3. Chemical Section

3.1. General

The large-scale synthesis of the phosphoric acid esters is a relatively young branch of industry, the development of which is clearly associated with their increasing significance as insecticides in crop protection, beginning after the Second World War. It is difficult to form an exact picture of the production figures because in general there is a delay of 5–10 years before these figures are made known by industry. Thus the last comprehensive compilation of American production was made in 1958 [1004]. According to this report, only 4% of phosphorus production was used for the manufacture of organophosphorus derivatives. However, this still means a total of over 45,000 tons of chemical products, one third of which were insecticides.

While the total production of organic insecticides of the earlier chemical types fluctuated, in 1951 to 1961, between 140,000 and 180,000 tons per year, the proportion of phosphorus insecticides (see Fig. 12) rose from 3,200 tons in 1951 (estimated) to 11,500 tons (1957), 17,300 tons (1960), 22,200 tons (1961), 25,500 tons (1962), 33,600 tons (1963), 37,000 tons (1964), 43,300 tons (1965) to 54,500 tons in 1966. These figures refer solely to American production and to active constituents.

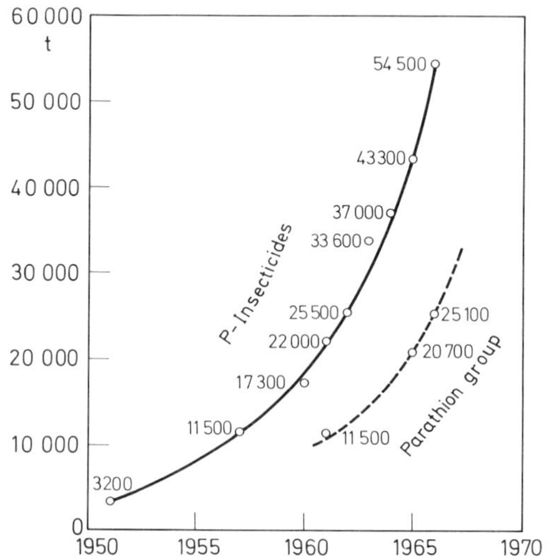

Fig. 12. US production of insecticidal phosphoric acid esters

The production of *parathion-methyl* and *parathion* in the USA in 1965 was about 20,700 tons and, in 1966, 25,100 tons [298].

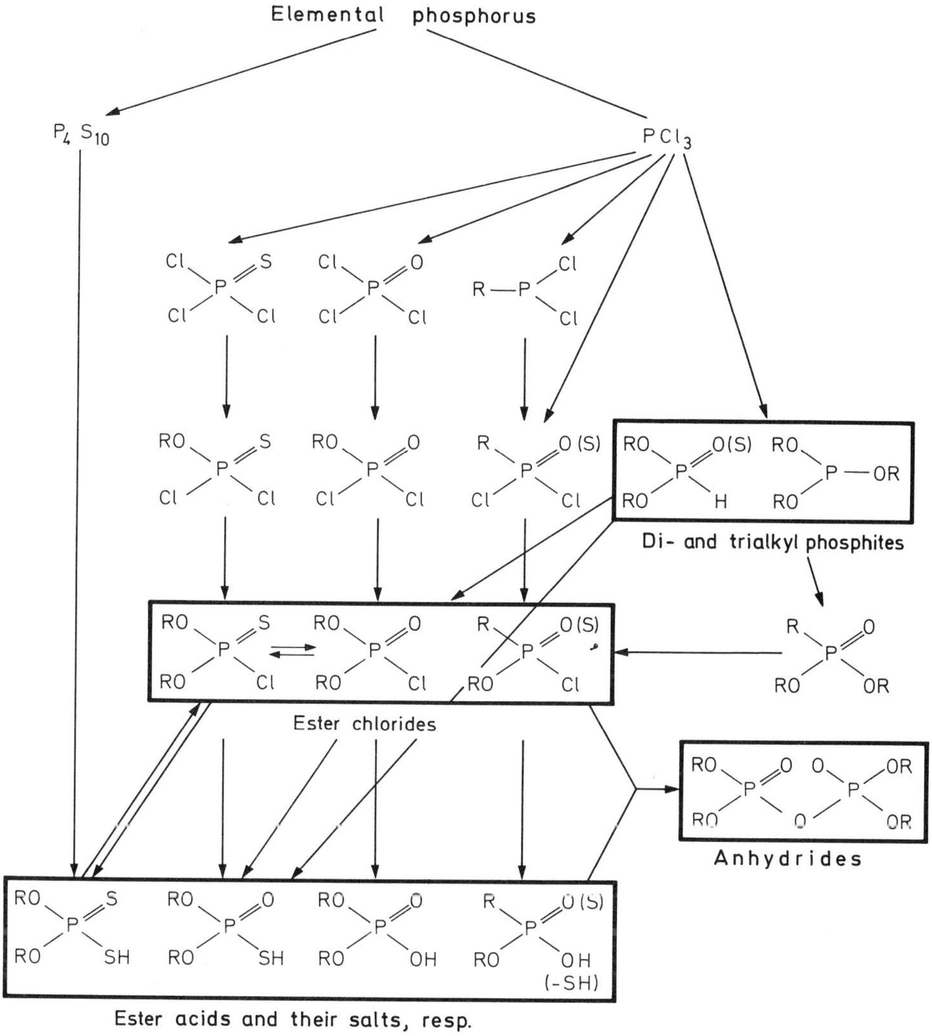

Fig. 13. Schematic synthesis of the important phosphorus intermediates for insecticide manufacture

The industrial synthesis of the more important phosphorus intermediates will be discussed in connection with the scheme of manufacture illustrated in Fig. 13. Elemental phosphorus is obtained by reduction of phosphates with carbon in an electrical furnace with the addition of silica. The overall reaction proceeds as follows:

$$Ca_3(PO_4)_2 + 3\,SiO_2 + 5\,C \xrightarrow[\text{[– ca. 730 kcal]}]{1300-1400\,°C} 3\,CaSiO_3 + {}^1/_2\,P_4 + 5\,CO \qquad (1)$$

The elemental phosphorus is further purified, if necessary, by distillation [1004]. For the manufacture of one ton of phosphorus about 6–10 tons of raw phosphate, 1–3 tons sand, 1.3–1.4 tons coke and 12,500–16,000 kwh. are required. As a rule raw phosphate and energy requirements are each responsible for 30–35% of the production costs. It was not until 1953 that elemental phosphorus was first manufactured in the Federal Republic of Germany, the firm Knapsack being the sole producer. In 1965, 70,000 tons were made. In 1968 the first furnace of a new plant with a yearly capacity of 30,000 tons was started. Accordingly, Knapsack's contribution to world production would be about 10% [492].
For example, in a new plant with a capacity of 2,000 tons per year for the manufacture of P_4S_{10} (vapour: P_2S_5), Knapsack AG employ a process in which the appropriate quantities of phosphorus and sulfur are reacted together at 350 °C in liquid form under a protective gas [223].
The second starting material, phosphorus trichloride, can also be obtained from the elements in an exothermic reaction when gaseous chlorine is passed over molten phosphorus. The resulting PCl_3 distills into the receivers and is redistilled with white phosphorus to remove the phosphorus pentachloride.
While phosphoryl chloride is relatively easy to obtain by various processes, for example by the controlled hydrolysis of PCl_5, by oxidation of PCl_3 or by reaction of phosphorus pentachloride with phosphorus pentoxide, the synthesis of the analogous sulfur compound $PSCl_3$ was, at first, far more difficult. Originally thiophosphoryl chloride was manufactured by heating PCl_3 with sulfur in a sealed tube at 130 °C. It was not possible to apply this method on a large scale because the required autoclave size was impractical, and corrosion by PCl_3 and resulting hydrochloric acid made economic production impossible. It was not until 1939 that a patent appeared [826] claiming a process in which PCl_3 in vapour form at 140 °C is passed over liquid sulfur of low viscosity. This permitted the continuous production of $PSCl_3$, which was later further improved by the use of catalysts such as charcoal [454], zinc chloride etc.

a) Phosphoric and Phosphonic Acid Ester Chlorides

Phosphoric and phosphonic acid ester chlorides can be prepared by relatively simple fundamental reactions. The phosphinic acid chlorides are not considered here because their manufacture involves a Grignard reaction which is difficult on an industrial scale (cf. [886]).
Phosphoryl chloride is used as starting material for the preparation of the phosphoric acid ester chlorides [669, 1018]. Uniform dialkyl ester chlorides are produced on reaction with alcohols, provided the resulting hydrogen chloride is removed from the reaction mixture. In the first stage the dichlorides are obtained (II).

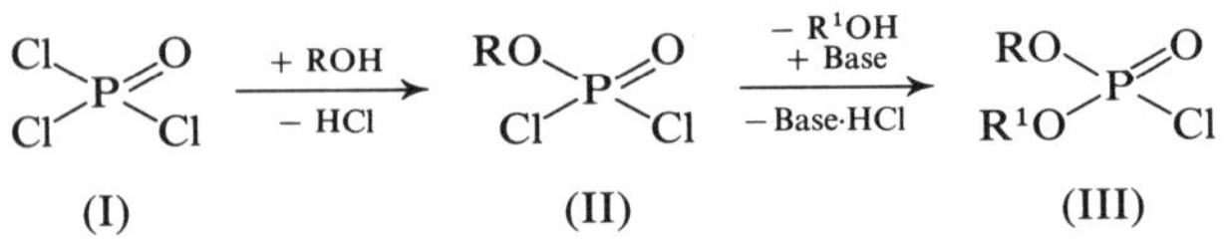

(2)

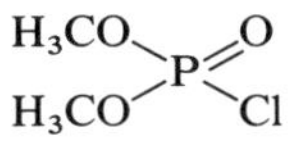

(IV)

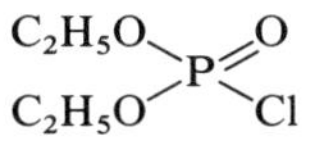

(V)

Under suitable conditions they can be isolated, so that at this point of the synthesis, mixed ester chlorides (II → III) can also be achieved [991]. The quantitatively most important of the ester chlorides are dimethyl and diethyl phosphorochloridate (IV) and (V) respectively.
With halogens, trialkyl phosphites (VI) undergo a type of Michaelis-Arbusov reaction [614] and also yield dialkyl ester chlorides:

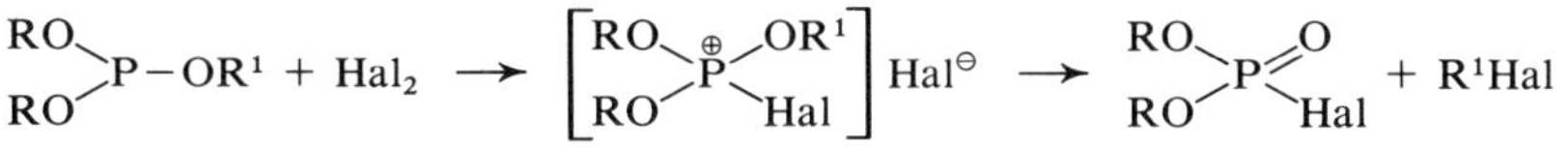 (3)

(VI) (III)

Dialkyl phosphites (VII), more conveniently prepared as trialkyl phosphites, react directly with chlorine to give dialkyl phosphorochloridates [370, 447, 614]:

 (4)

(VII) (III)

Diethyl ester chlorides can be obtained in one process from phosphorus trichloride, alcohol and subsequent reaction with sulfuryl chloride [285, 823]. An elegant synthesis is that of STEINBERG [921]:

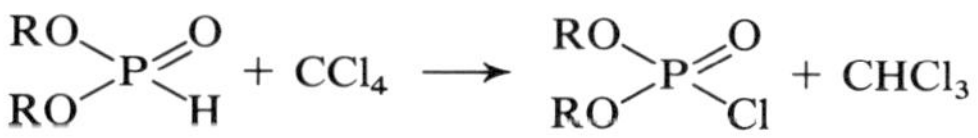 (5)

(VII) (III)

The reaction is carried out in the presence of tertiary amines. If primary or secondary amine is present, then dialkyl phosphoroamidates result. In the presence of phenolate, dialkyl phenyl phosphates are produced (see p. 105).
The reactions illustrated can also be applied to the thio-analogues. Thus thiophosphoryl chloride can be reacted in a corresponding manner with alcohol and alcoholate, respectively [288, 621]. As by-product, the corresponding tri-ester is always obtained in relatively large quantities.

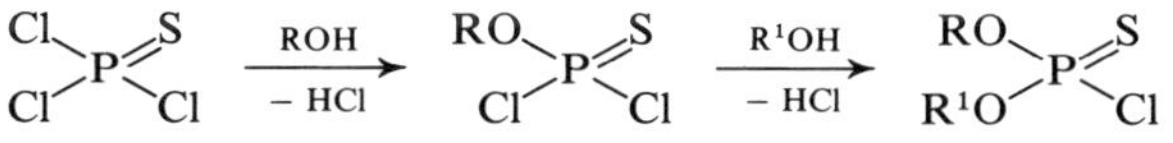 (6)

(VIII) (IX) (X)

The thiophosphoric acid chlorides are difficult to prepare by thionation of the corresponding oxygen compound. Thionation of the trivalent phosphorus compounds is to be preferred. Appropriate sulfur carriers are P_2S_5, elemental sulfur and $PSCl_3$ [51, 261, 343, 929, 955].
Groenweghe and Payne [355] have published results of systematic attempts with reorganization reactions between $X_3P{=}O$ and $X_3P{=}S$ compounds. They found that the reaction temperatures and times required correspond to the reaction conditions needed for halogen or alkoxy group exchange. One must always expect such reactions as side reactions when $P{=}O$ and $P{=}S$ compounds are present at the same time in a reaction mixture and when temperatures of about 150 °C are reached. The formation of $POCl_3$ from $PSCl_3$ is favoured. In some cases it is possible to use catalysts such as PCl_3 or $AlCl_3$.
Another route to the thionochlorides starts with P_2S_5 via the dialkyl phosphorodithioic acids (XI) which react with chlorine following Eq. (7) [289, 389, 678]: By this means and under carefully controlled conditions dialkyl thioesters are obtained which are very pure and contain very little trialkyl ester which, according to the method of Eq. (6), always appears as an impurity. The sulfur chlorides are hydrolyzed, and the resulting sulfur kept in solution by the addition of sulfite. Purification is achieved by vacuum distillation [389, 390].

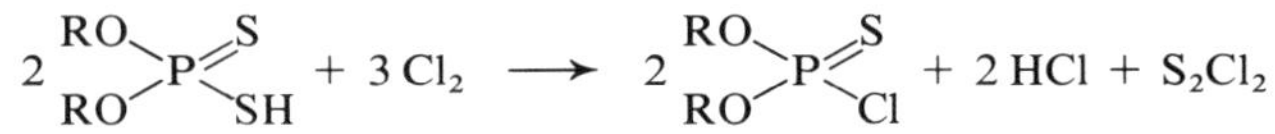
(7)

(XI) (X)

If PCl_5 is used as the chlorinating agent [965], thiophosphoryl chloride is obtained as a by-product, which can be further utilized by the method shown in Eq. (6).

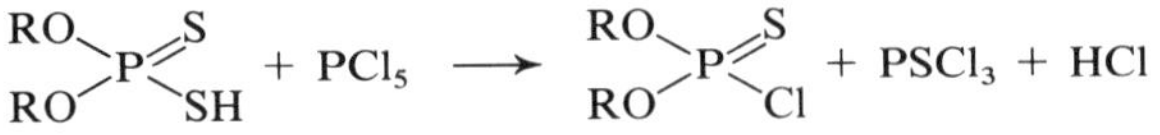
(8)

The chlorination of dialkyl phosphorothioites is also

(9)

suitable for the synthesis of the dialkyl phosphorothiochloridates. Completely triester-free ester chlorides can be obtained using the Steinberg reaction [571]:

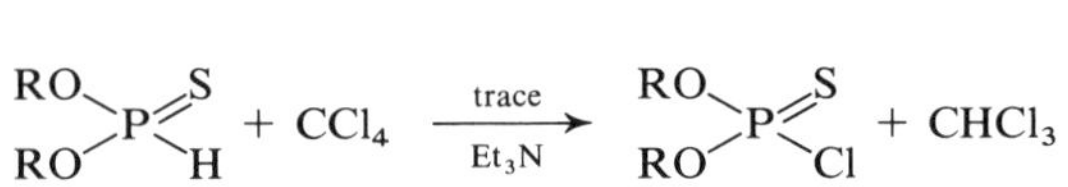
(10)

In addition to the ester chlorides, the amide chlorides, especially tetramethyl phosphorodiamidochloridate (XII), and the thiono-analogue (XIII) should be mentioned. $POCl_3$ is also used for their preparation, being reacted with secondary amines or their hydrochlorides:

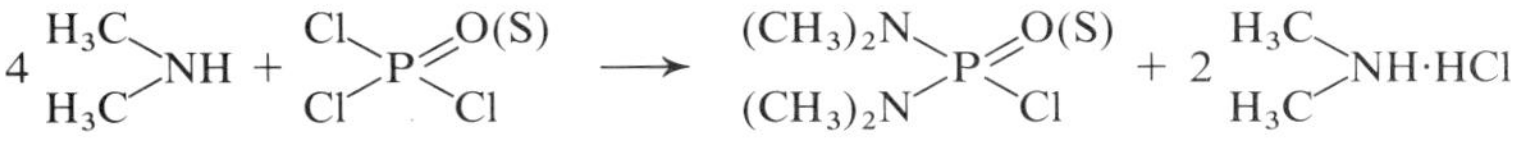

(XII) and (XIII) (11)

In order to obtain higher yields, the reaction is carried out in two stages [231, 417, 639, 644, 645, 324, 840, 998]:

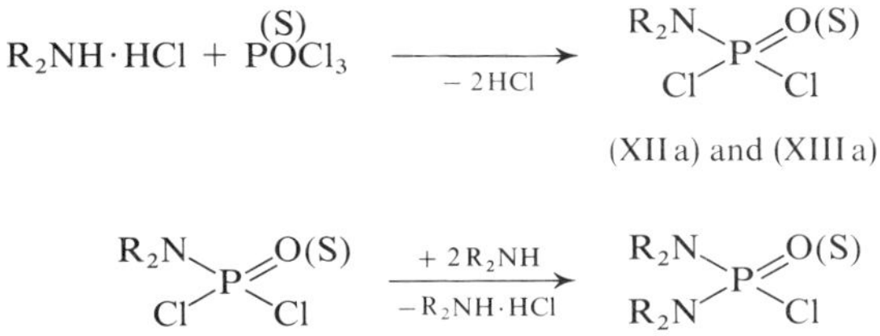

(XII a) and (XIII a) (12)

(XII) and (XIII)

The reaction with thiophosphoryl chloride proceeds analogously to (XIII a) and (XIII) [461, 793].
In addition to the phosphoric acid halides as intermediates for phosphoric acid esters and anhydrides, the phosphonic acid halides are becoming of increasing importance. Since here a direct phosphorus-carbon bond must first be formed, the syntheses are somewhat different. Nearly all methods procced via the phosphonic acid di-halides to the mono-halides, usually with trivalent phosphorus compounds as starting material [751].
A special method for the preparation of methanephosphonodichloridate is the pyrolysis of dimethyl phosphite [76, 211] (Eq. (13)). It proceeds via cleavage of dimethyl ether to methanepyrophosphonic acid (XIV), which can be isolated. Chlorination with thionyl chloride provides the dichloride (XV).

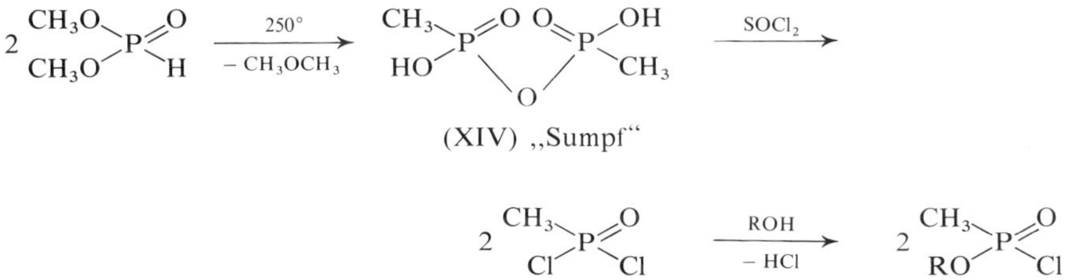

(XIV) „Sumpf"

(XV) (XVI) (13)

The synthesis preferred for large-scale manufacture originates from KINNEAR and PERREN [485, 486] (Eq. (14)). Phosphorus trichloride is reacted with alkyl or aryl halides and aluminium trichloride. The intermediate is alkyl or aryl phosphorus-tetrahalide, which can be processed in a variety of ways [191] (see p. 57).

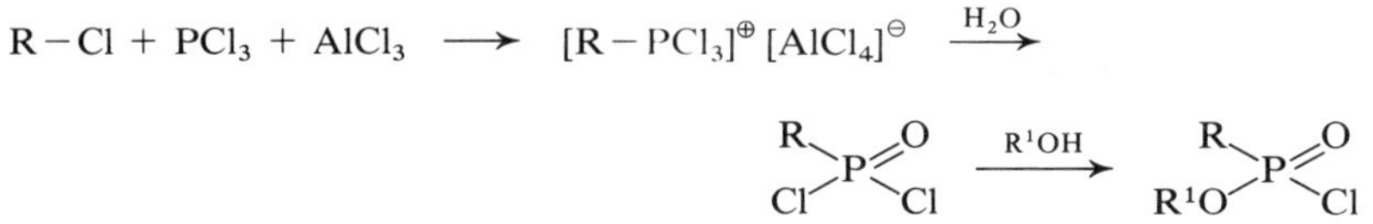

(14)

The Graf reaction is also worth mentioning [349] (Eq. (15)). Aliphatic and cyclo-aliphatic hydrocarbons are reacted with phosphorus trichloride and elemental oxygen:

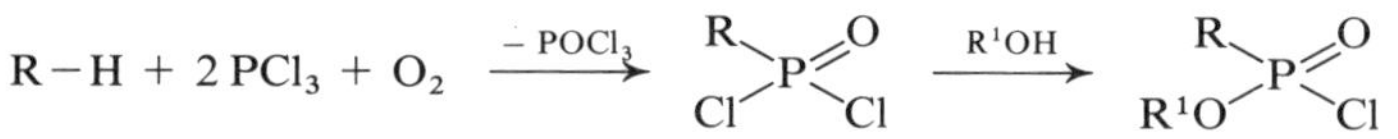

(15)

The phosphonates, which can be obtained by the Michaelis-Arbusov reaction (a) or by the Michaelis-Becker reaction (b), can be converted with phosgene to phosphonic acid ester chlorides [136, 198, 200, 235, 778, 822]:

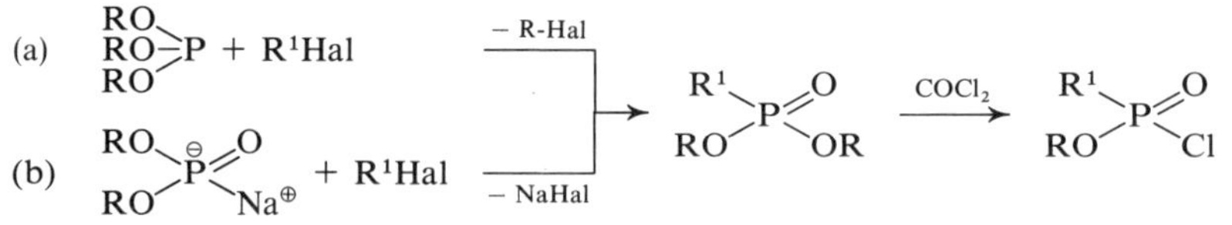

(16)

According to WARREN [1000], the last step in Eq. (16) starts with an electrophilic attack of the carbonyl C on the phosphoryl O, followed by nucleophilic replacement of the oxygen atom by a chloride ion.

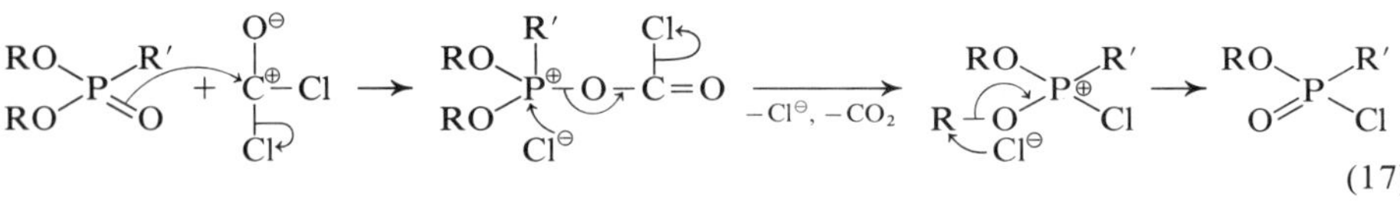

(17)

The phosphonic acid dichlorides synthesized by the method of KINNEAR and PERREN [485, 486] can be thionated to the thionophosphonic acid halides.
The most important compounds are obtained by the following syntheses:

1) Benzenephosphonodichlorothioate results from benzene and phosphorus trichloride in the presence of aluminium chloride and subsequent thionation of benzene phosphonodichloridite with $PSCl_3$ or elemental sulfur [139, 346, 448]:

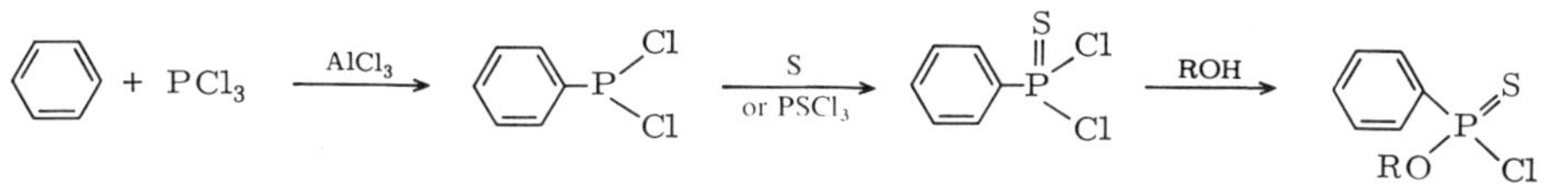

(18)

(Thionation can also be carried out with P_2S_5 [464]). Benzenephosphonodichloridite has been produced by passing phosphorus trichloride and benzene through glowing tubes [422].

2) For alkyl derivatives, variations of the Kinnear-Perren synthesis are known (Eq. (19a)) [470]:

$$R-Cl + PCl_3 + AlCl_3 \longrightarrow [R-PCl_3]^{\oplus}[AlCl_4]^{\ominus} \qquad (19\,a)$$

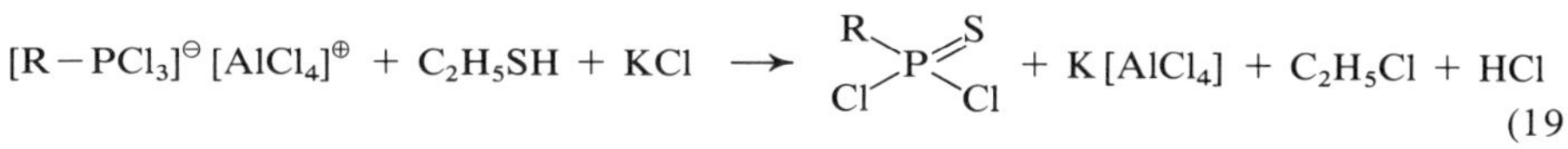

(19b)

If the complex is cleaved with potassium chloride and alkyl mercaptan (Eq. (19b)), then the phosphonodichlorothioate is likewise obtained and can be converted in the usual manner with alcoholate into the monochloride. A technically feasible process for the manufacture of the lower alkanephosphonodichlorothioate was developed by SCHLIEBS *et al.* [807, 808]. The advantage of this method is that normal pressures can be used, no solvent is required and a portion of the aluminium trichloride can be reclaimed. This is achieved by replacing the solid complex $(R-PCl_3)^+ (AlCl_4)^-$ by the liquid complex $(R-PCl_3)^+ (Al_2Cl_7)^-$ [810, 811, 813].

The process proceeds as follows:

$$R-Cl + PCl_3 + 2\,AlCl_3 \longrightarrow [R-PCl_3]^{\oplus}[Al_2Cl_7]^{\ominus} \qquad (20\,a)$$

$$[R-PCl_3]^{\oplus}[Al_2Cl_7]^{\ominus} + {}^2/_3\,Al \longrightarrow R-PCl_2\cdot AlCl_3 + {}^5/_3\,AlCl_3 \qquad (20\,b)$$

(The aluminium chloride which is not complex-bound can be reclaimed with suitable solvents such as methylene chloride).

(ref. [504]) (20c)

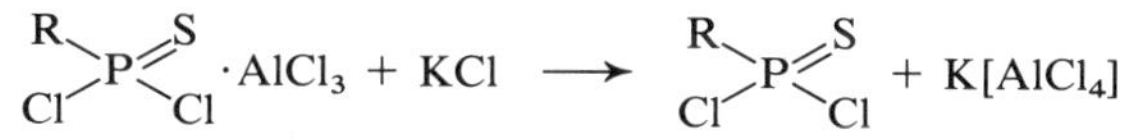

(20 d)

In stage (b) of the reaction aluminium can be replaced by phosphorus for the reduction of the complex [505]:

$$3\,[R-PCl_3]^{\oplus}[AlCl_4]^{\ominus} + 2\,P + 3\,KCl \longrightarrow 3\,R-PCl_2 + 2\,PCl_3 + 3\,K[AlCl_4] \qquad (20\,e)$$

An elegant laboratory process is illustrated by the following equations:

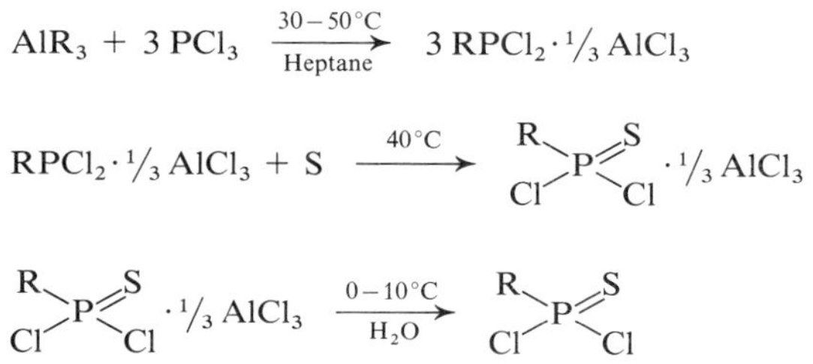

(21)

Trialkyl aluminium likewise reacts with phosphorus trichloride to an alkane-phosphonodichloridite-aluminium chloride complex, which can be thionated and then hydrolytically decomposed [626].
Tetra ethyl lead behaves in a similar manner [480]:

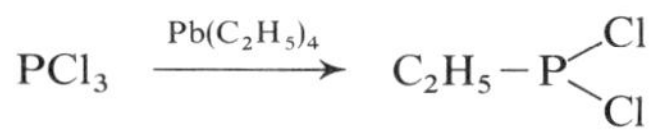

(22)

It has the advantage of not forming a complex such as trialkyl aluminium during the reaction.
Pentavalent phosphorus is also a useful starting material for a widely applicable synthesis of the phosphonyl chlorides. MARSH and GARDNER [598, 599] used the addition of PCl_5 to camphene; BERGMANN and BONDI [87, 88, 89] recognized the general principle of this reaction, in which phosphorus pentachloride is added mainly on the α-carbon atom to 1-olefines without substituents:

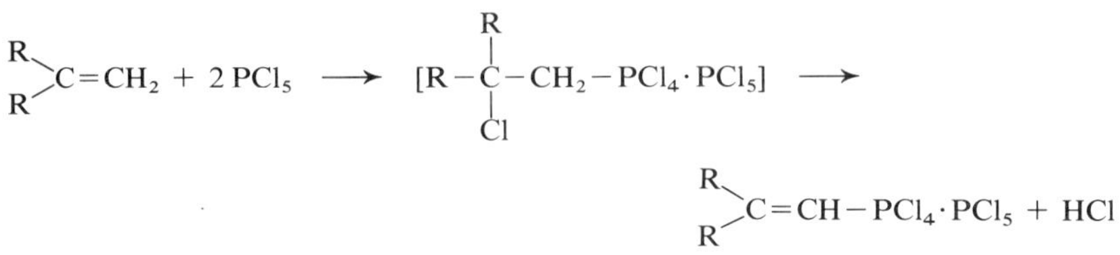

(23)

For example, this reaction offers the only route to "Phosphono DDVP" (XVII):

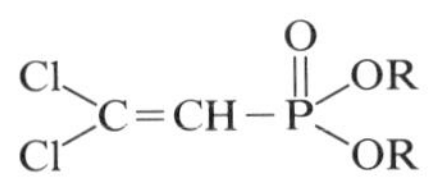

(XVII)

In the case of one β-substituent, the addition of phosphorus occurs in the β-position:

$$R-CH=CH_2 + 2\,PCl_5 \longrightarrow R-\underset{\displaystyle PCl_4\cdot PCl_5}{\underset{|}{CH}}-CH_2-Cl \qquad (24)$$

With an excess of olefine the resulting PCl_5 complexes are cleaved at high temperatures, if necessary in a solvent [423].

If the ester chlorides described in Section 3.1.a. are now reacted with appropriate molecules containing hydroxyl or mercapto groups in the presence of proton acceptors, the corresponding esters as possible insecticides are obtained in a normal acylation reaction [626, 939]:

(25)

R = Alkyl
R^1 = Alkyl, Aryl, Alkoxy
X = Oxygen or Sulfur

R^1 and RO may also represent NH_2-, monoalkyl or dialkylamine. The significance of the function R^2 will be discussed in Section 4.2 (p. 187).

b) Phosphoric and Phosphonic Ester Acids

The synthesis shown on page 51 illustrates two possible ways of obtaining dialkyl phosphoric acids. One route starts with P_4S_{10} and leads by way of reaction with appropriate quantities of alcohol under defined reaction conditions to O,O-dialkyl phosphorodithioic acids [185, 590, 591].

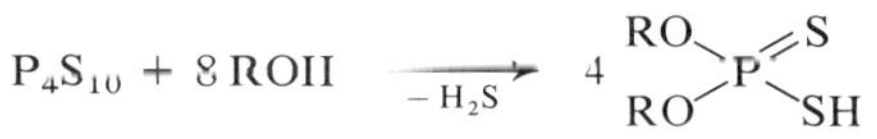

(26)

The compounds manufactured on the largest scale by this reaction are O,O-dimethyl and O,O-diethyl phosphorodithioic acid. They are, in some cases, alkylated directly to insecticidally active esters (see p. 113) and also converted to O,O-dimethyl and O,O-diethyl phosphorochlorothioate (see p. 54). (Large quantities of the homologues with long alkyl groups are used as flotation agents in the mining industry; amounts used being in the order of 10–1000 g per ton of ore) [29].

Analogously, mono- and dialkyl phosphoric acids are formed from phosphorus pentoxide with alcohols. By thionation of dialkyl phosphites, O,O-dialkyl phosphorothioic acids are obtained in almost quantitative yield [453, 791].

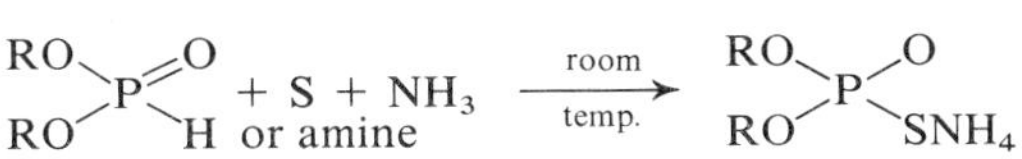

(27)

The second route by which variously substituted ester acids are also obtainable starts with thiophosphoryl chloride and phosphoryl chloride respectively, which then yield the corresponding ester chlorides.

(28)

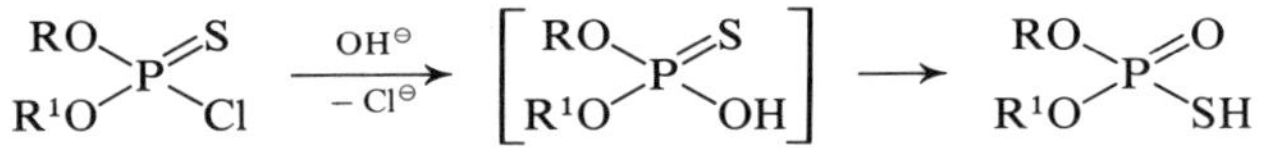

(29)

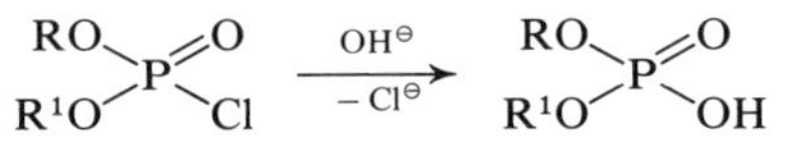

(30)

Phosphonic ester acids or their salts, which are industrially of secondary importance, are prepared by hydrolysis or by using potassium sulfide following the Scheme on p. 51.
In comparison to the thiophosphoric acids, the phosphonic acid series offers a chemical speciality (Eq. (31)). If phosphonodichlorothioates are treated with hydrogen sulfide, then the so-called "dithioanhydrides" (XVIII) are achieved [208, 809]:

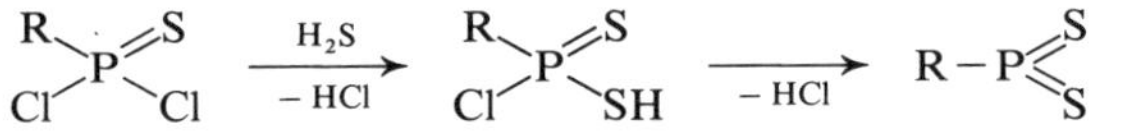

(XVIII)

(31)

Wheatley [1012] determined the structure of methanedithiophosphonic acid anhydride by X-ray analysis. It was found to be a dimeric compound with the following bond lengths and angles:

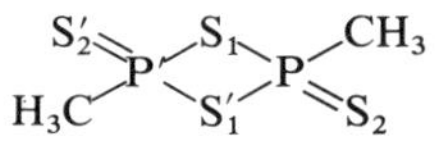

$P-S_1$: 2.144 Å (equiv. single bond)

$P-S_2$: 1.940 Å (equiv. double bond)  84.2° 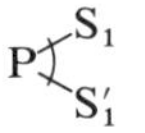 95.8°

$P-C$: 1.828 Å

The dithiophosphonic acid anhydrides react readily in a formal addition of salts of the formula MeX such as KF, NaN_3, KCN, and KSCN to dithiophosphonic acid salts [868]:

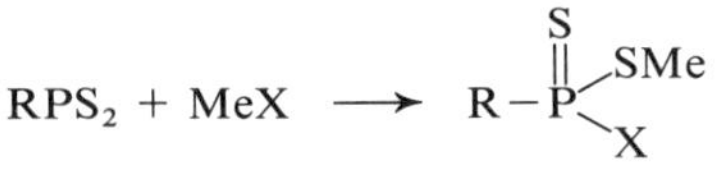

(32)

With alcohols, O-alkyl alkanephosphonodithioic acids are formed [871]:

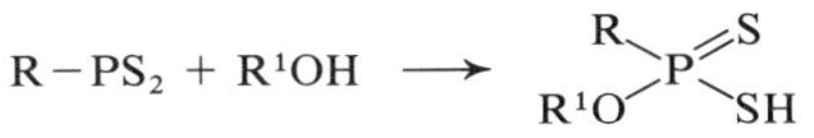
(33)

If secondary alcohols are used, compounds of the same type are obtained, they cannot, however, be achieved with acid chlorides [209]:

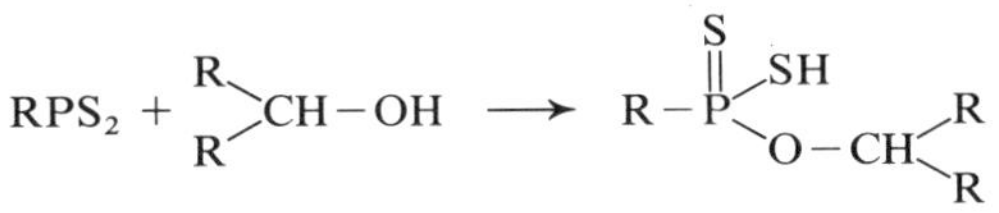
(34)

Dithioester acids also result by the addition of amino-alcohols and they exist in the form of their inner salts [210]:

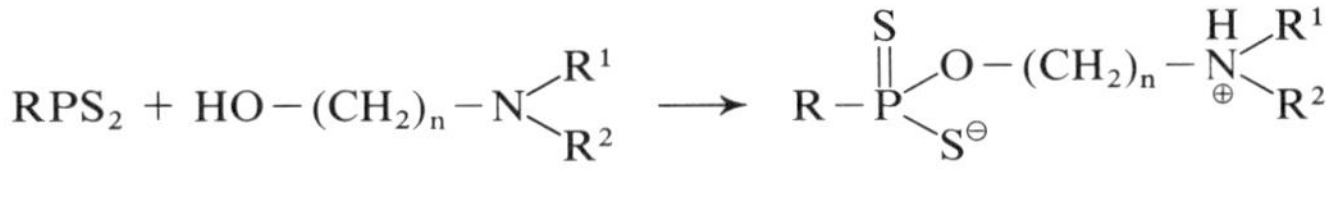
(35)

When the dithioanhydride is reacted with water [486], esters of the thiol acids are obtained via the oxysulfide and reaction with alcohol:

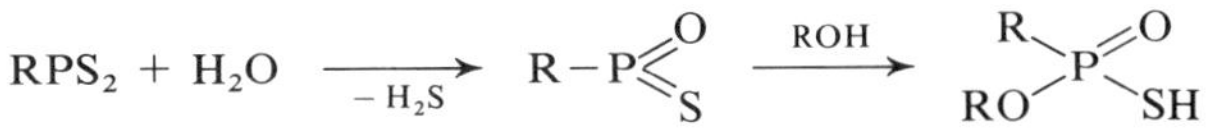
(36)

The addition of mercaptans to give trithio acids was described in 1959 [858]:

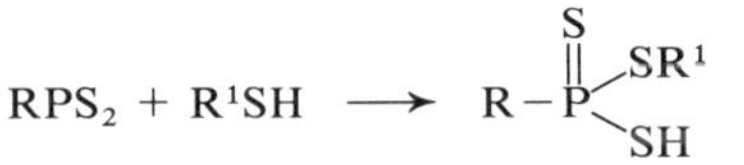
(37)

The dithio- and trithiophosphonic acids so obtained can in some cases be further alkylated to insecticidally active esters [868]. A new type of synthesis for halides of dithiophosphonates [291, 857] is the addition of alkyl halides to dithioanhydride:

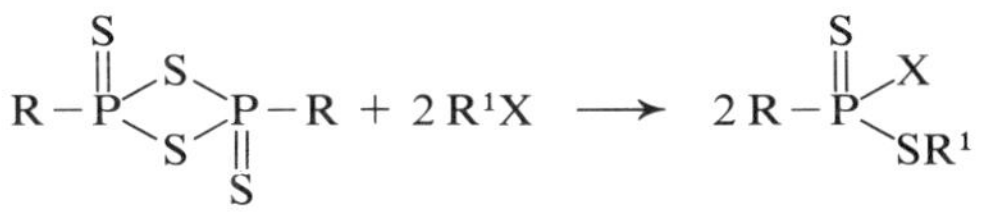
(38)

The acidity of phosphoric ester acids, as well as the structure of thiophosphoric acids (thiono-thiolo-equilibrium) has been investigated in particular by Russian authors in a Hammett treatment [462, 465, 524].

While JAFFÉ *et al.* [436] determined the ionization constants of benzenephosphonic acids (XIX),

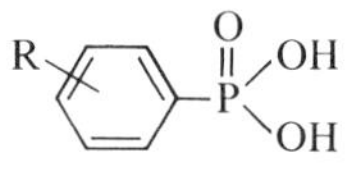

(XIX)

KABACHNIK *et al.* went a step further and assumed that the readily polarizable P=O and P=S groups, like the benzene ring, were able to confer substituent effects on the reaction center, i.e. OH or SH groups of acids of the following formulae [462]:

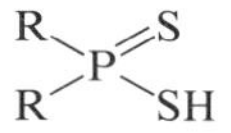
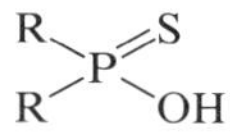
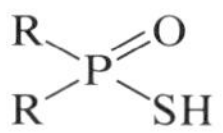
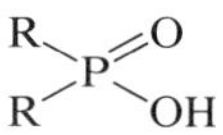

where R may represent H, OH, alkyl, aryl, alkoxy, or aryloxy groups. The reference acid was hypophosphoric acid, ϱ was chosen as the unit for the first ionization stage in aqueous solution. The calculation was made by the following formula:

$$pK = pK^0 - \varrho \cdot \Sigma\sigma \qquad (39)$$

where $\Sigma\sigma$ stands for the sum of the σ constants of the substituents R.
The σ values deviate somewhat from those found for the benzene ring [616]; however, about a hundred different phosphoric acids admirably fitted the linear Eq. (39). By a stepwise approach, characteristic ϱ, σ and pK^0 values were determined for each substituent or for each series. The measurements were made in 7% and 80% alcohol.
For the experimental details and their evaluation, we would refer to the original work. The acidities of the phosphoric ester acids conform in a qualitative sense to the following rules:

1) Dissociation in alcohol is less than in water.
2) Alkylation of phosphoric acids increases the acidity (pK values for the first dissociation step in 7% alcohol):

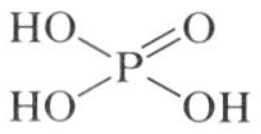

1.97

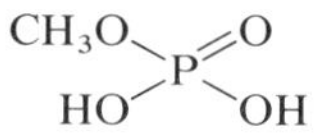

1.54

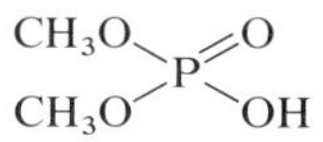

1.25

3) The acidity decreases with increasing size of the alkyl substituents:

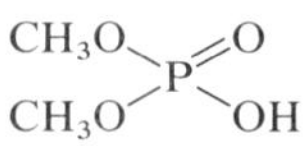

1.25

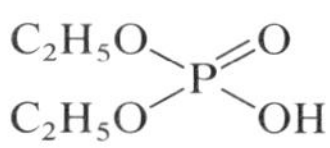

1.37

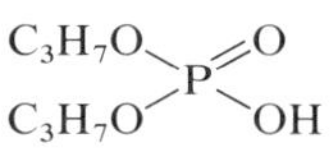

1.52

4) The acidity falls in the order phosphoric-phosphonic-phosphinic acids:

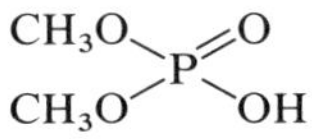

1.25

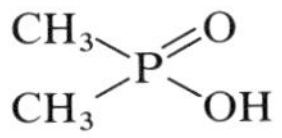

3.13

5) Phenyl compounds are strongly acidic:

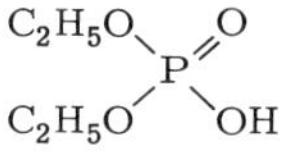

1.37

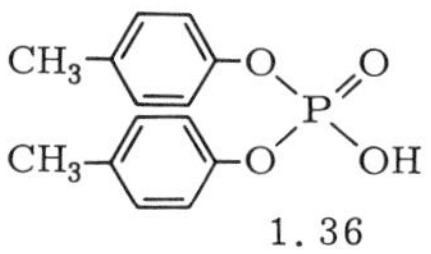

1.36

6) Increasing sulfur content decreases the acidity in water, but increases acidity in alcoholic solution:

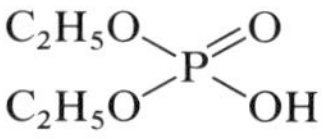

7% Alk.: 1.37
80% Alk.: 3.15

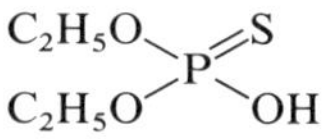

1.49
2.84

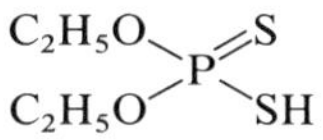

1.55
2.64

The state of the thiono-thiol equilibrium in thioester acids is given by the following rules:

7) The thiono-form becomes increasingly stable in the order phosphoric- phosphonic- phosphinic acid:

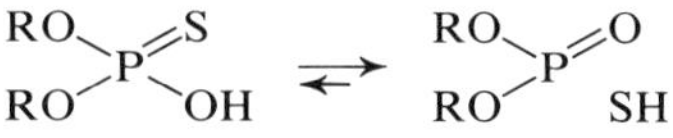

80%

Thiol form present in 7% alcohol:

5 − 20%

0 − 1%

8) When 80% alcohol is used as solvent, the thiono-form is stabilized. At least 98% of the thiophosphonic acid is present in the thiono-form.

These facts are qualitatively in agreement with the bonding conditions discussed on p. 22f.

Insecticidally active esters of phosphoric and phosphonic acids are obtained by alkylating the acids according to the following general scheme:

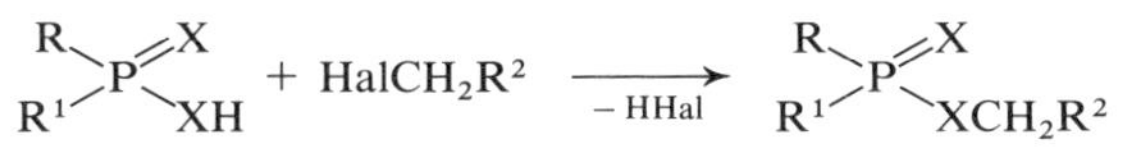

(40)

Examples of these alkylation reactions will be discussed in Chapter 3.2.c (p. 113).

c) Di- and Trialkylphosphites

As was shown on p. 51, a third group of phosphorylating agents of technical importance is formed by the reaction of phosphorus trichloride with alcohol.
Trialkyl phosphites can be obtained from stoichiometric quantities of phosphorus trichloride, alcohol and bases such as pyridine [331, 456, 655, 777]:

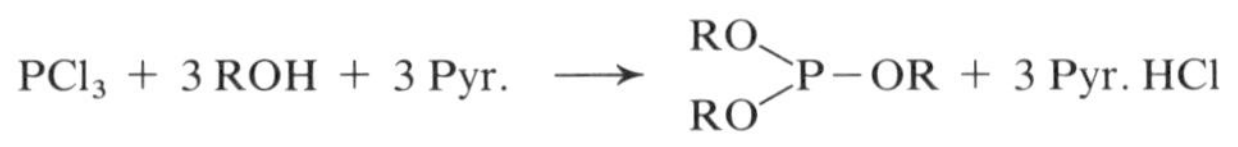

(41)

Separation of the resulting pyridine hydrochloride and the intense cooling required limit the reaction economically. These disadvantages are overcome in a process by Farbenfabriken Bayer AG [806], in which the phosphorous acid trisamide is reacted with any desired alcohol (Eq. (42)):

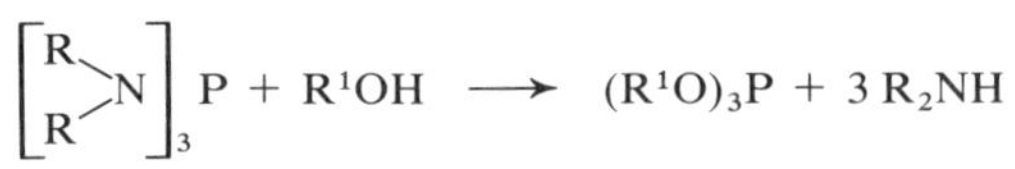

(42)

A condition, however, is that the secondary amine used must boil at a lower temperature than the corresponding alcohol and can, therefore, be continuously removed from the reaction equilibrium by distillation. Without the addition of base, the reaction takes another course than Eq. (41), yielding dialkyl phosphites [323, 592, 614, 684, 759, 1018, 1036]:

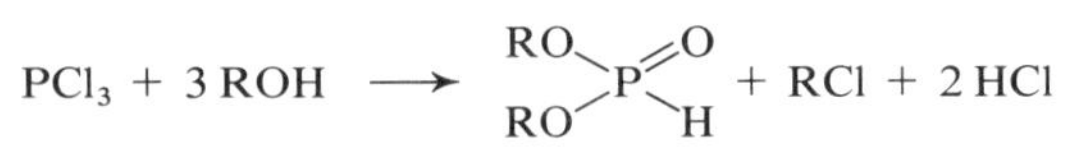

(43)

Therefore, only two-thirds of the alcohol used is recovered in the dialkyl phosphite. On the other hand the alkyl halide serves as coolant and diluent. In particular, dimethyl and diethyl phosphites are prepared by this process. Dipropyl phosphite and the higher homologues can be prepared by a method of Jonas and Thraum [455] in which 1 mol phosphorus trichloride is reacted with only 2 mol alcohol and 1 mol water.

There is also a transesterification process for the preparation of dialkyl phosphites [997] in which no undesirable impurities arise:

(44)

An indication of the mechanism involved in dialkyl phosphite formation in Eq. (43) is given by the reaction of trialkyl phosphites with hydrochloric acid to dialkyl phosphites and alkyl halides:

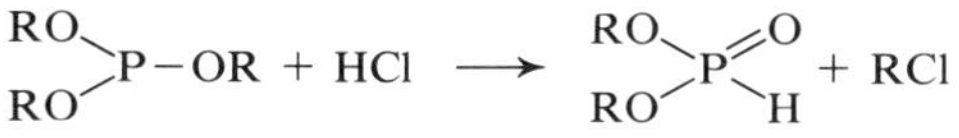

(45)

Reaction (45) can therefore be regarded as the simplest case of an Arbusov reaction (Eq. (46)).

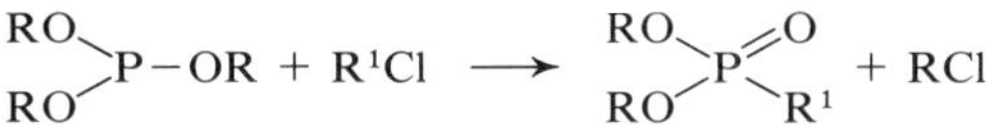

(46)

It is, however, too expensive for the industrial manufacture of dialkyl phosphites because, for the synthesis of the trialkyl phosphites, bases are also necessary. Reaction (45) proceeds by way of quasi-phosphonium compounds (XX) and (XXI) as intermediates which, on alkylation of the chloride ion, decompose to dialkyl phosphite and alkyl halide (Eq. (47)):

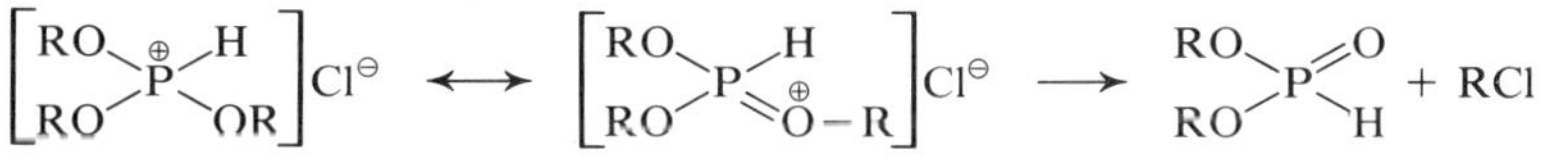

(47)

(XX) (XXI)

With sodium alcoholate or with sodium metal in inert solvents, the dialkyl phosphites are converted to their sodium salts which can be used in the Michaelis-Becker reaction.

Thionation of dialkyl phosphites with phosphorus pentasulfide [1038] yields O,O-dialkyl phosphorothioites (thiol phosphites).

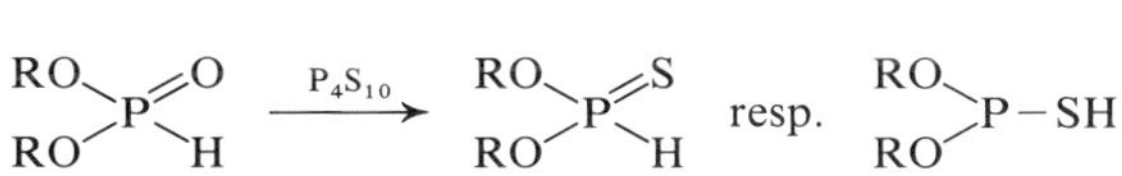

(48)

The same type of compound is accessible by way of O,O-dialkyl phosphorochloridites and hydrogen sulfide [515, 647] with the exception of the dimethyl compound for which only small yields are obtainable [648, 790].

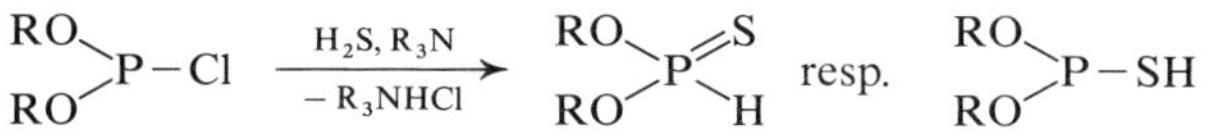

(49)

Dialkyl phosphites with different alkyl groups can be obtained by trans-esterification [207, 366, 895] but these products are not uniform. Stepwise reaction of phosphorus trichloride with alcohols gives satisfactory results only with secondary alcohols [594, 861, 514]. One clear synthesis is by way of mixed anhydrides [650] and their alcoholysis as follows:

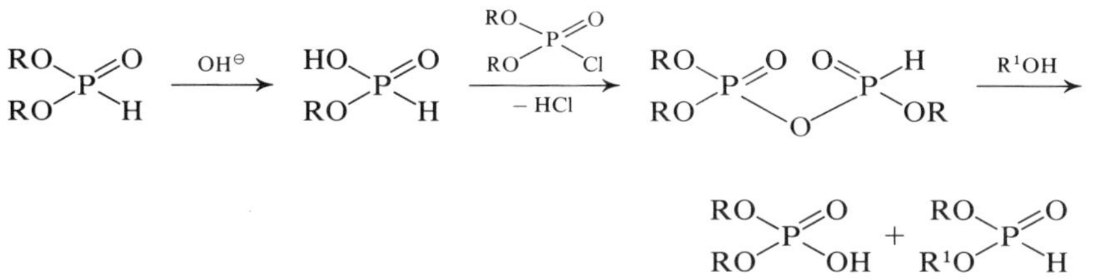

(50)

According to CÖLLN [205], phosphorochloridates can be replaced by alkyl chloroformates:

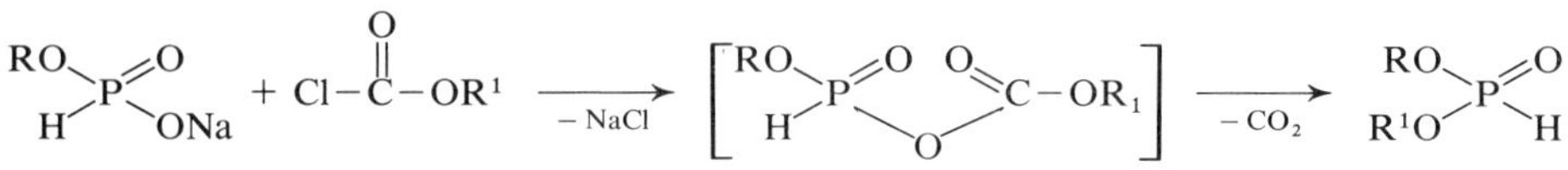

(51)

Michaelis-Arbusov and Perkow Reactions

Esters of trivalent phosphorus have the tendency to change into compounds of pentavalent phosphorus. In many reactions, including that of the dialkyl phosphites, this is formally equivalent to electrophilic substitution on the phosphorus. There are several papers on the mechanisms of such S_E reactions [1000, 488, 431].

One of the most important examples of the transformation of trivalent to pentavalent phosphorus compounds is the conversion of trialkyl phosphites to the thermodynamically more stable phosphonates; the reaction is accompanied by the formation of a phosphorus-carbon bond. This is known as the *Michaelis-Arbusov reaction* and is effected by the action of alkyl halides:

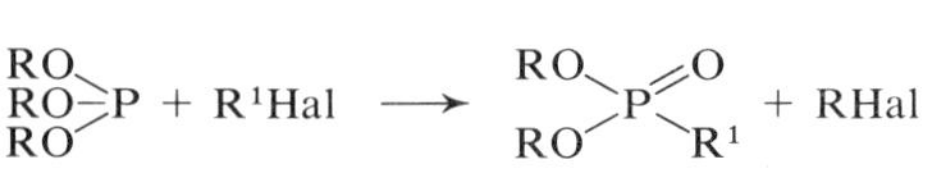

(52)

If the alkyl groups of phosphite and halide are identical, then isomerization of the phosphite to phosphonic ester occurs (intrinsic *Arbusov reaction*). In this case,

catalytic quantities of alkyl halide are sufficient. When the alkyl substituents are different, then the alkyl halide added to the reaction competes with that being released. The ratio of resulting products can be adjusted in the required direction by distillation or by an excess of alkyl halide [217].
Sodium salts of dialkyl phosphites react with alkyl halides to yield the same final products.

(53)

This synthesis is termed the *Michaelis-Becker-Nylén reaction* and may be regarded as an extension of the *Michaelis-Becker reaction*. ARBUSOV suggested two steps – an addition followed by removal of the alkyl halide:

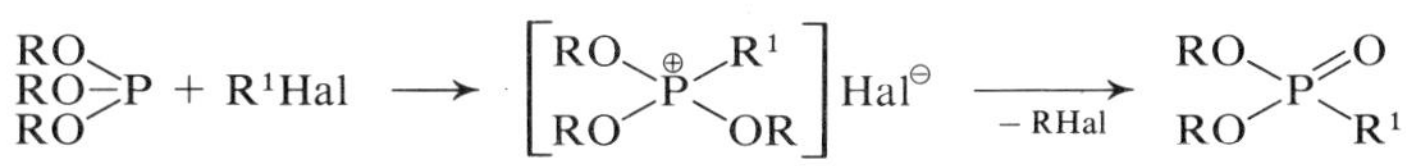

(54)

When R represents a phenyl group, the addition product can be isolated. This intermediate, and the fact that compounds of trivalent but not of pentavalent phosphorus readily form complexes with copper (I) halides, served ARBUSOV as evidence for the validity of this interpretation [685].
The reaction of the trialkyl phosphites with hydrogen chloride to give dialkyl phosphites (Eq. (55)) is easily understood if the role of alkyl halide can be attributed to the hydrochloric acid:

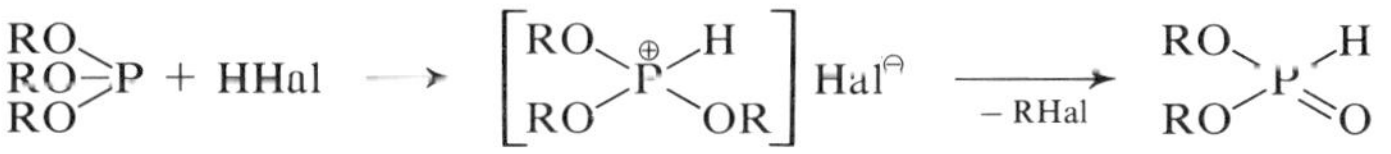

(55)

The Michaelis-Arbusov reaction is not limited to esters of phosphorous acid but can also be applied to esters of phosphonous and phosphinous acids, i.e. to all trivalent phosphorus compounds possessing at least one OR-group [53]:

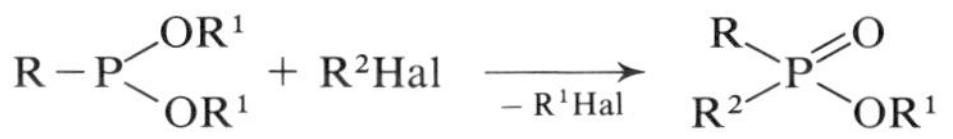

(56)

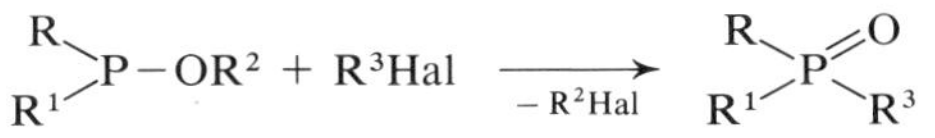

(57)

Like the alkyl halides, halogen-substituted acids, alcohols, acid anhydrides etc., yield Arbusov products:

(58)

Unsaturated or secondary alkyl halides can also take part in this reaction. The secondary compounds, however, tend to give an undesired side reaction, the formation of olefines, by analogy to Hofmann's olefine synthesis. The reactivity in the Michaelis-Arbusov reaction decreases in the order [53]:

Acyl > prim. Alkyl > sec. Alkyl

S-Alkyl dialkanephosphinothioites give a normal Michaelis-Arbusov reaction:

(59)

whereas the trithiol phosphites are transformed by another mechanism into S,S-dialkyl phosphorochloridites

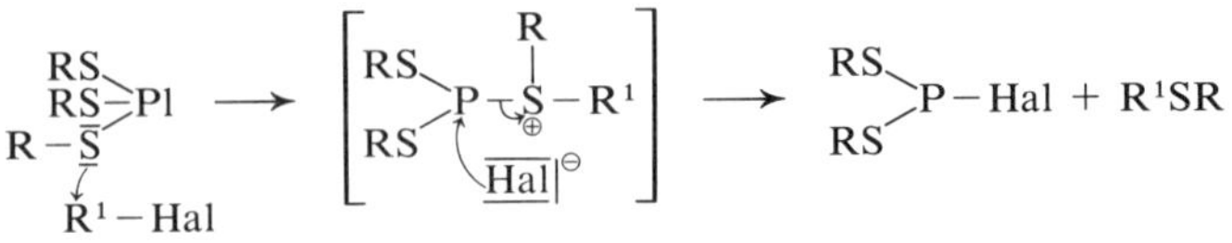
(60)

Contrary to the Michaelis-Arbusov reaction, alkyl halide does not result, but the corresponding thioether [53].
Various mechanisms have been suggested for the course of the Michaelis-Arbusov reaction. It is now clear that the first step involves the formation of a quasi-phosphonium intermediate (XXII) which, in some cases, could be isolated (e.g. R = phenyl) or demonstrated by conductivity measurements [217, 548].

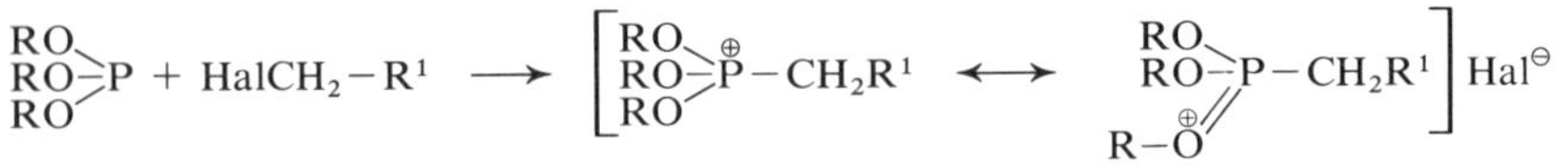
(XXII a) (XXII b)

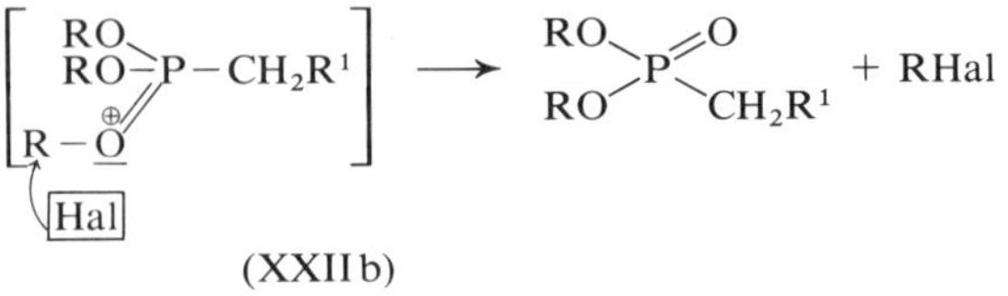
(61)

(XXII b)

The phosphonium compound (XXII) is mesomerically stabilized, because oxonium forms (XIIb) are involved. The last step consists in alkylation of the halide ion accompanied by formation of a P=O double bond. This reaction proceeds by expanding the octet, a property of great importance to phosphorus and a decisive factor in many reactions. This is ensured by the low-energy, empty 3*d* orbitals entailing an interaction between these *d* orbitals of phosphorus and the *p* electrons of oxygen ($d_\pi - p_\pi$ bonding) which is favoured energetically. Evidence for the oxonium structure as an intermediate (XXIIb) is the fact that groups bound to phosphorus which accept a positive charge more readily than oxygen, yield more stable onium structures, for example when $-OR$ is replaced by $-NR_2$ (XXIII):

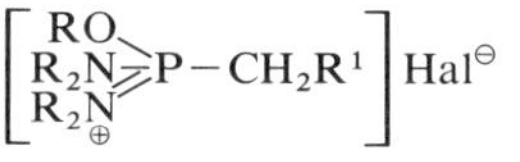

(XXIII)

Increasing polarizability of the anion should favour the Arbusov reaction. This can really be observed in the order fluorine, chlorine, bromine to iodine.
A very important synthesis related to the Michaelis-Arbusov reaction is the *Perkow reaction* (Eq. (62)) by which well-known commercial products are obtained, e.g. dichlorovinyl phosphates:

(62)

During treatment of α-halogen carbonyl compounds in place of alkyl halides with trialkyl phosphite, ARBUSOV and RAZUMOV [55] observed the release of alkyl halide, but were unable to prepare the sodium salt of the presumed "acyl" dialkyl phosphonate (XXIV). The reaction must, therefore, have taken another course. It was PERKOW who first discovered that α-halogen carbonyl compounds did not yield Arbusov products with trialkyl phosphites but a new type of compound, the enol phosphates (XXV).

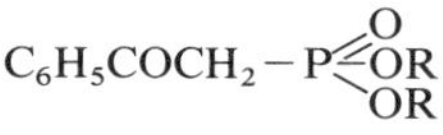

(XXIV)

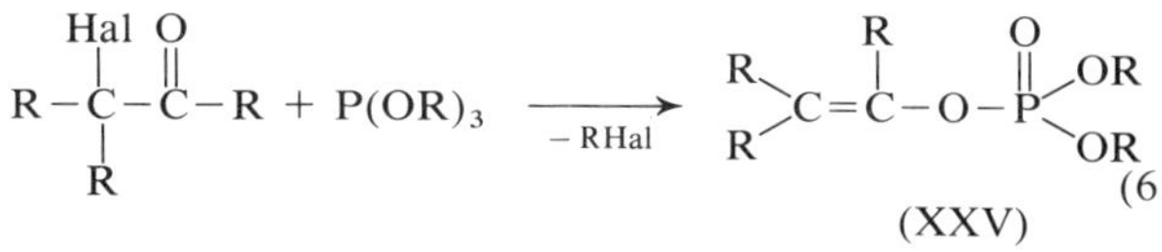

(XXV) (63)

As evidence for their structure, he demonstrated [724] that chlorine or bromine can be added, that the products show no carbonyl reactions, and the IR spectra are in agreement with the structure suggested. It is not possible in all cases to distinguish clearly between these types of reaction for Arbusov and Perkow products are often produced simultaneously.

As a mechanism for both reactions, it has been postulated that the nucleophilic attack of phosphorus might be directed against the oxygen or carbon of the carbonyl group, possibly against the α-carbon atom, or against the halogen atom.

PUDOVIK suggested attack on the carbonyl oxygen [756]:

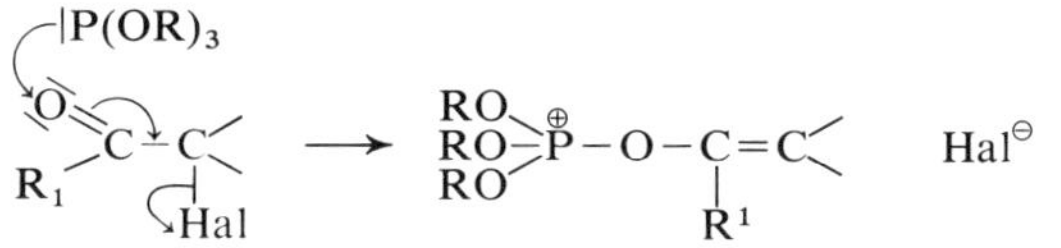

(64)

"Enol phosphonium salt" (XXVI)

where synchronously a halide ion leaves with formation of an enol phosphonium salt (XXVI) (according to ALLEN and JOHNSON [15], KHARASCH and BENGELSDORF [479]). Attack on the carbonyl carbon could be formulated as follows:

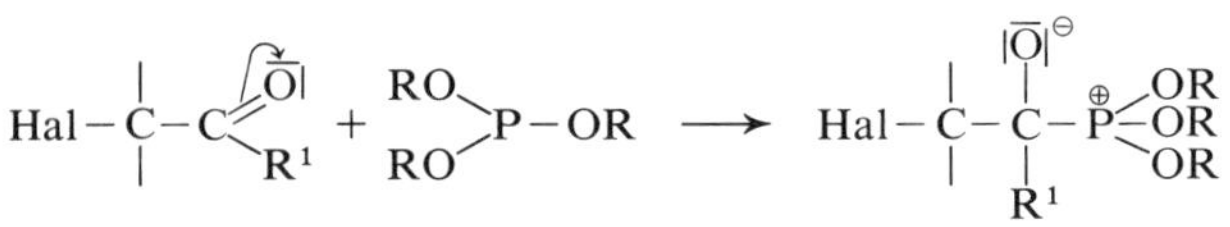

(65)

If this formulation is correct

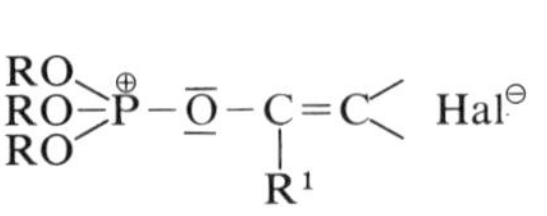

(XXVII)

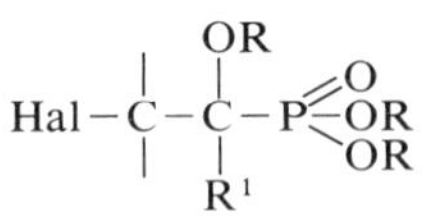

(XXVIII)

then an enolphosphonium salt (XXVII) and an α-alkoxy phosphonate (XXVIII) should be formed, which to date has not been found. Attack on the α-carbon atom might be expected to produce an "oxophosphonium" compound (XXIX) which, by rearrangement, would yield an enol phosphonium salt (e.g. XXX).

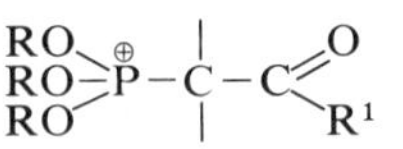

"oxophosphonium" compound

(XXIX)

Since the halogen is in the α-position to an electron-withdrawing group, the attack of the phosphite might be the last possibility here. The greater the tendency of the halogen to accept a positive charge and the more stable the enolate ion, the more readily the enol phosphonium salt (XXX) should be formed.

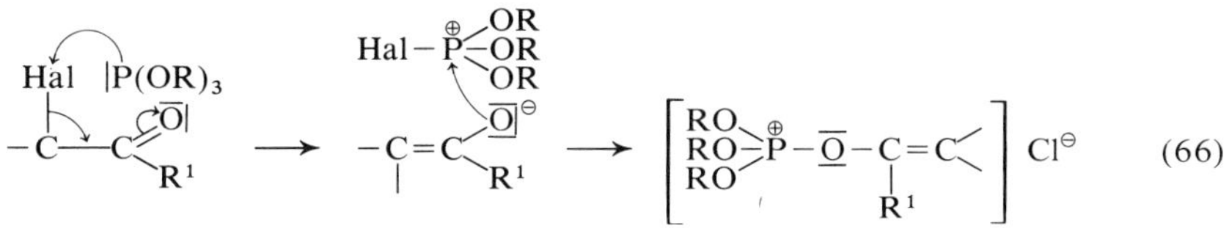

enol phosphonium salt (XXX)

The release of the halogen atom and formation of the phosphorus-oxygen bond may be synchronous, although some authors [217] consider a stepwise cleavage to be more probable.

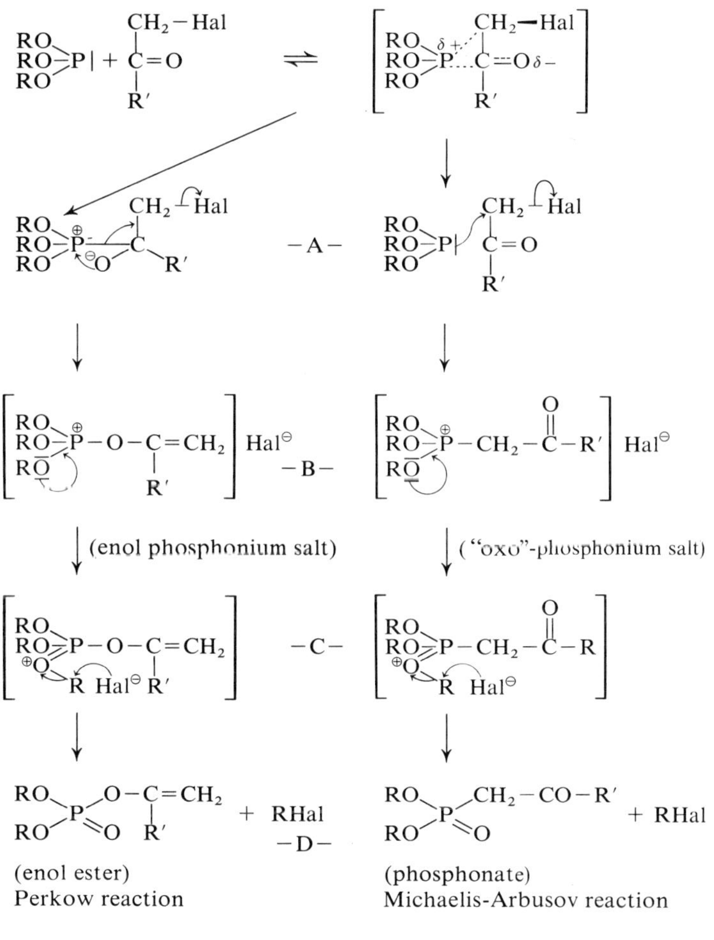

Scheme 4. Mechanisms of the Perkow and Michaelis-Arbusov reactions according to CRAMER [217]

CRAMER [217, see also 548] has attempted to refer the mechanism of the Michaelis-Arbusov and the Perkow reactions to a common intermediate state (Scheme 4) in which the reaction is influenced in one direction or another by factors such as:

1) Electronegativity of the halogen atom
2) Polarizability of the halogen atom
3) Number of halogen atoms
4) Substituents on the α-C atom
5) The double bond character of the carbonyl group (influenced by R)
6) Reaction conditions (polarity of the solvent, temperature etc.)

Although the Michaelis-Arbusov reaction can take place with suitable α-halogen carbonyl compounds, compounds with a strongly polarized carbonyl group should, however, yield enol phosphates. "Acyl" phosphonates should result from compounds with a polar carbon-halogen bond. When the carbonyl group and carbon-halogen bond show comparable reactivity, both products ought to occur together, which can also be confirmed experimentally. In the case of the halogen carboxylic acid chlorides, both Perkow and Arbusov reactions proceed simultaneously in the same molecule. Further systematic investigations are, however, prerequisites for the final establishment of the mechanism involved in the Arbusov and Perkow reactions. Various mechanisms suggested for the Perkow reaction are summarized by CHOPARD, CLARK, HUDSON and KIRBY [181].
A particularly important example would be an explanation of the rearrangement of ®Dipterex to DDVP (Eq. (67)), which in agreement with CRAMER can be formulated as follows:

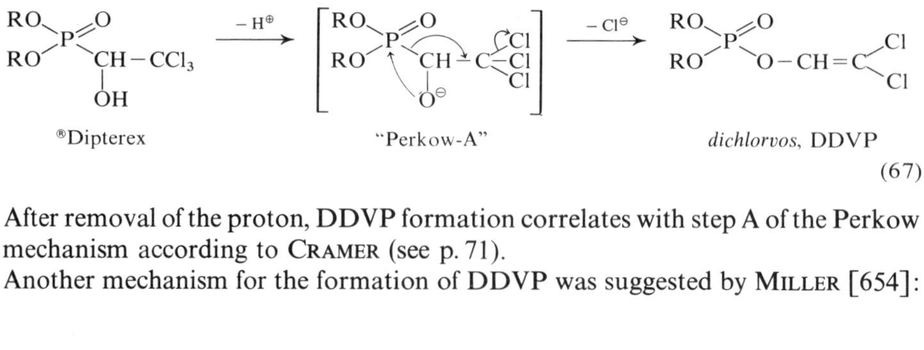

®Dipterex "Perkow-A" *dichlorvos*, DDVP

(67)

After removal of the proton, DDVP formation correlates with step A of the Perkow mechanism according to CRAMER (see p. 71).
Another mechanism for the formation of DDVP was suggested by MILLER [654]:

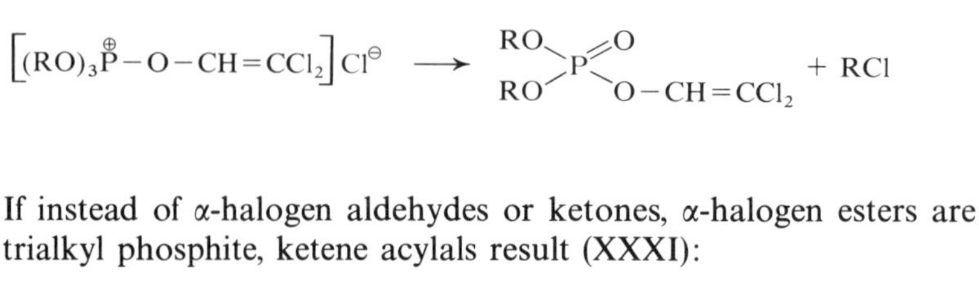

(68)

If instead of α-halogen aldehydes or ketones, α-halogen esters are reacted with trialkyl phosphite, ketene acylals result (XXXI):

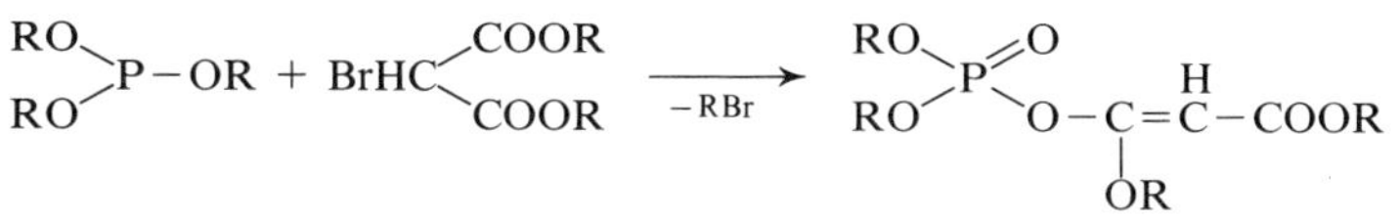 (XXXI) (69)

The very reactive ketene acylals [217] react with a large number of carboxylic acids, HOAc,to malonic ester and the corresponding anhydrides (Eq. (70)), which result from the phosphorylium cation and the acid anion.

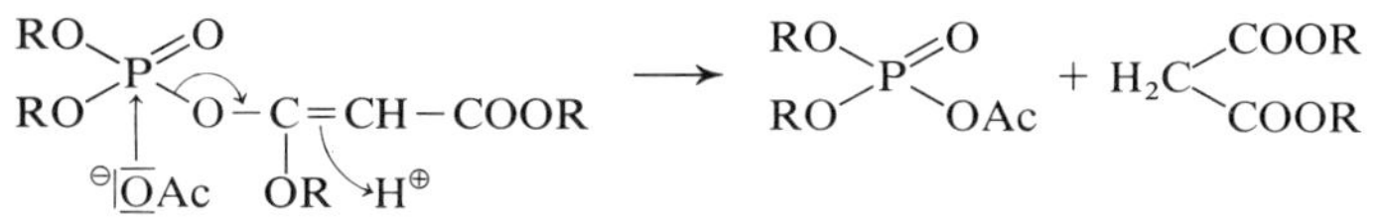 (70)

When HOAc represents phosphoric ester acids, pyrophosphates are obtained.
One important application of the Perkow reaction is the synthesis of phosphoenol pyruvic acid.

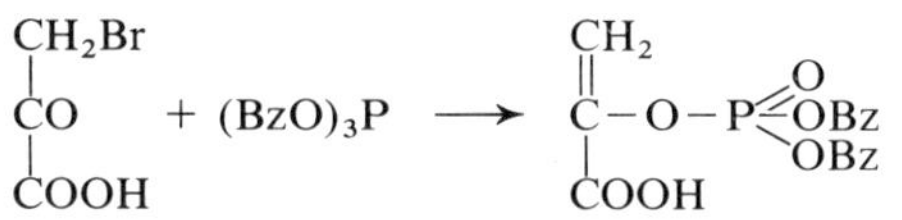 (71)

Benzyl phosphite is used because the benzyl groups are readily removed by hydrogenation.
The preparative importance of the di- and tri-alkyl phosphites for the manufacture of insecticidally active compounds can be seen from following review of the more important reactions:

Dialkyl phosphites:

a) Michaelis-Becker reaction:

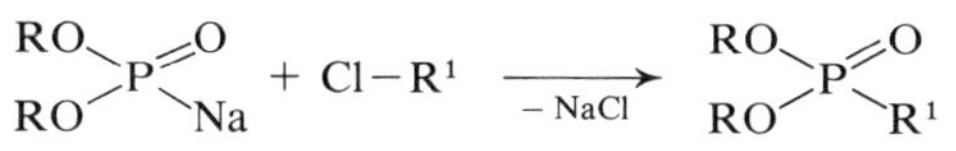

b) Addition to carbonyl groups:

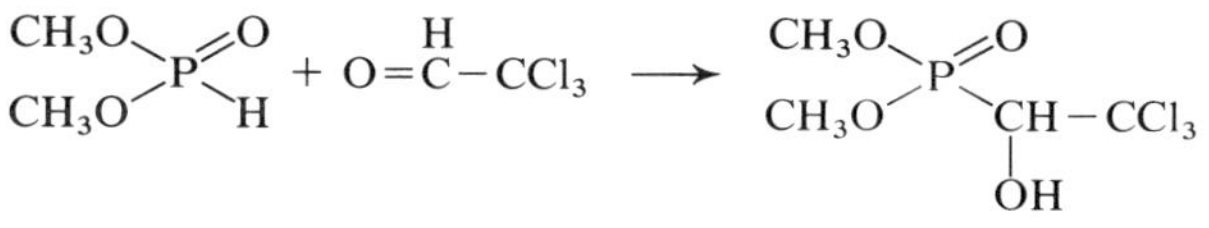

trichlorfon

c) Reaction with sulfenyl chlorides to thiol phosphates [797]:

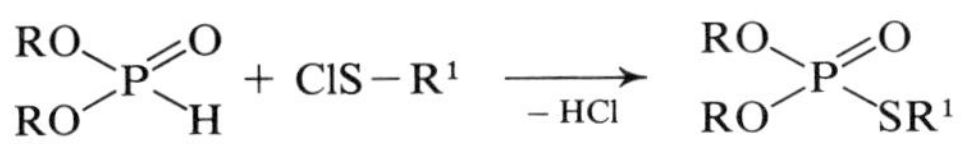

This reaction replaces the direct acylation of thiophenols with diester chlorides to thiol esters, which is not successful, but leads to alkyl thioethers:

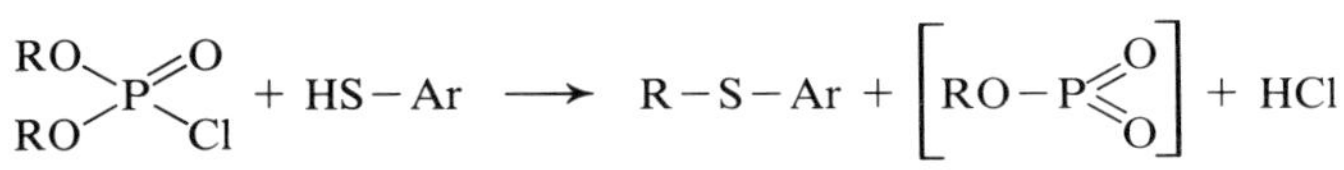

d) Reaction with disulfides [575]:

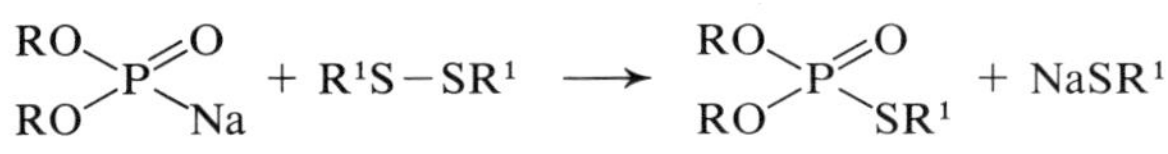

Trialkyl phosphites:

a) Michaelis-Arbusov reaction

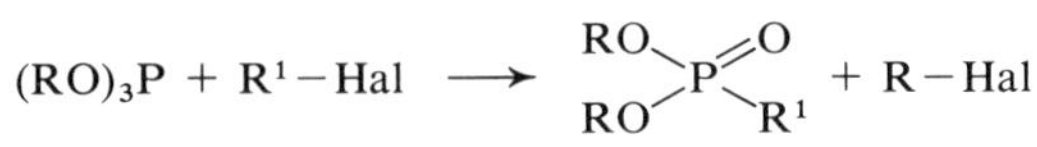

b) Perkow reaction with suitable α-halogen carbonyl compounds to enol phosphates:

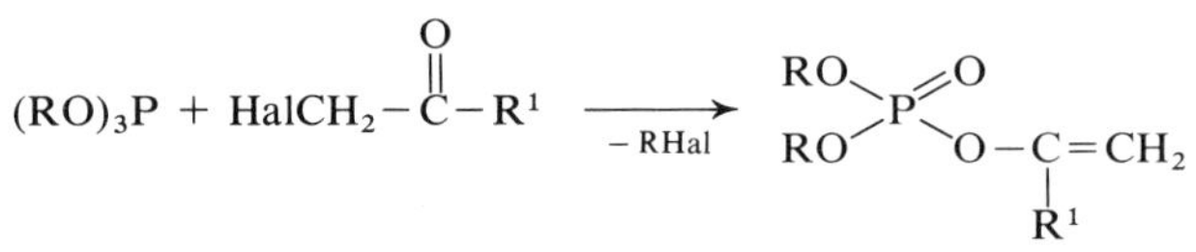

c) Reaction with sulfenyl chlorides [295, 664]:

d) Reaction with isothiocyanates [649]:

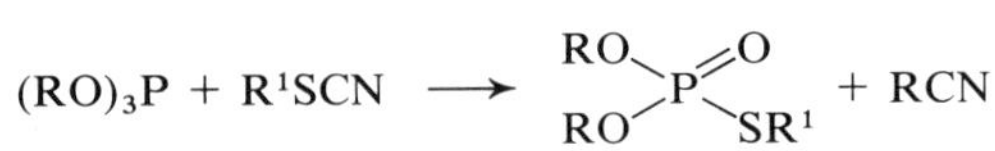

e) Reaction with disulfides [435]:

f) Reaction with sulfonyl chlorides [412]:

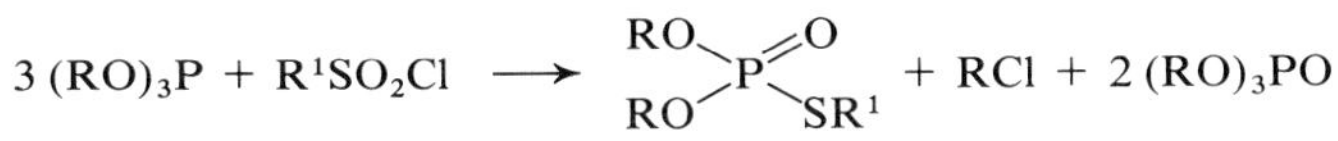

g) Reaction with aldehydes [217, 548, 721] (no reaction with CH_3CHO)

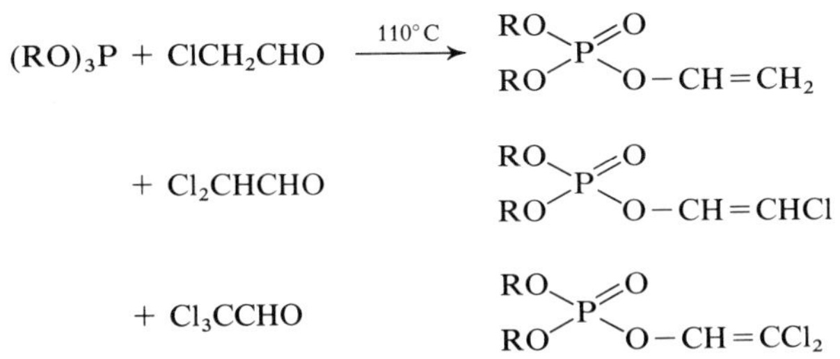

dichlorvos

The reactivity increases with the number of halogen atoms at the α-carbon atom.

h) Reaction with unsaturated aldehydes [488]:

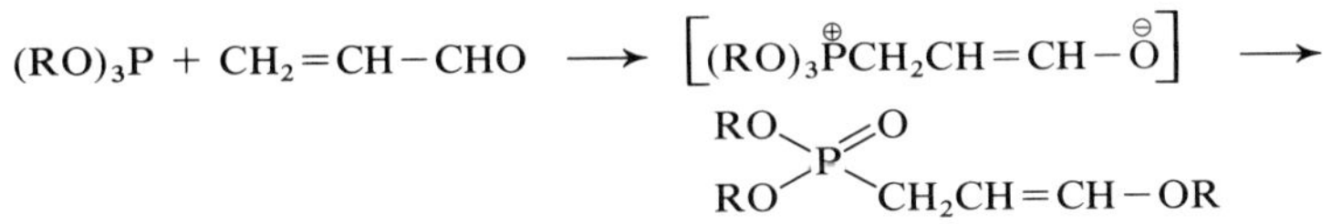

i) Reaction with ketones [516, 217, 548, 755]:

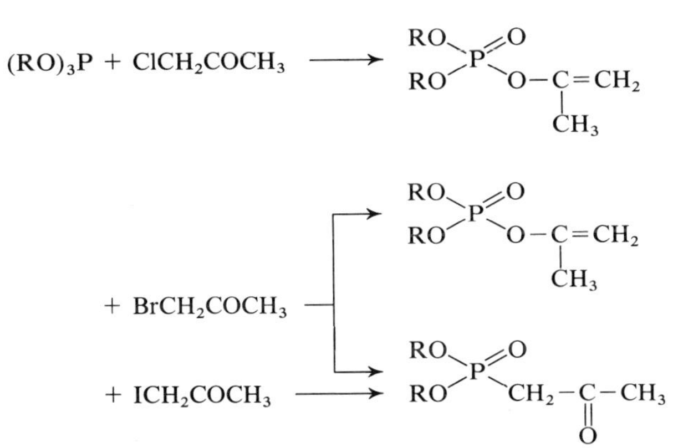

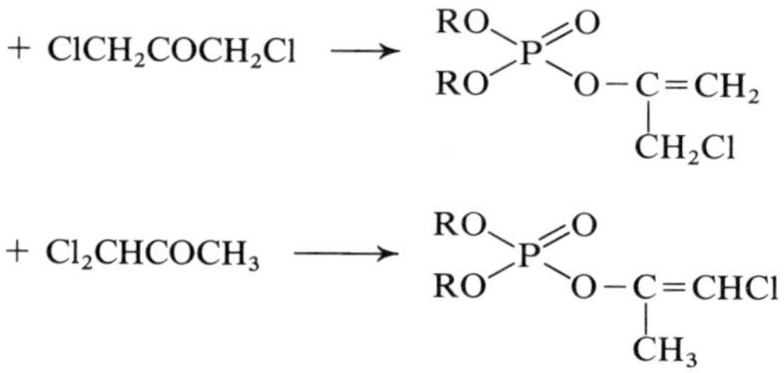

Ketones react less readily than aldehydes, the α-halogen carboxylic acid esters being considerably less reactive than ketones.

j) Reaction with carboxylic acid derivatives [217, 548]:

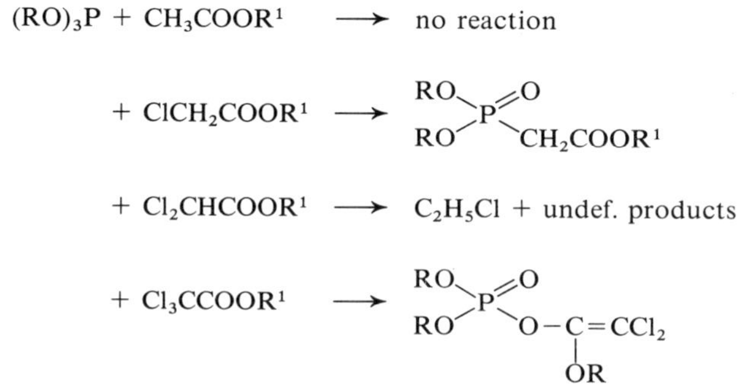

k) Reaction with unsaturated carboxylic acid derivatives [488]:

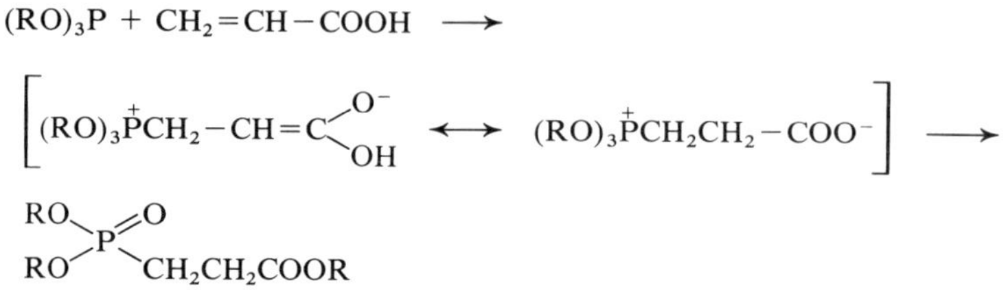

l) Reaction with carboxylic acid amides [217, 548]:

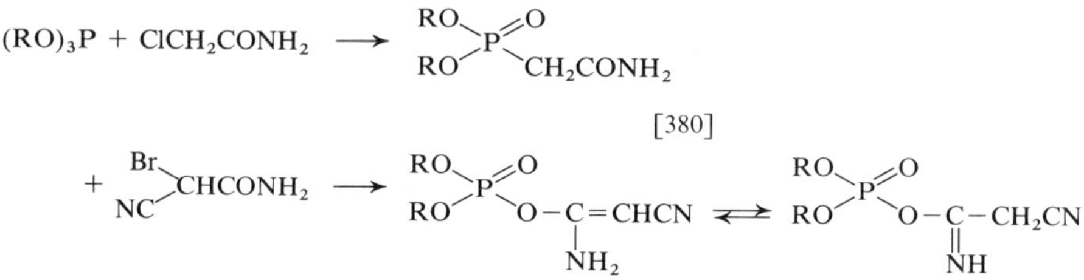

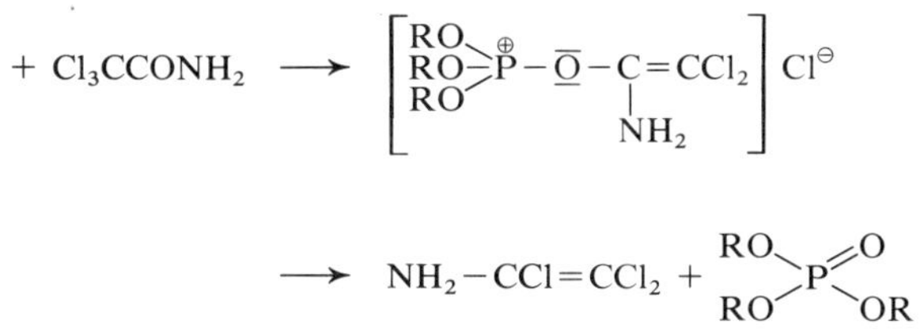

m) Addition to β-dicarbonyl compounds [762]:

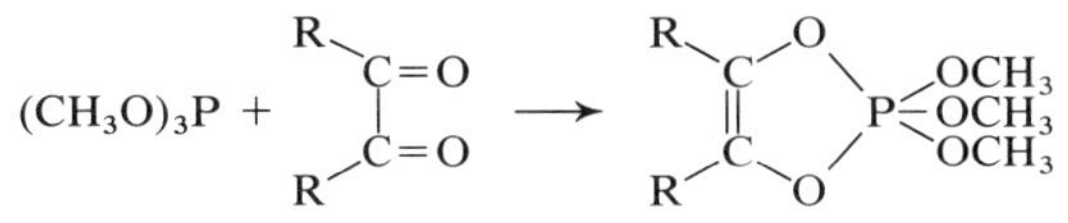

n) Reaction with α-halogen dicarbonyl compounds [217, 548]:

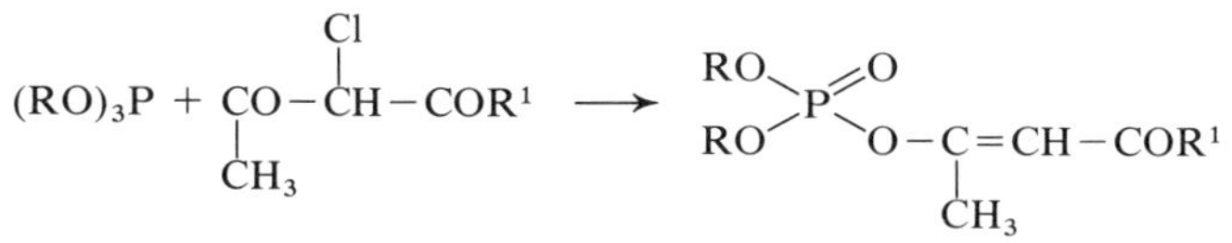

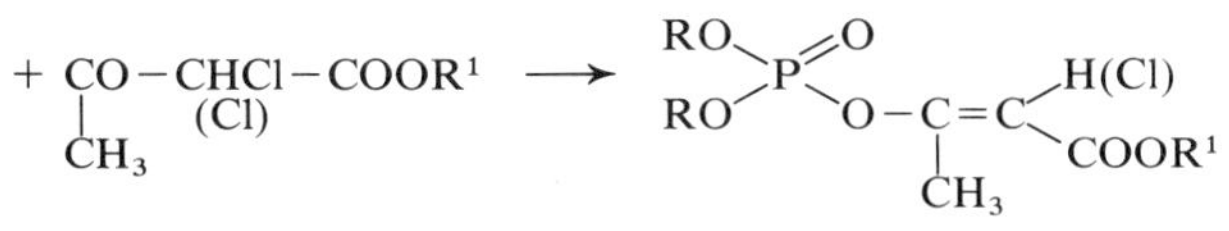

mevinphos

o) Reaction with acid chlorides [217, 548]:

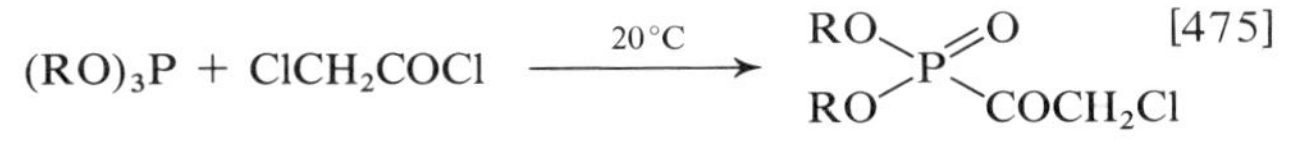

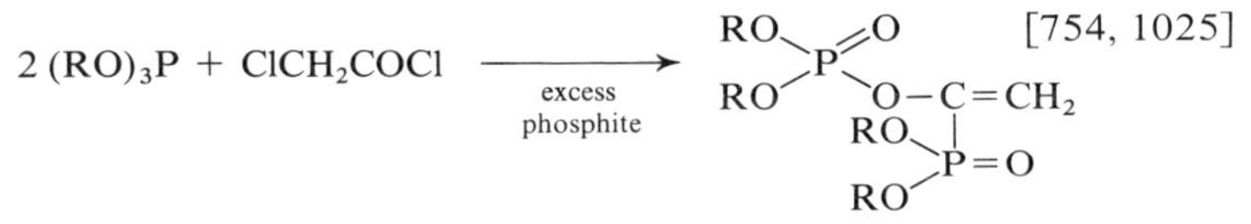

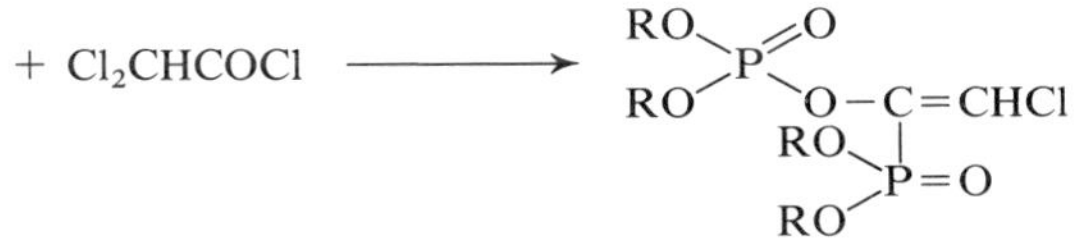

p) Reaction with nitro-compounds [14]:

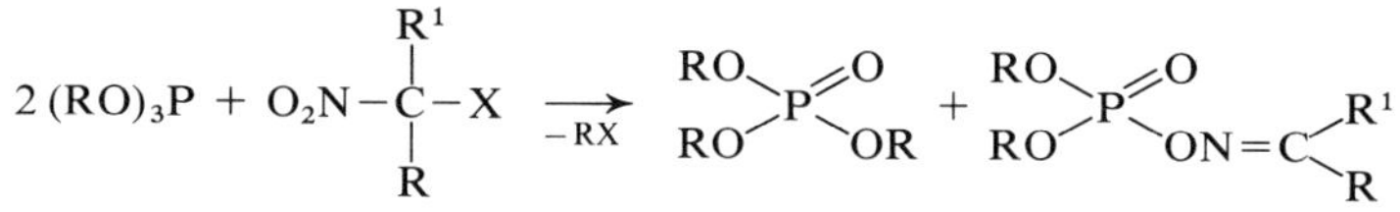

3.2. Single Compounds

Note: The trade and common names used in the following text refer only to the pure active substance and not to specific formulations.

a) Trans-Halogenation of Ester Chlorides

Examples of acid halides and pseudo-halides were the phosphoric acid fluorides, cyanides and phosphonic acid fluorides. They are some of the early important phosphorus compounds and can be obtained by trans-halogenation of simple starting materials (see Section 3.1.a).
For their synthesis, SCHRADER [824] 1937 started with N,N-dimethyl phosphoroamidodichloridate, which reacts with alkali cyanides in the presence of alcohol to give Tabun (I) [O-ethyl N,N-dimethyl phosphoroamidocyanidate] [345, 828]:

(1)

(I)

SAUNDERS *et al.* [795] chose a synthesis via diethyl phosphoryl chloride with a subsequent Michaelis-Arbusov reaction:

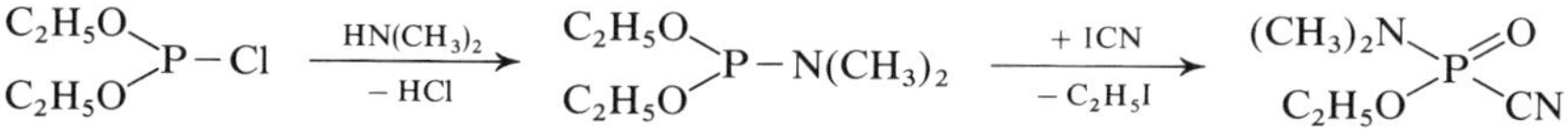

(I) (2)

Iodine cyanide can be replaced by bromine cyanide [794]. Tabun is a highly toxic compound and one of the most potent cholinesterase inhibitors known.
Dimefox (II) [N,N,N′,N′-tetramethyl phosphorodiamidofluoridate] [418, 843] is also derived from N,N-dimethyl phosphoroamidodichloridate. It was synthesized by SCHRADER in 1940 and in 1949 was introduced commercially by Fisons Pest Control Ltd.

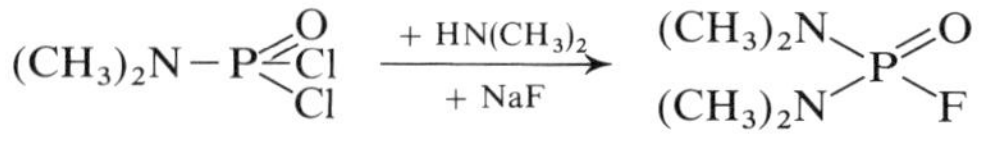

(3)

(II)

Phosphoryl chloride (II) [384, 613] can be replaced by phosphoryl fluoride in the synthesis:

$$4\,(CH_3)_2NH + POF_3 \longrightarrow II + 2\,(CH_3)_2NH \cdot HF \qquad (4)$$

Dimefox is very toxic, but has only weak contact action. Its LD_{50} for the rat after oral administration is 3 – 5 mg/kg. It is used mainly in hop culture as a systemic agent against aphides and the red spider mite.
Another member of this group is the isopropyl analogue called *mipafox* (III) [N,N-Diisopropyl phosphorodiamidofluoridate] [376, 747].

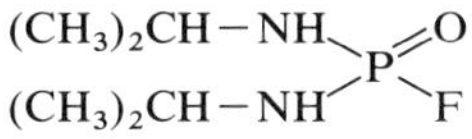

(III)

The compound was first described by HARTLEY *et al.* in 1950 and introduced by Fisons Pest Control Ltd. Compared with *dimefox* it has a lower toxicity with an LD_{50} of 25 – 50 mg/kg orally for the rat. The product would have been suitable as a systemic insecticide and acaricide, but after absorption through the human skin it produced paralysis and was, therefore, withdrawn [96, 97] (see pp. 264, 266).
Another dialkyl fluorophosphate which must be mentioned is DFP (IV) [O,O-diisopropyl phosphorofluoridate] [534, 843, 877].

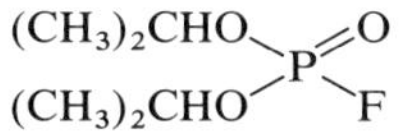

DFP (IV)

In 1932 LANGE and v. KRUEGER developed the method of alkylating fluorophosphoric acid silver salts with alkyl iodides.

(5)

It was later replaced by other syntheses. In 1938 SCHRADER [877] prepared DFP as follows:

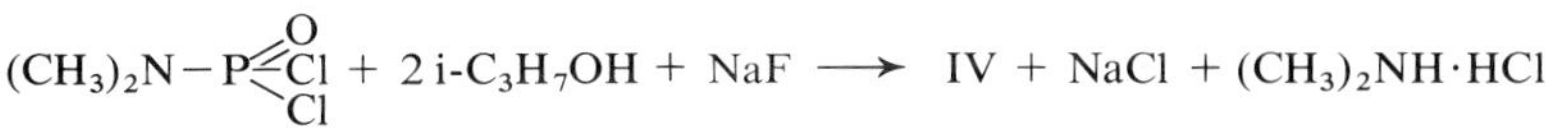

(6)

SAUNDERS (1941) started with diisopropyl phosphite [794] which he chlorinated and then reacted with sodium fluoride.

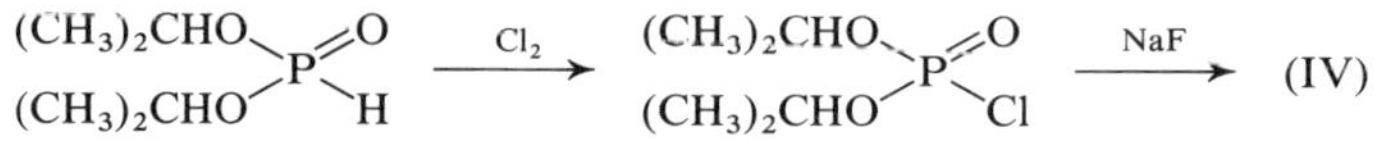

(7)

Phosphoryl dichloride fluoride is only of academic interest as starting material (SAUNDERS *et al.* (1944) [107, 613]):

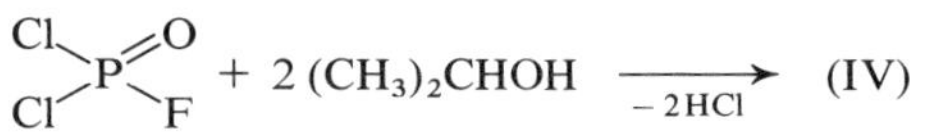

(8)

Because of its high toxicity (LD_{50} 5 – 13 mg/kg for the rat orally) DFP cannot be used as an insecticide, although it has a good contact action.
Fluorides derived from phosphonic acids have become known because of their unusually high toxicity, particularly the two compounds Sarin (V) (O-isopropyl methanephosphonofluoridate) and Soman (VI) (O-(1,2,2-trimethyl propyl) methanephosphonofluoridate).
The most convenient method of preparation starts with the phosphonic acid ester chlorides, which are reacted with alkali fluoride in the presence of alcohol [845]:

(9)

(V)

The dichloride required was obtained in a Michaelis-Becker reaction from diethyl phosphite [845]:

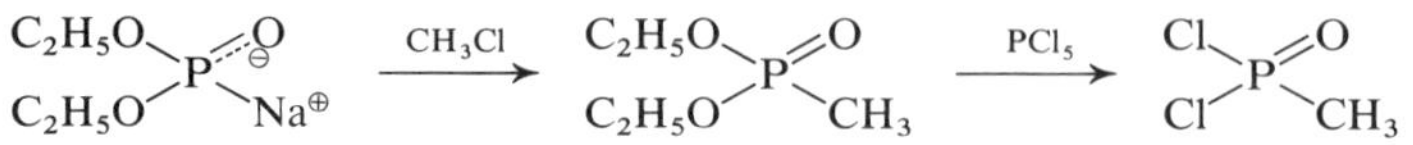

(10)

or by an Arbusov reaction from trimethyl phosphite:

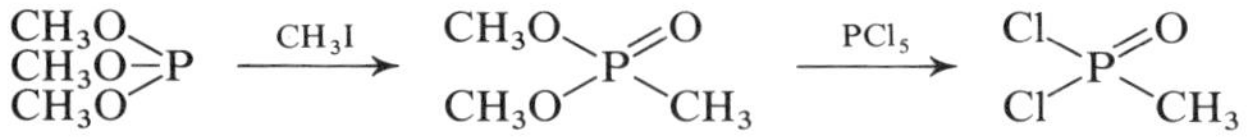

(11)

A third possibility was the pyrolysis of dimethyl phosphite [76, 211]

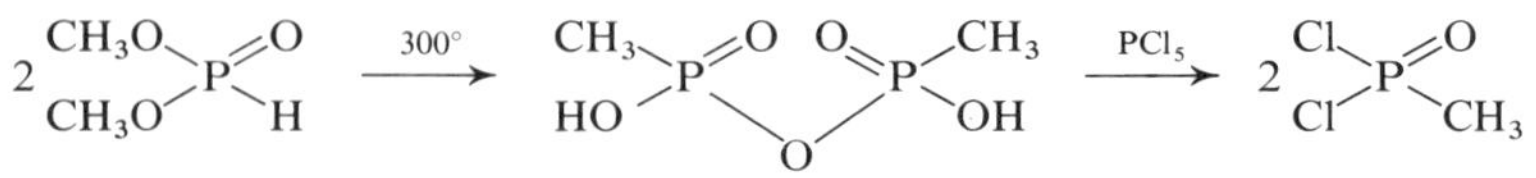

(12)

As BOTER, OOMS, VAN DEN BERG and VAN DIJK [119] were able to show, acetylcholinesterase (E.C. 3.1.1.7) is preferentially inhibited by the levorotatory

enantiomer of Sarin. From the results of CHRISTEN *et al.* [182] it may be concluded that (+)-sarin is preferentially hydrolyzed by sarinase in plasma, while (−)-sarin is the more active inhibitor of AChE. (Fluoride ions catalyze the racemization of optically active Sarin [85]).
In the case of Soman the isopropyl group is replaced by a pinakolyl group [555]:

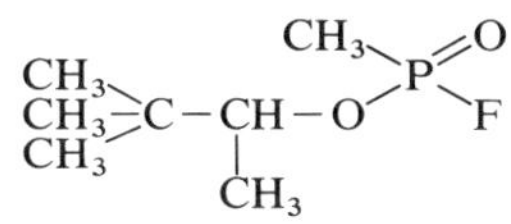

(VI)

This highly toxic compound was discovered shortly before the end of the Second World War in Heidelberg, but not by SCHRADER [775].
Although it had always been SCHRADER'S objective to find insecticidal esters, the fluorophosphonates were screened by the German authorities as military weapons.
Research with this class of substance is extremely dangerous, even when the most stringent precautions are taken. A special technique is required in handling such neurotoxic compounds.

b) Acylation with Ester Halides

1. Preparation of Anhydrides

In general, ester anhydrides are prepared by the condensation of phosphoric ester acids or by the acylation of phosphoric acid salts with phosphoryl ester chlorides. Suitable condensation agents are carbodiimides, trichloroacetonitrile, imidochlorides, dialkyl cyanamides [424] etc. They form a so-called "energy-rich phosphate" compound as intermediate.

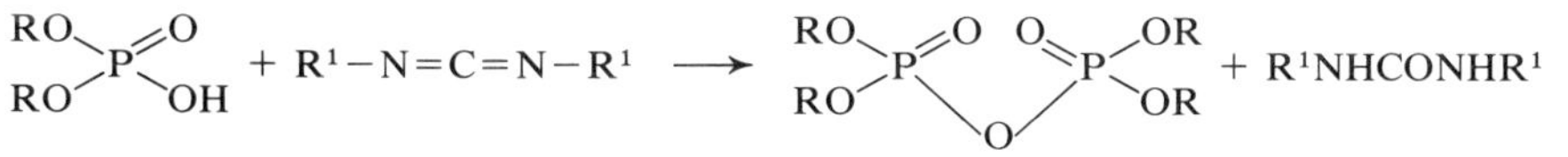

(1)

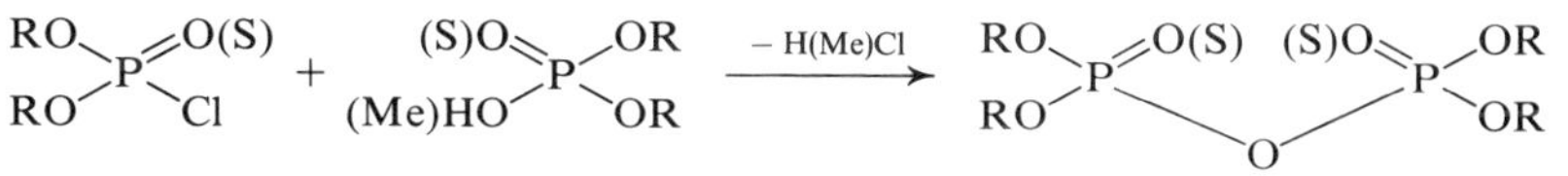

(2)

In practice, the pyroester is synthesized by treating 1 mol acyl chloride with the appropriate quantity of tertiary amine and 0.5 mol water. By that means the chloride is partly hydrolyzed to acid and acylated to the pyroester by excess chloride [200, 416, 425, 843, 963].

In a similar way trialkyl phosphates can be reacted with diester chlorides [363, 426, 508]:

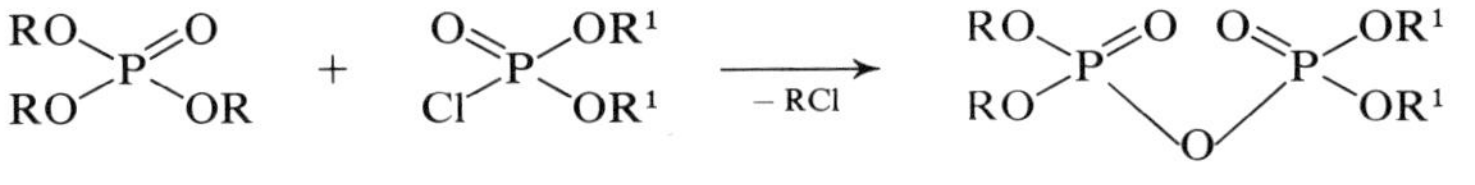

(3)

When thiol acids are acylated, thionopyroesters are formed irrespective of the method used [284, 286, 427, 511, 512, 887, 890]:

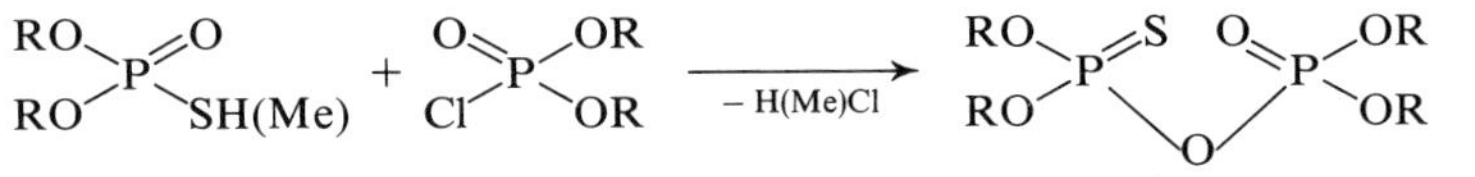

(4)

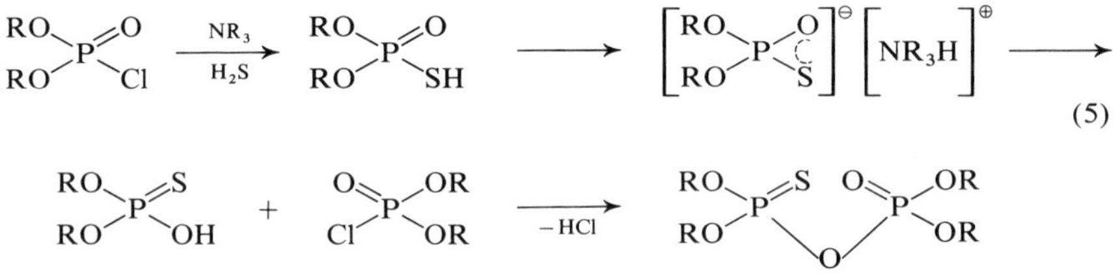

(5)

or [371, 567, 735, 887]:

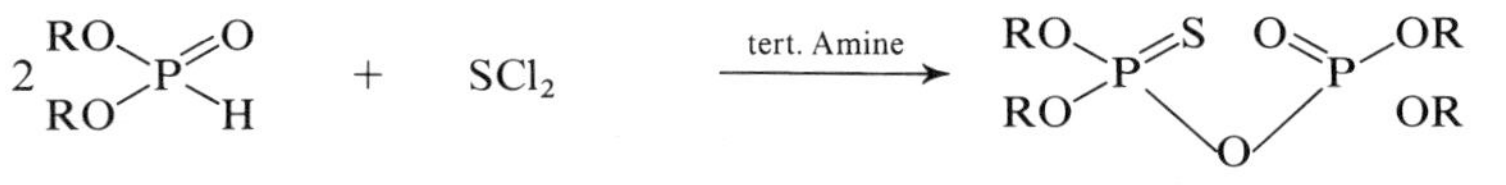

(6)

These can also, of course, be obtained from the thionochloride and the corresponding alkali salt [428, 618]:

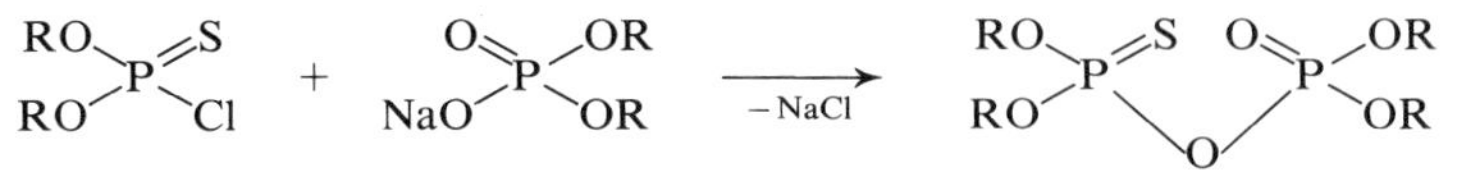

(7)

In contrast to the alkylation, acylation does not take place on the sulfur atom. This was substantiated by MILLER [653] who found that the displacement of groups bound to phosphorus was not determined by the nucleophilic properties but rather by the basicity of the attacking ambident ions (acylation on the oxygen atom):

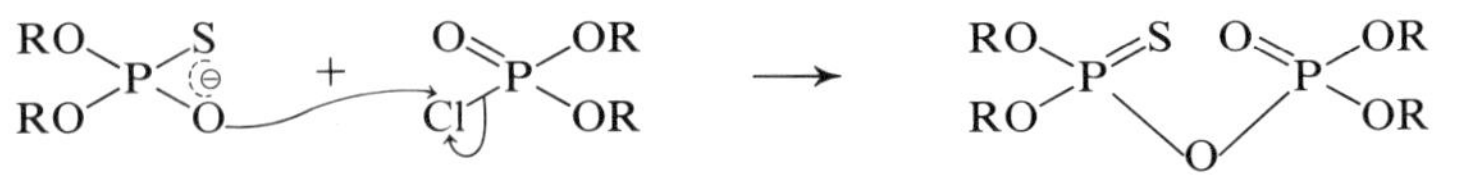

(8)

In contrast, an electrophilic carbon atom reacts with the most nucleophilic center of the ambident thiophosphate anion, i.e. alkylation occurs at on the sulfur atom:

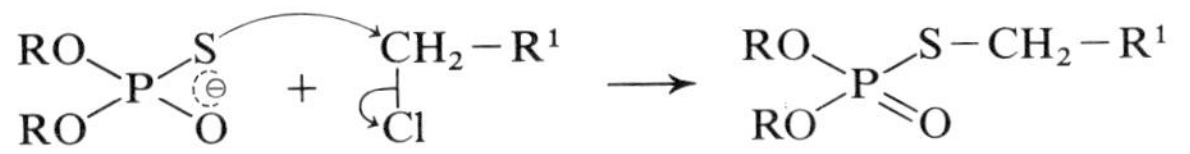
(9)

PEARSON's concept [713, 714, 715, 716] provides an explanation as to why the thermodynamically stable thionopyroesters are combinations of

hard acids 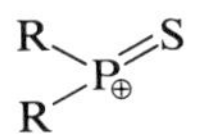 with hard bases 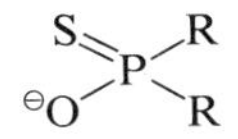 .

Acylation on the sulfur is only successful if no oxygen atom is competing [590]:

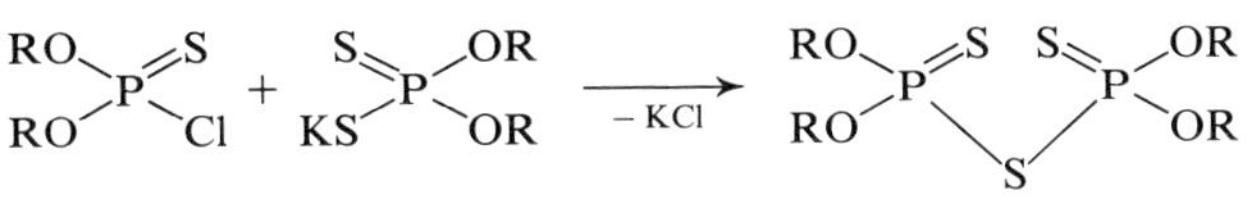
(10)

Another possible way of preparing trithiopyrophosphates is by the reaction of the alkali salt of dithiophosphoric acid with chlorocyanide [585].

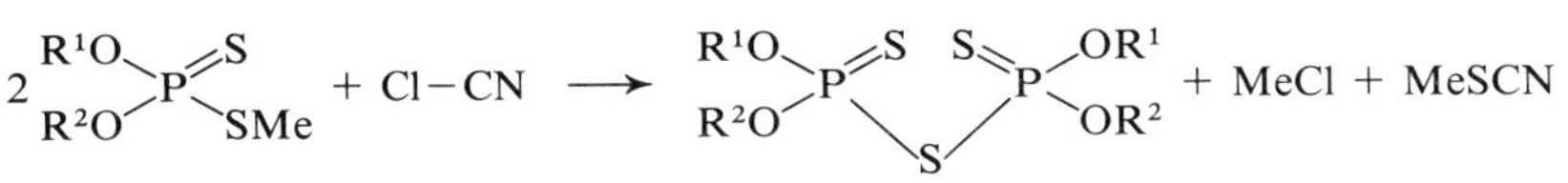
(11)

It is preparatively possible to isolate thionothiol pyroesters as kinetically controlled reaction products when sulfenamides are reacted with dialkyl phosphites [16]:

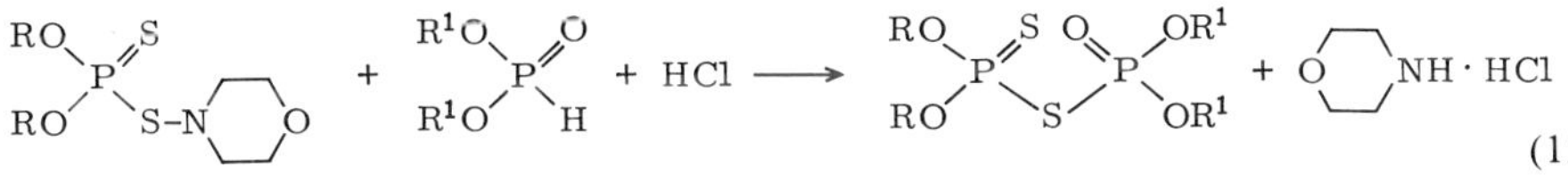
(12)

The very pure asymmetric dithiopyrophosphates obtained by this method are converted, at elevated temperature to the thermodynamically more stable bisthionocompounds.

In view of the amount of work done with tetraethyl pyrophosphate (I), it is surprising that its toxicity and insecticidal properties remained undiscovered for a long time.

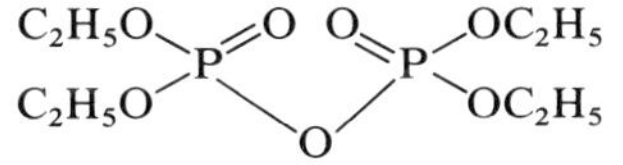

(I)

There are numerous syntheses for this ester which must be mentioned here, because they are not only of historical significance, but also of technical and scientific interest. The English abbreviation TEPP has become established for the compound (I).
The first synthesis carried out by CLERMONT consisted in a classical alkylation of the silver salt of pyrophosphoric acid with ethyl iodide [195].

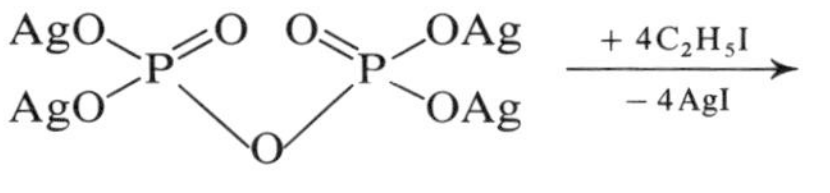
(13)

NYLÉN [685] synthesized the pyroester in the following way (1930):

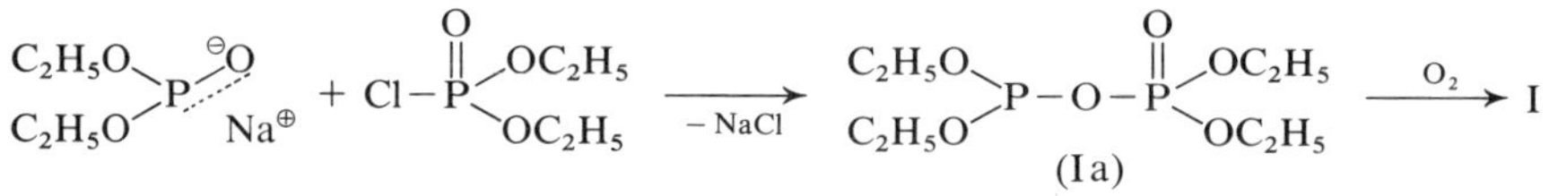
(Ia)
(14)

The oxydation can also be carried out with chlorine [510] in place of oxygen.

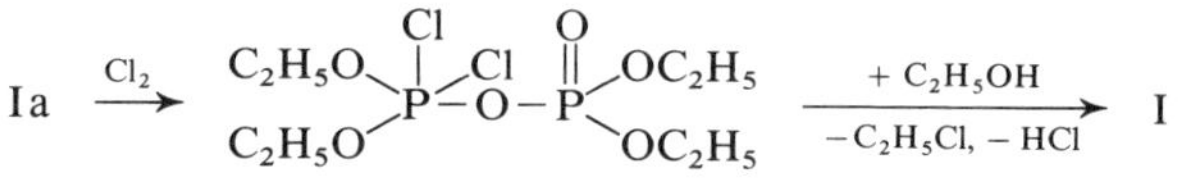
(15)

In addition to the general synthesis for pyroesters, there are special industrial-scale processes applicable to TEPP.
SCHRADER [827, 842] reacted phosphoryl chloride with triethyl phosphate in a mol ratio of 1 : 3. This reaction is known as the "Schrader process".

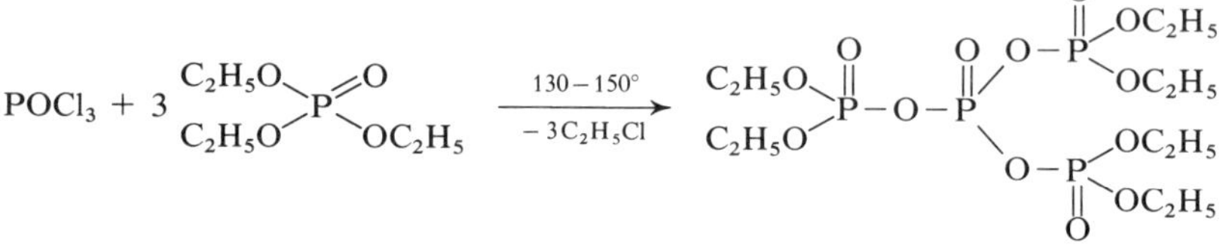
(16)

"HETP"

The reaction product was formulated as hexaethyl tetraphosphate "HETP". The phosphoric acid ester mixture prepared by this process was introduced in 1943 under the name ®Bladan (Farbenfabriken Bayer) as the first contact insecticidal organophosphate and was intended as a substitute for nicotine and other natural insecticides which were in short supply. JANNING [438] was the first to discover a synthesis of "HETP" from $POCl_3$ and alcohol, which was later described also by THURSTON [952].

$$4\,POCl_3 + 9\,C_2H_5OH \longrightarrow 9\,HCl + 3\,C_2H_5Cl + (C_2H_5O)_6P_4O_7 \qquad (17)$$

If the "Schrader process" is carried out with a mol ratio of 1 : 5, then TEPP [1024] is obtained as the main product.

$$5\,(C_2H_5O)_3PO + POCl_3 \longrightarrow 3\,C_2H_5Cl + 3\,(C_2H_5O)_4P_2O_2 \qquad (18)$$

In a method described by WOODSTOCK [1033] phosphorus pentoxide may be used in place of phosphoryl chloride. According to the ratio of substances used in the reaction, varying proportions of TEPP result.

$$2\,(C_2H_5O)_3PO + P_2O_5 \longrightarrow (C_2H_5O)_6P_4O_7 \qquad (19)$$

$$4\,(C_2H_5O)_3PO + P_2O_5 \longrightarrow 3\,(C_2H_5O)_4P_2O_3 \qquad (20)$$

Another synthesis is given by Eq. (21):

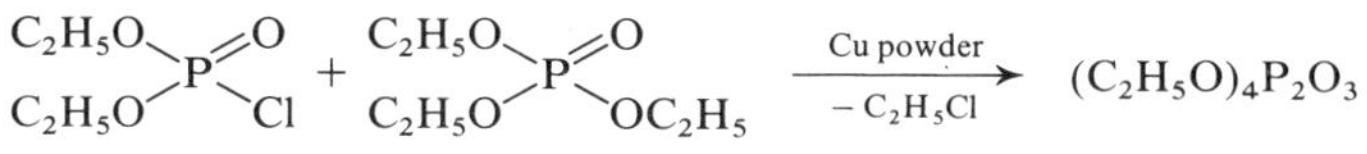

(21)

The products obtained by both processes have substantially the same properties. McCOMBIE [614] reports that the following reactions are involved in the "Schrader process":

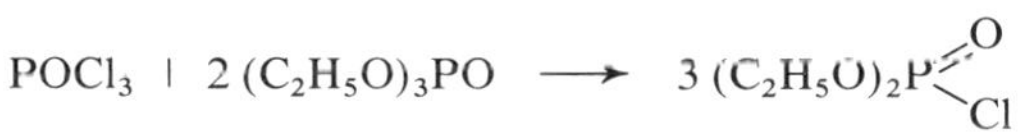
(22)

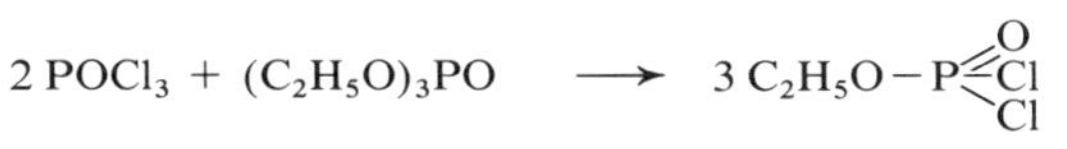
(23)

Other authors [363] suggest that "HETP" is not a uniform product but rather a mixture of "HETP", TEPP, triethyl phosphate and ethyl metaphosphate, the ratio of which depends upon the quantities of phosphoryl chloride and triethyl phosphate used. In all product mixtures, TEPP would appear to be the actual active constituent.

The compound possesses insecticidal and acaricidal activity, recognized by KÜKENTHAL in 1938. Due to its low stability towards hydrolysis, it is nowadays seldom used. Occasionally, however, a rapid degradation is desirable, for such

preparations may be applied until shortly before harvest. The oral LD_{50} for the rat is 1.12 mg/kg.
Replacement of the two oxygen atoms in TEPP by sulfur results in *sulfotepp* (II) [O,O,O′,O′-tetraethyl pyrophosphorodithionate]. The syntheses proceed in the same manner as for the oxygen compound. Also the desired product is obtained by the addition of sulfur to O,O,O′,O′-tetraethyl pyrophosphite [839] (1944).

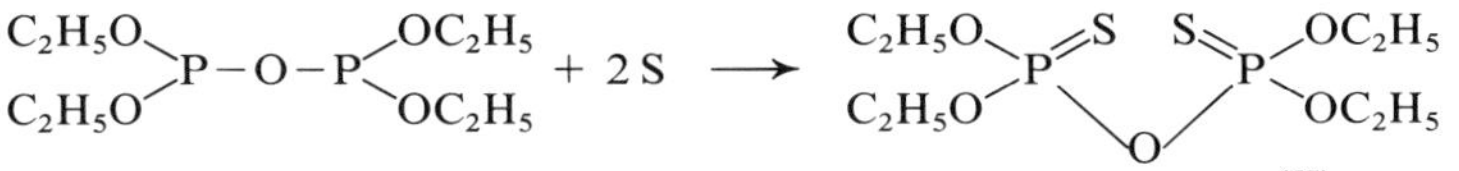

(II)

(24)

For industrial synthesis another process is used, i.e. that described by SCHRADER *et al.* in 1950 [618].
It is the partial hydrolysis of O,O-diethyl phosphorothiochloridate [855, 890, 964].

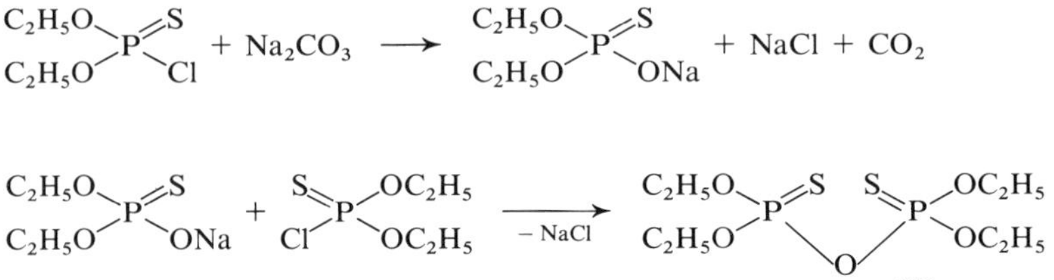

(II)

(25)

This compound known as ®Bladafum (Farbenfabriken Bayer) has contact insecticidal, acaricidal activity und an oral LD_{50} of 5 mg/kg for the rat. On account of its high vapour pressure and thermic stability, *sulfotepp* is used in greenhouses as a fumigant. Because of the thiono-group, the compound is more stable towards hydrolysis than TEPP.
When the ethyl groups are replaced by n-propyl groups, a considerably less toxic compounds results: ®NPD (III) [O,O,O′,O′-tetrapropyl pyrophosphorodithionate] (E. I. du Pont de Nemours & Co.) [890].

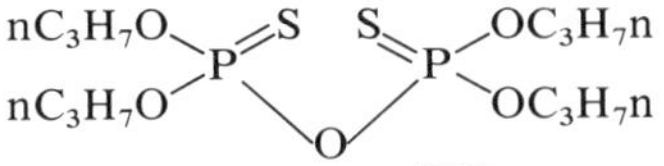

(III)

The oral LD_{50} for the rat is 1400 mg/kg. The ester is prepared by partial hydrolysis of the corresponding thionophosphoryl chloride.
Pyrophosphoric acid amides are also known. Thus the same tetramethyl phosphorodiamidochloridate, from which the phosphoric acid fluorides are obtained, is the starting material for OMPA (IV) [Octamethyl pyrophosphoroamidate], first described by SCHRADER in 1941. The synthesis is carried out by well-known procedures [509, 748, 843, 961]:

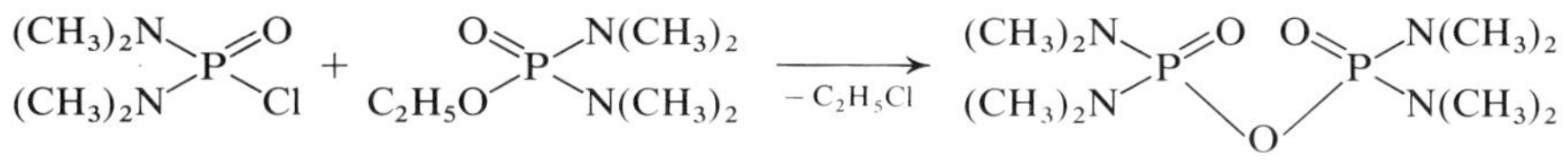

(26)

schradan, OMPA (IV)

For large-scale manufacture, the partial hydrolysis of tetramethyl phosphorodiamidochloridate [377, 881] was selected.

(27)

In this way OMPA can be obtained in a one-step reaction from phosphoryl chloride and dimethylamine without first isolating the dichloride. The LD_{50} of this compound for the rat after oral administration is 8 mg/kg; it thus belongs to the more toxic phosphoric acid esters. OMPA is well-known for its systemic insecticidal properties but has now been replaced by the more economical ®Systox series (see p. 118f.) which are better tolerated by plants. In honour of the discoverer, this compound has also been named *schradan* and is of historical significance in so far as it was the first systemic phosphoric ester insecticide to be recognized. OMPA was introduced commercially by Pest Control Ltd. under the name ®Pestox III.

2. Preparation of Phenol Esters

By far the greatest number of insecticidally active phosphoric acid esters are obtained by acylation of phenols or heterocyclic hydroxy compounds with phosphoryl and phosphonyl chlorides:

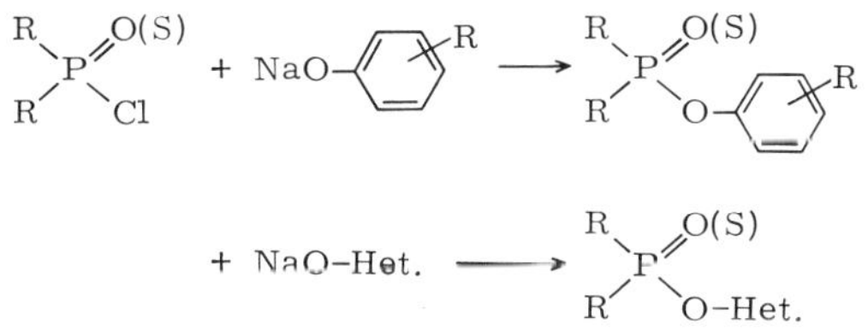

(1)

(2)

$R = R^1, R^1O, R_2N$

The fluorides and anhydrides discussed in the last paragraph are to be regarded as forerunners in the development of the ®E 605 group discovered by SCHRADER in 1944, which includes the most important insecticides both in quality and quantity.

The first member of the ®E 605 series is O,O-diethyl O-*p*-nitrophenyl phosphate with the laboratory number ®E 600 and today known as *paraoxon* (I).

The insecticidal and acaricidal properties of *paraoxon* are exceptional, but due to this high acute mammalian toxicity (oral LD_{50} for the rat 3 mg/kg) ®E 600 was hardly ever used as a systemic and contact insecticide. It is, however, used in ophthalmology as a miotic agent under the name ®Mintacol.

It may be prepared by two processes: firstly by the reaction of diethyl phosphorochloridate with *p*-nitrophenolate [830, 961]:

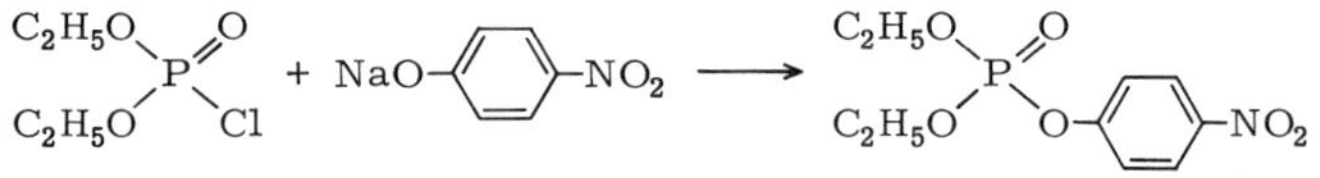
(3)

paraoxon, ®E 600 (I)

or secondly by phosphorylation of phenol with subsequent nitration [863]:

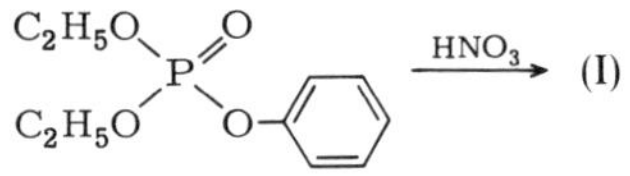
(4)

Paraoxon is fairly soluble in water and able to penetrate into the plant, exerting a systemic insecticidal action (see p. 125).

The corresponding phosphonate (II) is used in human medicine for glaucoma, and under the name Armin [864] in gynaecology. With an oral LD_{50} of 1 mg/kg for the rat, Armin or O-ethyl O-*p*-nitrophenyl ethanephosphonate is even more toxic than *paraoxon*. Its cholinesterase-inhibiting activity (I_{50}) is 2×10^{-9} mol/liter, i.e. about ten times that of *paraoxon*, and of the same order as for TEPP. *Paraoxon*, Armin and TEPP are, therefore, among the most potent cholinesterase inhibitors.

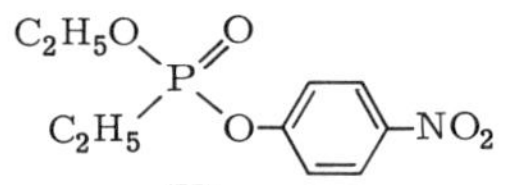

(II)

The most important insecticide with the broadest spectrum of activity in the phosphate series is *parathion* (III) [O,O-diethyl O-*p*-nitrophenyl phosphorothioate] which SCHRADER synthesized in 1944 [879]. The biological activity was found by KÜKENTHAL in the same year [863a]. The trade name ®E605 is the current number in SCHRADER's laboratory note book.

In 1948 it was suggested by UNTERSTENHÖFER for the control of orchard pests [972].

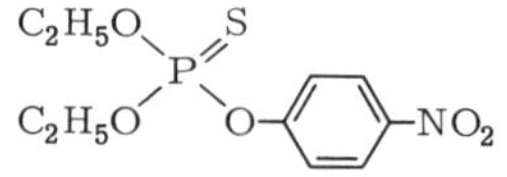

parathion, ®E 605 (III)

There are two methods for its industrial synthesis [829], which differ only in the preparation of the O,O-diethyl phosphorothiochloridate (cf. p. 53f.). The German process starts with thiophosphoryl chloride, the American with P_4S_{10}. The last step, phosphorylation of the *p*-nitrophenol sodium salt (Eq. (5)) is the same in both routes. Numerous variations are suggested for the special reaction conditions [138, 185, 288, 965].

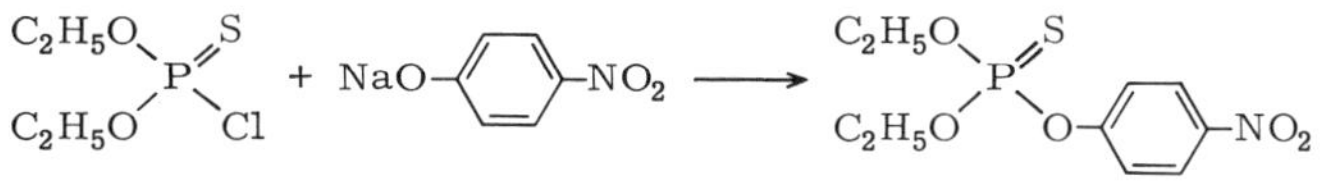

(III) (5)

Parathion has an oral LD_{50} for the rat of 6.4 mg/kg and 40 mg/kg (rabbit cutaneously) [24]. It belongs, therefore, to the rather toxic phosphates and can cause severe poisoning in inexperienced hands. ®E 605 is stable in aqueous solution at pH 4–8, but is rapidly hydrolyzed at pH 9–11. In order to give an idea of the insecticidal activity of *parathion*, a number of its active concentrations are listed. (LC_{95} values for different insects) [24]:

Table 4. Biological activity of *parathion*

Mosquito (*Aedes aegypti*)	0.000005 %
Red spider mite (*Tetranychus urticae*)	0.001 %
Potato Aphid (*Macrosiphon solanifolii*)	0.0005 %
Citrus fruit spider (*Paratetranychus citri*)	0.0001 %
Greenhouse thrips (*Heliothrips haemorrhoidalis*)	0.0001 %
House fly (*Musca domestica*)	0.0001 %

Parathion works as a non-systemic contact and stomach insecticide, as well as ovicide [691]. It is noteworthy, however, that *parathion* shows a "depth action", i.e. it penetrates into the plant to some extent but without any true translocation [300]. On the living plant the substance is broken down relatively rapidly (within 2–8 days) which is largely due to the effect of sunlight and enzyme systems of the plant. On inanimate materials it persists much longer.

On the one hand both *paraoxon* and *parathion* possess excellent insecticidal properties, yet on the other they are relatively toxic to mammals. SCHRADER therefore sought to synthesize esters with as low a toxicity as possible to ensure maximum safety for all users. He came considerably closer to this goal with the change from ethyl to methyl esters.

Parathion-methyl (IV) or O,O-dimethyl O-*p*-nitrophenyl phosphorothioate is a much-used contact and stomach insecticide [833] (trade names: ®Folidol-M, ®Dalf, ®Metacide etc.).

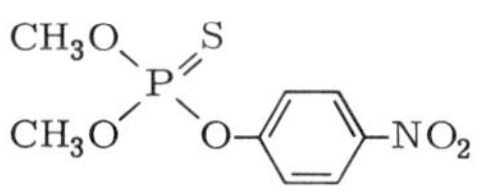

parathion-methyl, ®Folidol-M (IV)

The oral LD_{50} for the female rat is 15–20 mg/kg, for the rabbit 300–400 mg/kg cutaneously. The skin toxicity is, therefore, ten times more favourable than that

of the diethyl compound, i.e. the rabbit skin is considerably less permeable to the dimethyl esters. Their manufacture follows the standard procedures.
At elevated temperatures rearrangement of *parathion-methyl* to the thiol ester (V) can be effected [620].

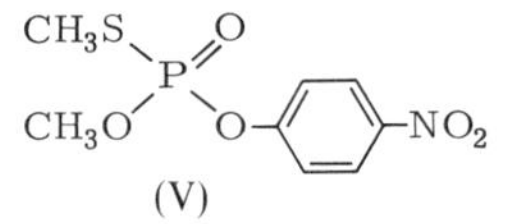

(V)

In practice, however, isomerization is best carried out with a base such as sodium hydroxide, so that a salt of the ambident anion (VI) is obtained, which can be alkylated on the sulfur with dimethyl sulfate [863].

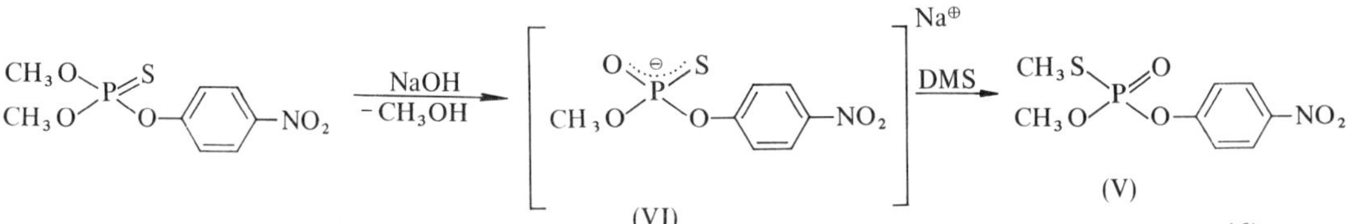

(6)

The thionoesters can be oxidized to the corresponding P=O compounds using oxidizing agents such as chlorine, bromine or nitric acid. This step increases the serum cholinesterase inhibiting activity from $I_{50} = 10^{-2}$ mol/liter to the methyl-paraoxon value of $I_{50} = 10^{-6}$ mol/liter [62].
Surprisingly, a less toxic compound was found in the series of phosphonates. EPN (VII) [O-ethyl O-*p*-nitrophenyl benzenephosphonothionate] was described by JELINEK in 1948 and was the first phosphonic acid ester to be put on the market in 1949 [77, 446].

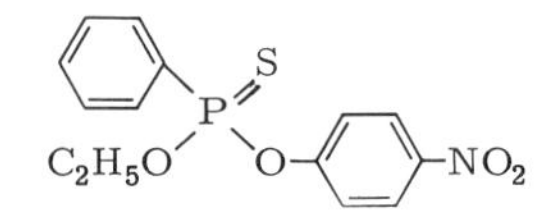

EPN (VII)

The compound is an insecticide and acaricide, its synthesis follows in principle the same route as that depicted for *parathion*. The ester chloride required may be obtained by the method of KINNEAR and PERREN described on p. 56.
The LD_{50} for the male rat orally is 36 mg/kg. Because it has been shown that the toxicity in the rat generally increases about tenfold when a phosphate is replaced by its corresponding phosphonate, EPN would appear to be somewhat of an exception. Benzenephosphonic acid esters are, however, in general less toxic than analogously composed alkanephosphonic acid esters.

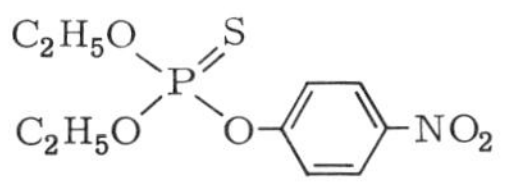

LD_{50}: 6.4 mg/kg orally for the rat

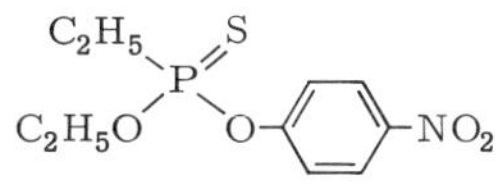

LD_{50}: 2.5 mg/kg orally for the rat

A more recent example is ®Colep (VIII) [O-phenyl O-*p*-nitrophenyl methanephosphonothioate], which was filed for a patent in 1962. Colep is a contact insecticide with a specific spectrum of activity [187].

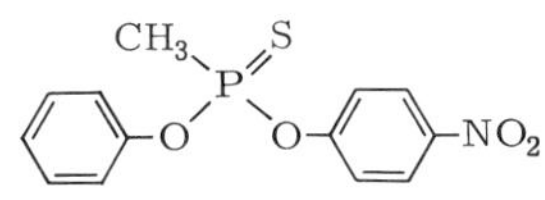

(VIII)

If an alkoxy group in the *parathion* molecule is replaced by a dimethyl aminogroup, then toxicity and biological activity are reduced. Both are also lowered when the thionoester is replaced by the alkylthio-ester (see p. 194). Methyl substitution in the phenyl ring results in a decreased toxicity without a significant alteration in biological activity. Stability, however, towards hydrolytic influences is improved. *Fenitrothion* (IX) [O,O-dimethyl O-(3-methyl 4-nitrophenyl) phosphorothioate] as an example of the *parathion* series, was first described by DRÁBEK and PELIKÁN [249] and later named *metathion.* Quite independently, both Farbenfabriken Bayer [561] (®Folithion) and Sumitomo Chemical Co., [47, 936] (®Sumithion) were working on the same compound.

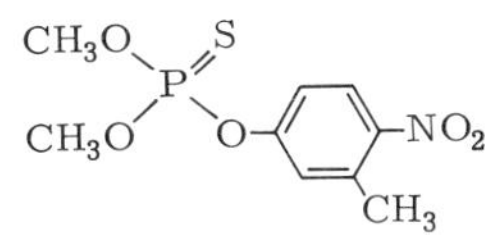

®Folithion, ®Sumithion (IX)

Fenitrothion has an oral *LD_{50}* of 500 mg/kg for the female rat. It was introduced onto the market in 1961. The spectrum of activity as contact and stomach insecticide is comparable to that of *parathion-methyl.* Furthermore, in the field of hygiene, it is suitable for controlling flies and mosquitoes which are resistant to chlorohydrocarbons. The synthesis follows the standard procedures.
Introduction of a chlorine atom into the *o*-position to the nitrogroup also markedly diminishes the toxicity. For instance, the oral *LD_{50}* of (X) in the male rat is 880 mg/kg. ®Chlorthion (X) [O,O-dimethyl O-(3-chloro-4-nitrophenyl) phosphorothioate] [847] was introduced in 1952 by Farbenfabriken Bayer.

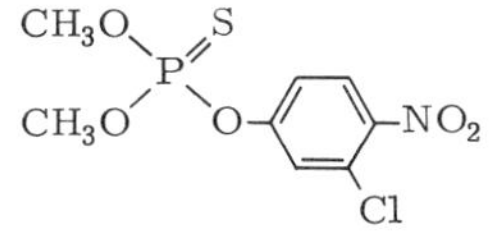

(X)

At first, the synthesis of 3-chloro-4-nitrophenol proved difficult. During normal nitration of *m*-chlorophenol, 3-chloro-6-nitro-phenol is produced in addition to the desired 3-chloro-4-nitrophenol. If *m*-chlorophenol is esterified, e.g. phosphorylated [628], and subsequently nitrated, then substitution occurs exclusively in the 4-position.

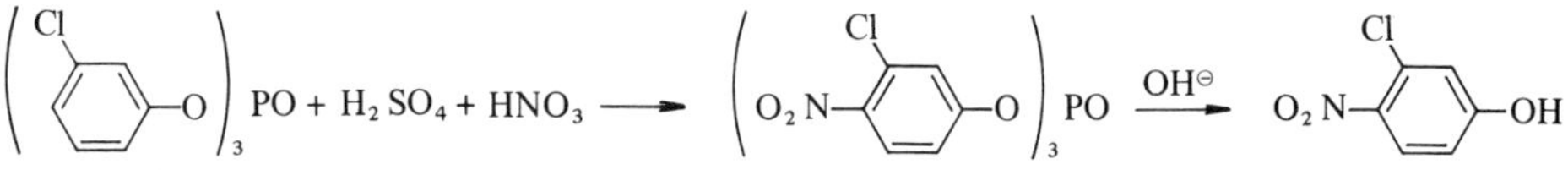

(7)

The ester is then hydrolyzed again and the 3-chloro-4-nitrophenol phosphorylated in the usual manner.

The insecticidal properties remain almost intact, but sensitivity to hydrolytic influences increases somewhat. Chlorthion is effective as a contact insecticide, especially against mosquito larvae and DDT-resistant flies and has become established in cotton cultivation. Metabolism in the insect organism results in the corresponding compounds as found with the other nitrophenyl esters (see pages 226, 234).

Possible explanations for the diminished toxicity on *m*-substitution are, for instance:

a) partial dearrangement of the planar position of the nitro group to the ring weakens the mesomeric effects, hence, the tendency to hydrolysis indicated by the σ values according to Hammett (physicochemical explanation) and/or
b) steric hindrance to the approach of the nitro group to a biochemical surface at the site of action in the organism.
c) the *m*-substituent may be involved in selective interactions with different enzymes (cf. p. 179).

Placing the chlorine atom in the *o*-position to the hydroxyl group decreases somewhat both insecticidal and toxic properties. This compound was introduced by American Cyanamid Company under the name *dicapthon* (XI) [O,O-dimethyl-O-(2-chloro-4-nitrophenyl) phosphorothioate] in 1954 [229, 287, 321, 329, 852].

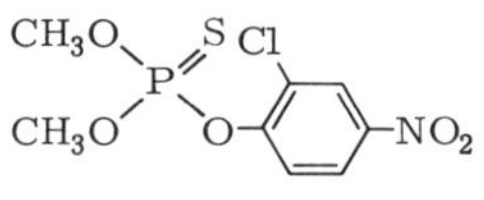

(XI)

The oral LD_{50} in the male rat is 400 mg/kg. *Dicapthon* is used mainly against flies. Its synthesis follows the methods mentioned above. If the positions of the chlorine and nitro groups are interchanged, compounds are obtained with poor insecticidal properties and average toxicity.

Exclusive chlorine substitution in the phenyl ring leads to ®VC 13-Nemacide (XII) [O,O-diethyl O-2,4-dichlorophenyl phosphorothioate] [123], which was developed by Virginia-Carolina Chemical Corporation in 1955.

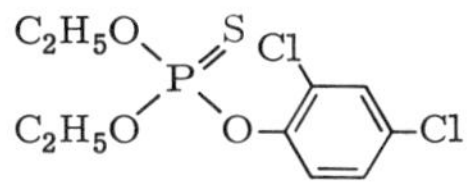

(XII)

The oral LD_{50} for the rat is 270 mg/kg. In comparison to the compounds so far described, it is rather ineffective as an insecticide, but possesses remarkable nematicidal activity. VC-13 was the first organophosphate to control soil nematodes. Synthesis is carried out in the usual manner by phosphorylation of 2,4-dichlorophenol. A closely related compound is ®Zytron (XIII) [O-methyl O-(2,4-dichlorophenyl) N-isopropyl phosphoroamidothioate] [538], originally introduced by the Dow Chemical Company as a systemic insecticide and now used as a herbicide.

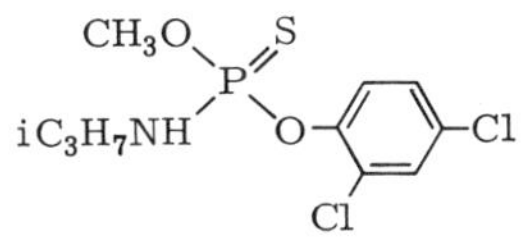

(XIII)

A benzenephosphonate of 2,4-dichlorophenol was marketed recently by the Japanese Nissan Kagaku Company [186, 854] under the name ®S-Seven (XIV) [O-ethyl O-(2,4-dichlorophenyl) benzenephosphonothioate] with the following structural formula:

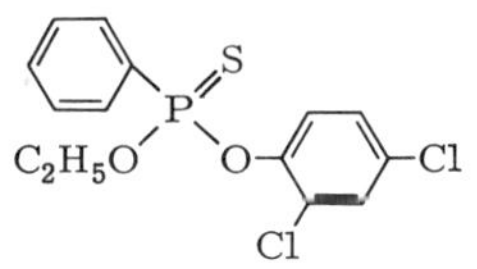

(XIV)

This product may be regarded as a combination of the phenol of ®VC-13 and the ester radical of EPN. It has been assigned as an acaricide and soil insecticide for the control of maggots, flea-beetles, potato or root mites.

The trichloro-derivative Ronnel or *fenchlorphos* (XV) [O,O-dimethyl O-(2,4,5-trichlorophenyl) phosphorothioate] [600, 683, 940] has found wide application in the veterinary medicine under the name ®Nankor.

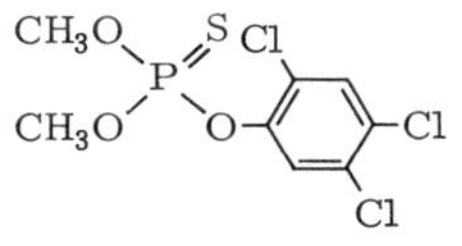

(XV)

Because of its low mammalian toxicity, it was introduced in 1954 on an experimental basis in crop protection by Dow Chemical Company but on account of its phytotoxicity it has never become established as an agricultural insecticide. The oral LD_{50} for the male rat is 1250 mg/kg. The ester amides of Nankor have also been described [100, 101]; their phytotoxicity, however, only permits their use as insecticides for hygiene.
The diethyl phosphonate of 2,4,5-trichlorophenol was recently suggested by Farbenfabriken Bayer as agent against soil insects, wire worms and miner flies [419]. It is on the market under the name ®Agritox, *trichloronate* (XVI) [O-ethyl O-(2,4,5-trichlorophenyl) ethanephosphonothioate] [812].

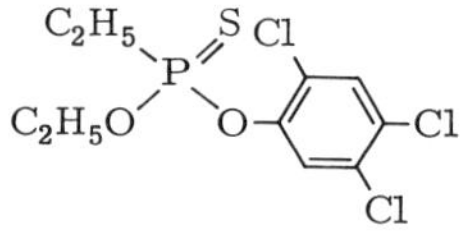

(XVI)

The oral LD_{50} for the male rat is 16 mg/kg.
If the chlorine atom in the 4-position is replaced by bromine, *bromophos* (XVII) [O,O-dimethyl O-(4-bromo-2,5-dichlorophenyl) phosphorothioate] is obtained [896]. *Bromophos* is an effective agent against ectoparasites in mammals, well tolerated by the skin and mucous membranes, and possessing low mammalian toxicity with an oral LD_{50} for the rat of 3750 – 6100 mg/kg. In crop protection, *bromophos* serves not only as an insecticide but also as an acaricide.

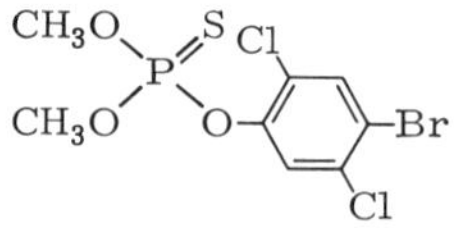

(XVII)

For the synthesis of the corresponding trihalo-phenol there exist numerous laboratory procedures which start from *p*-bromophenol [296], from 2,5-dichloro-4-nitrophenol [296] or from 2,5-dichlorophenol [682] which, for its part, can be obtained from 1,2,4-trichlorobenzene with sodium hydroxide under pressure [22]. These processes are hardly feasible in practice, the starting materials being presumably polyhalogen benzenes.
The iodine analogue of *bromophos* was introduced under the name ®Nuvanol N, *iodofenphos* (XVIII) [O,O-dimethyl O-(2,5-dichloro-4-iodophenyl) phosphorothioate] by Ciba as an insecticide and since 1965 has been protected in Germany by patent [91].

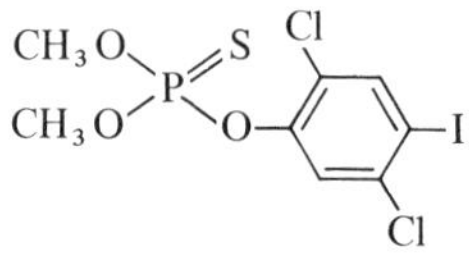

®Nuvanol N (XVIII)

The phosphonate analogue of *bromophos*, ®Phosvel (XIX) [O-methyl O-(2,5-dichloro-4-bromo-phenyl) benzenephosphonothioate] was filed as patent by Velsicol Chemical Co. in 1965 in the USA [770]. Its synthesis can clearly be seen from the structural formula.

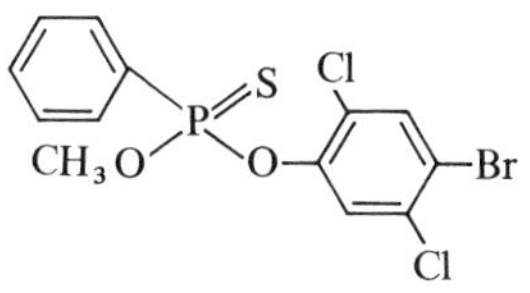

®Phosvel (XIX)

Phosvel has a broad spectrum of activity against insects in corn, cotton, vegetables, fruit and sugar cane, and has also been used to control the rice stem borer and *Prodenia litura.*

In general, the esters of halogen-substituted phenols are said to be usually of low mammalian toxicity. They are insecticides of average toxicity, often acting systemically and therefore suitable for the control of ectoparasites in cattle, [292], in some cases usable as herbicides. The chlorophenyl phosphates, therefore, possess a noteworthy spectrum of biological activity (cf. p. 191).

Cyanophenols are related to the halogen phenols, the phosphoric acid derivatives of which possess, as might have been expected, good insecticidal properties. In 1961 ®Cyanox or *cyanophos* (XX) [O,O-dimethyl O-(4-cyanophenyl) phosphorothioate] was claimed as patent in USA by Sumitomo Chemical Co. [525].

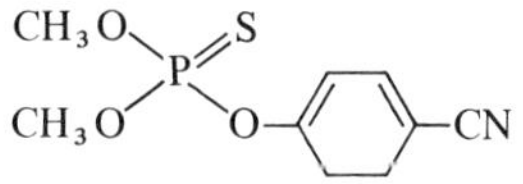

®Cyanox (XX)

The oral LD_{50} for mice is 920 mg/kg. Cyanox kills the rice stem borer, paddy borer, purplish stem borer and insects injurious to health, especially house flies.

A phosphonate of *p*-cyanophenol was also marketed by Sumitomo Chemical under the name ®Surecide (XXI) [O-ethyl O-(4-cyanophenyl) benzenephosphonothioate]; as in the case of Cyanox, its synthesis is evident from the structural formula [526].

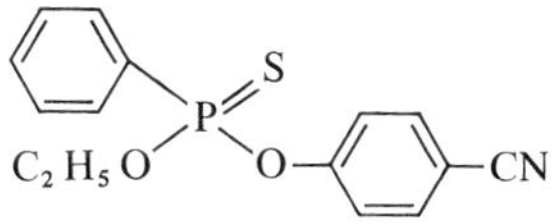

cyanofenphos, ®Surecide (XXI)

The compound has insecticidal properties, particularly against rice stem borer (*Chilo suppressalis* Walker etc.) and house flies. The oral LD_{50} for mice is 46 mg/kg.

The esters of alkylthiophenols which have been investigated by Farbenfabriken Bayer since 1956, are apparently at variance with SCHRADER's rule (see p. 40). Although one might expect a reduction in insecticidal activity in view of the fact that the methylthio-group, in comparison for example to the nitro group, decreases the hydrolysis of the P–O bond, one does, in fact, find a surprisingly high insecticidal activity in esters of *p*-methyl thiophenol such as (XXII) [O,O-dimethyl O-(*p*-methylthio)-phenyl phosphorothioate] [338]. The mammalian toxicity of the compound is also unexpectedly high. The oral LD_{50} for the rat is 10 mg/kg.

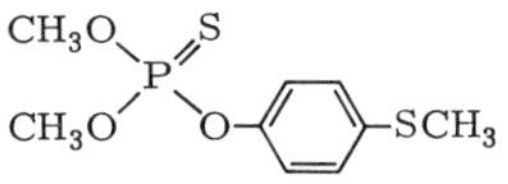

(XXII)

The principle of introducing a methyl group into the *o*-position to the substituent, which has already been mentioned in connection with Folithion and Chlorthion, leads in this series to a considerable reduction in mammalian toxicity. The resulting compound (XXIII) is known under the names *fenthion* and ®Baytex [O,O-dimethyl O-[3-methyl 4-methylthio-phenyl] phosphorothioate] [799]. The active constituent was synthesized in 1956 and was described as an hygiene insecticide (®Baytex) for the first time by JUNG, KÜKENTHAL and TECHNAU in 1959 [459, 460]. UNTERSTENHÖFER reported on the same compound as an agricultural insecticide (®Lebaycid) [859, 983].

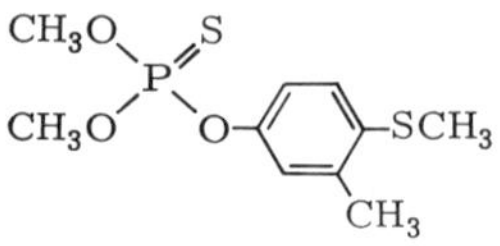

fenthion, ®Baytex, ®Lebaycid (XXIII)

The oral LD_{50} of this compound for the male rat is 313 mg/kg. If the methyl group is in the *m*-position to the methylthio-group, then mammalian toxicity and insecticidal action are both diminished. Elegant procedures for the synthesis of the necessary 3-methyl 4-methyl thio-phenol (XXIV) were developed by DELFS and WEDEMEYER: these involved the reaction of *m*-cresol with dimethyl disulfide and Lewis acids [237], with methyl sulfene chloride [238] (Eq. (8)),

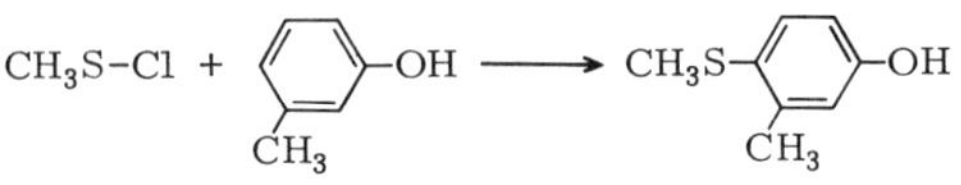

(XXIV) (8)

or with dimethyl sulfoxide and hydrochloric acid [1006] (Eq. (9)).

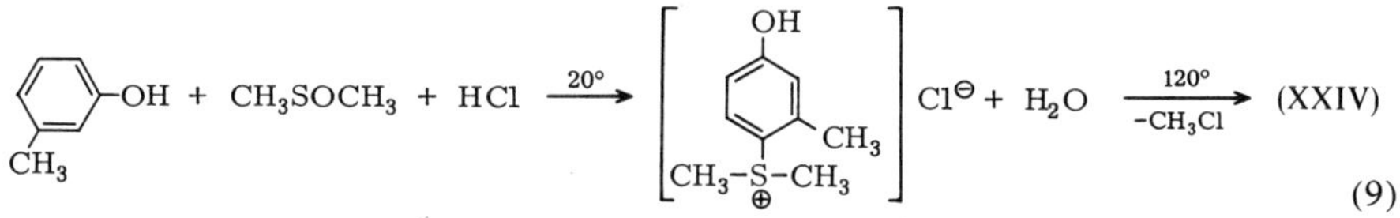

(9)

The usual methods are used for phosphorylation. *Fenthion* is a versatile product on account of its contact and stomach insecticidal properties. In the field of hygiene it is known under the name of ®Baytex and as ®Tiguvon in veterinary medicine as a broad-spectrum insecticide suitable for controlling insects that have become resistant to chlorinated hydrocarbons. Because of its stability towards hydrolysis a considerable residual effect is obtained. Some insecticidal data for *fenthion* are given in the following table [863]:

Table 5. Biological action of *fenthion*

	Conc.		Mortality
Calandra granaria	0.004 %		100 %
Sitona spp.	0.01 %		100 %
Phytonomus variabilis	0.01 %		80 %
Apion pisi	0.01 %		80 %
Phyllobius piri	0.1 %		100 %
Rhizotrogus solstitialis	0.001 %		100 %
		LC_{50}	
Plutella maculipennis		0.003 %	
Tortrix viridana		0.0004 %	
Doralis fabae on Vicia faba		0.0007 %	
Myzodes persicae on Brassica spp.		0.001 %	

The substance is of particular interest in the control of fruit flies and the rice stem borer (®Lebaycid), of flies and mosquitoes (®Baytex), and larvae of warble fly (®Tiguvon). The corresponding ethyl ester is used in veterinary medicine under the name ®Lucijet (XXV) [O,O-diethyl O-(3-methyl 4-methylthiophenyl) phosphorothioate] against blowfly larvae [79, 800].

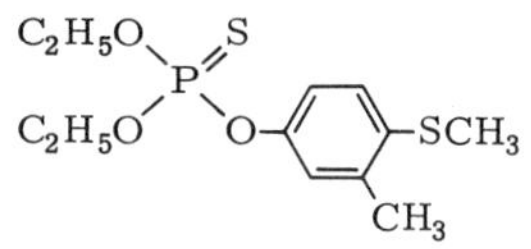

(XXV)

The oral LD_{50} of the compound for the rat is 100 mg/kg.

The amido-derivative ®Nemacur P (XXVI) [O-ethyl O-(3-methyl 4-methylthiophenyl) N-isopropyl phosphoroamidate] (sugg. common name: *phenamiphos*) was developed by Farbenfabriken Bayer AG [472] for the control of soil and leaf nematodes. Additionally, it acts as a contact insecticide with systemic properties. The compound is obtainable, for instance, by reacting 3-methyl 4-methylthiophenol with phosphoryl chloride to the ester dichloride. Nucleophilic displacement of the chlorine atoms by sodium ethylate and isopropyl amine yields Nemacur P with a melting point of 48 °C:

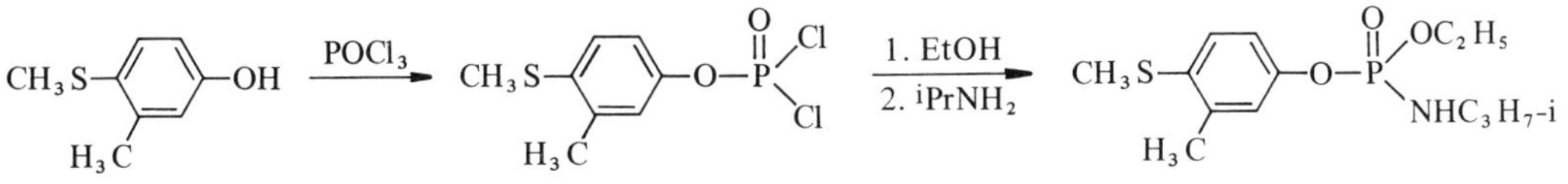

®Nemacur P (XXVI)

The oral LD_{50} for the rat is about 15–20 mg/kg. As a result of its highly nematocidal activity and favourable partition in the soil, Nemacur P can also be applied in the form of granules ensuring low costs of soil and plant treatment.
From a comparison of the *p*K values of phenol (9.99), *p*-methoxyphenol (10.2) and *p*-methylthiophenol (9.53) it might be expected that the biological activity of *fenthion* is only of the same order as is shown by the corresponding esters of phenol or *p*-methoxyphenol, both of which are rather inactive. This contradiction to the "acyl" rule is explained by the fact that the thioether group in the *p*-position is rapidly converted in the organism into the sulfoxide (see p. 236, 237) (cf. [631]).

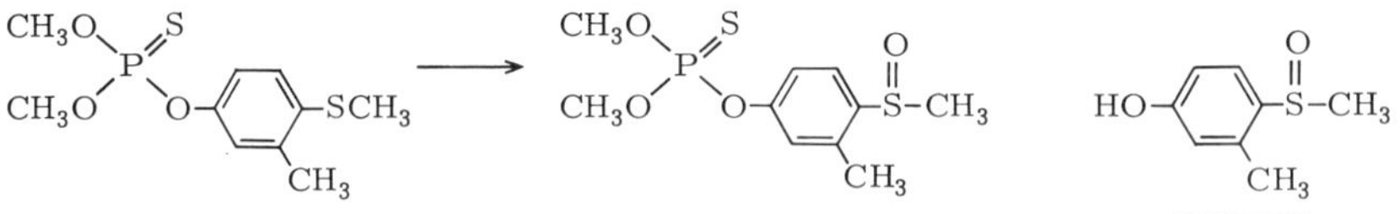

(XXVII) (XXVIII)

The *p*K value of the *p*-methylsulfinylphenol (XXVIII) is 8, the esters of this phenol therefore fit well into the scheme referred to: the actual active constituent is not the thioether (transport form), but the product resulting from the "lethal synthesis" (see p. 227, 230), i.e., the sulfoxide (active form).
By oxidation the mammalian toxicity is enhanced in the same manner (oral LD_{50} of the sulfoxide is 125 mg/kg for the rat). The *p*-methylsulfinyl ester as such exhibits the same insecticidal properties as the corresponding thioether and

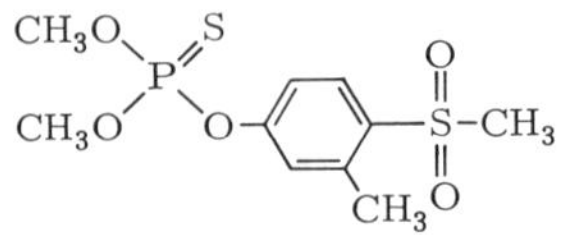

(XXIX)

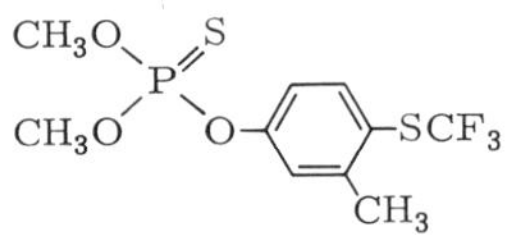

(XXX)

therewith offers an indirect confirmation of the theory. The next oxidation product, the sulfone (XXIX) has the same LD_{50} of 125 mg/kg but the biological activity is considerably reduced.
The mammalian toxicity of the trifluoromethylthio compound (XXX) resembles that of the methyl compound, its insecticidal activity is, however, somewhat lower. The phosphonates of this series are without exception good insecticides but possess a high mammalian toxicity (oral LD_{50} 1–2 mg/kg for the rat) so that they have received little attention.
If a second methyl group is introduced into the *fenthion* molecule in the *m*-position, then the toxicity falls further (oral LD_{50} 1000 mg/kg rat) (XXXI), as does unfortunately the biological activity. Surprisingly, after oral application its activity against the blowfly remains completely intact, as was found by BEHRENZ [79] who had also suggested the diethyl ester for use against ectoparasites in sheep (XXXII):

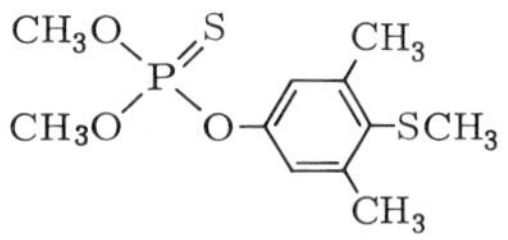

(XXXI)

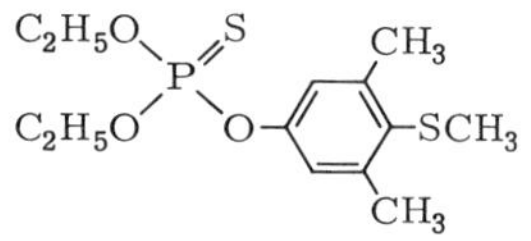

(XXXII)

Its oral LD_{50} for the male rat is 150 mg/kg.
The diethyl ester of the sulfoxide possesses both insecticidal and nematicidal activity. It has an LD_{50} of 4 mg/kg orally for the male rat.
Its trade name is ®Terracur P (common name: *fensulfothion*) (XXXIII) [O,O-diethyl O-(4-methylsulfinyl phenyl) phosphorothioate] [798]. The pesticide was put on the market in 1965 by Farbenfabriken Bayer.

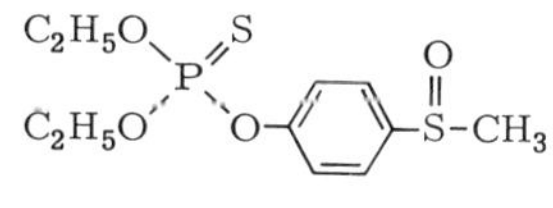

(XXXIII)

Recently phosphoric acid esters of sulfonamides have also appeared. Under the name *famphur* (XXXIV) American Cyanamid [93, 996] introduced O,O-dimethyl O-(4-dimethyl sulfamoyl phenyl) phosphorothiate.

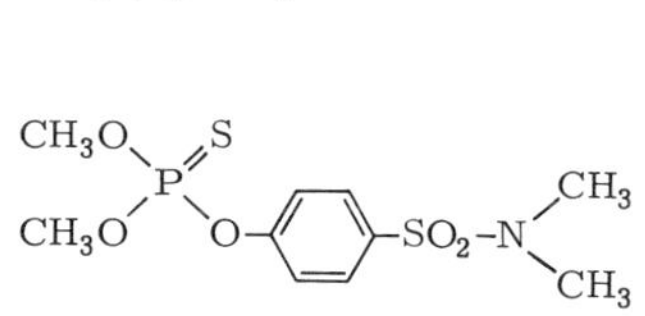

(XXXIV)

This class of compound shows systemic, nematicidal and anthelmintic activity with relatively low mammalian toxicity. The synthesis follows the usual methods from 4-hydroxy-benzenesulfonyl dimethylamide and thionophosphoryl chlorides in non-aqueous solvents. A noticeable loss in biological activity results if the sulfamoyl group stands in the *o*- or *m*-position to the phenolic group. A loss in activity is also incurred if the nitrogen is substituted with alkyl groups of increased size, with aryl groups, or belongs to a heterocyclic system. The diethyl derivative is also to be introduced commercially as a nematicide (XXXV) [680].

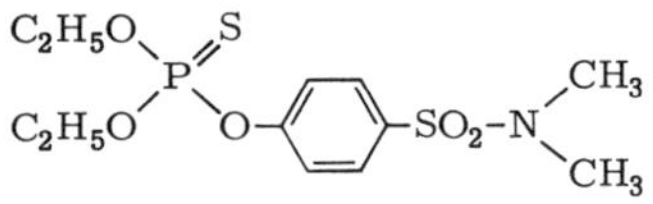

(XXXV)

The oral LD_{50} for the mouse is 23 mg/kg.
®Ruelene (XXXVI) [O-methyl O-4-tert.butyl 2-chlorophenyl N-methyl phosphoroamidate] was suggested in 1959 by Dow Chemical as a less toxic substitute for phenothiazine in the eradication of intestinal parasites [32, 532].

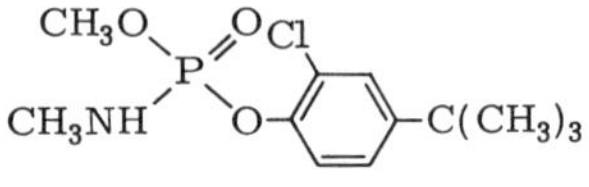

(XXXVI)

The oral LD_{50} for the rat is 1000 mg/kg. It can be synthesized from the corresponding substituted phenol, which is acylated with phosphoryl chloride:

(10)

The halogen atoms are replaced successively by alkoxyl and methyl amino groups.

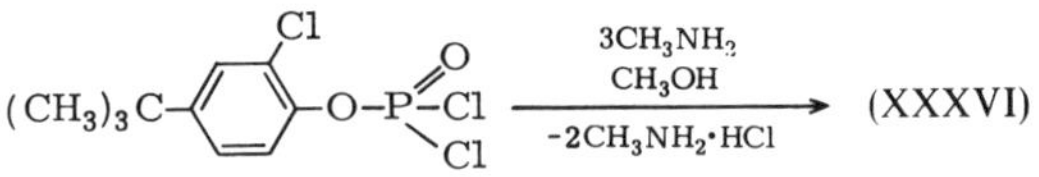

(11)

The thiono-analogue of ®Ruelene, ®Dowco 109 or ®Narlene (XXXVII) [O-methyl O-(2-chlorophenyl 4-tert.butyl) N-methyl phosphoroamidothioate] was also

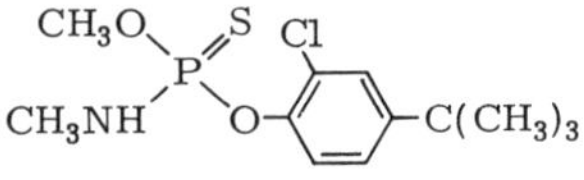

(XXXVII)

proposed as an anthelmintic, but compared with ®Ruelene had the disadvantage of accumulating in adipose tissue [137, 471, 532, 954].
In 1958 Dow Chemical offered ®Dowco 109 as an experimental compound. It is synthesized in a similar manner to ®Ruelene, with the exception that the sulfur is introduced afterwards (Eq. (12)).

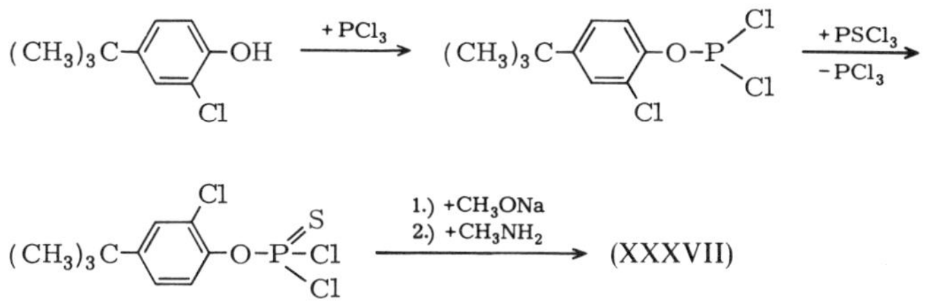

(12)

The oral LD_{50} for the rat is 1000 mg/kg.
The simplest compound of this series obtained by phosphorylation of phenol is ®Nellite (XXXVIII) [O-phenyl N,N-dimethyl phosphorodiamidate].

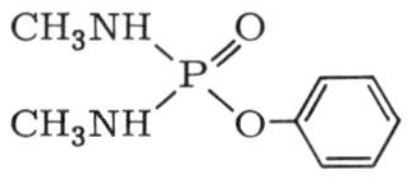

(XXXVIII)

It was also marketed by Dow Chemical in 1959 [1042] and has an oral LD_{50} of 140–200 mg/kg for the rat. Nellite fails to act as an insecticide but is peculiarly suited to the control of root nematodes.
A somewhat unusual type of phosphoric acid ester is ®Gophacide (XXXIX) [O,O-bis-(4-chlorophenyl)-N-acetimidoyl phosphoramidothioate].

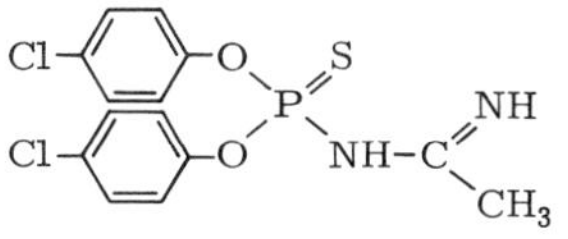

(XXXIX)

It is offered by Chemagro Corporation as a selective rodenticide having an oral LD_{50} for the male rat of 7.5 mg/kg [593].
®Fujithion (XL) [O,O-dimethyl S-(4-chlorophenyl) phosphorothioate] was put on the market by Ihara Chemical Co. in 1969 as an weather-resisting insecticide for controlling rice stem borer and leaf hopper. Monsanto Chemical Co. described this compound in a patent [98] as destroying undesired plants. Its synthesis may be carried out according to a patent of Schrader [860] (Eq. (13)):

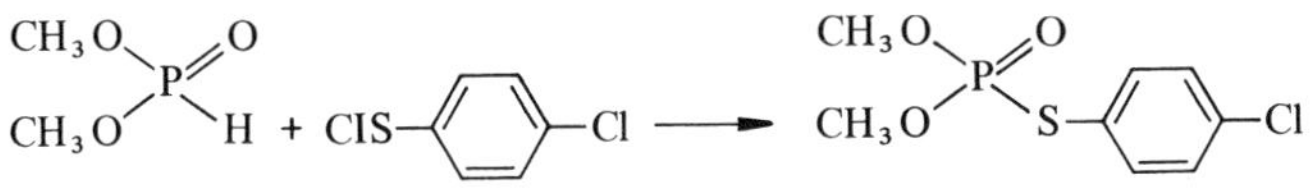

(13)

®Fujithion (XL)

In 1967 Stauffer Chemical Co. introduced ®Dyfonate (XLI) [O-ethyl S-phenyl ethanephosphonodithioate] as insecticide against soil insects [939]. The compound has an oral LD_{50} for the male rat of 8 – 17 mg/kg and is synthesized from thiophenol and O-ethyl ethanephosphonothiochloridate with triethylamine.

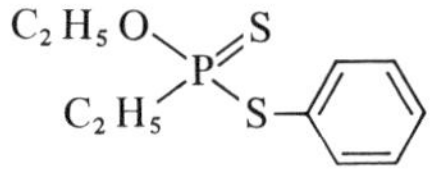

fonofos, ®Dyfonate (XLI)

Due to high manufacturing costs, the bis-phosphoryl compounds have not yet become established. Recently, however, ®Abate (XLII) [4,4′-bis-(O,O-dimethyl thionophosphoryloxy) diphenyl sulfide] was proposed for controlling malathion-resistant mosquito larvae [946].

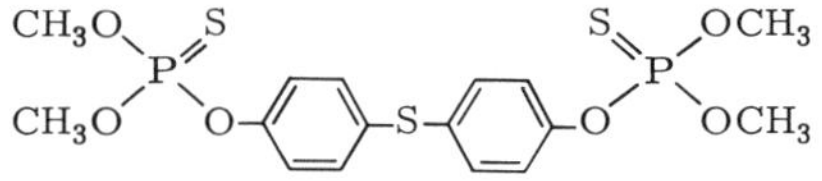

®Abate (XLII)

It has a notably low toxicity with an oral LD_{50} of the order of 2000 mg/kg for albino rats. ®Abate and its analogues were developed in 1960 by Farbenfabriken Bayer [904] and in 1963 by American Cyanamid [580]. It is prepared from the corresponding phenol and phosphoryl chloride in the presence of aqueous sodium hydroxide. The bis-(*p*-hydroxyphenyl) sulfide itself is obtained by reaction of phenol with sulfur dichloride.

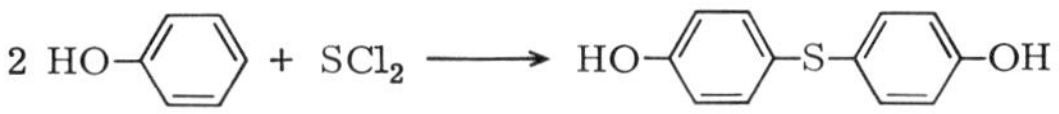

(14)

The closely related disulfide (XLIII) [805] (4,4′-bis-(O,O-dimethyl thionophosphoryloxy) diphenyldisulfide) acts as an excellent larvicide, killing 100 % mosquito larvae in a dilution of 0.000.000.000.01 %. This compound was synthesized by SCHICKE of Farbenfabriken Bayer in 1963.

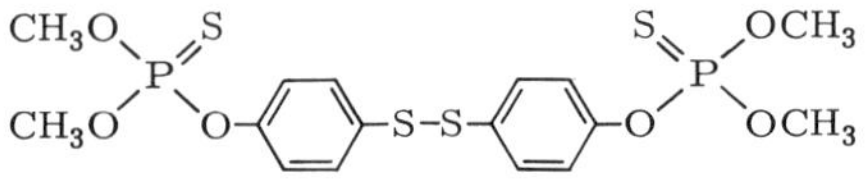

(XLIII)

The oral LD_{50} for the rat is 10–25 mg/kg. The bis-(*p*-hydroxy-phenyl)-disulfide may be prepared by reacting phenol with disulfur dichloride:

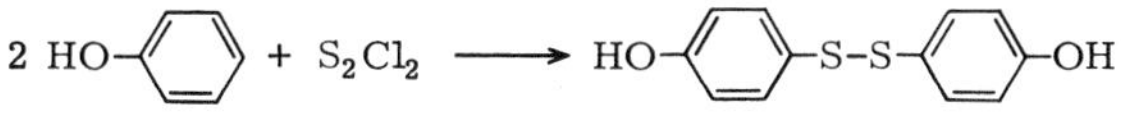
(15)

A bridged diphenyl compound of related structure is ®Slam C (XLIV) [O,O-dimethyl O-(4-(4′-chlorophenylazo)phenyl) phosphorothioate] for which, in 1956, a patent was filed by Montecatini Company [41]. The compound is a practically non-toxic phosphoric acid ester suggested for use as a solution in oil for the control of fly larvae.

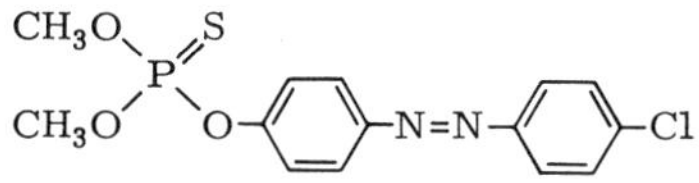
(XLIV)

Phosphorylation of salicyl alcohol, described by Eto, Eto and Oshima [272, 701], results in ®Salioxon (XLV) [2-methoxy-4H-1,3,2-benzodioxa-phosphoran-2-one] or the thiono-analogue ®Salithion (XLVI).

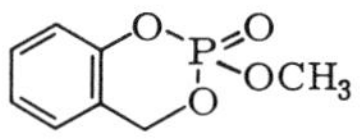
(XLV)

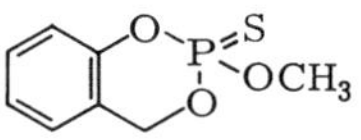
(XLVI)

Both compounds are effective insecticides, whose toxicity in mice is more favourable in comparison to *parathion* [269] (oral LD_{50} of Salioxon: 30 mg/kg, of Salithion: 91 mg/kg) ®Salioxon acts synergistically with *malathion* against resistant house flies and against the silk worm. The size of the group attached to phosphorus (XLV compared to SM-1) exerts a considerable influence on the biological properties of the compound. Aryl groups reduce insecticidal activity and increase synergistic properties; small alkyl groups enhance insecticidal activity and diminish the synergestic action with *malathion*. Steric effects are held mainly responsible for this phenomenon.
The synthesis of ®Salioxon and ®Salithion may be traced to the work of Eto, Casida and Eto [268] who were able to identify the *o*-tolyl ester "SM-1" as a toxic metabolite of tri-*o*-cresyl phosphate which might have resulted from hydroxylation of a methyl group ("hydroxy-TOCP") and intramolecular transesterification:

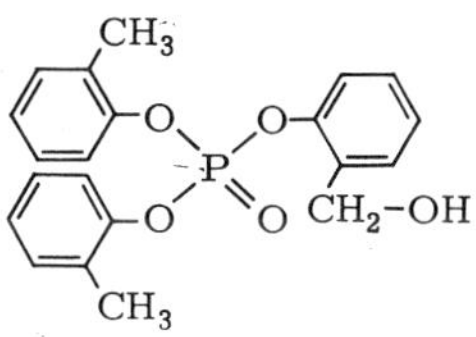
"Hydroxy-TOCP"

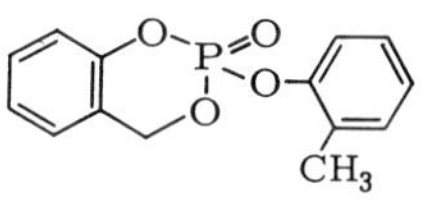
"SM-1"

The corresponding phosphonic acid esters were also synthesized [270]. Their toxicity (oral LD_{50} as mg/kg for mice) is, as might be expected, higher than that of the phosphates. Their hydrolytic stability is lower, which also correlates with the rules. Their activity against various arthropod species is of the same order as that of the phosphoric acid esters, so that overall there are no advantages of this series of compounds.
The phosphorylation of salicylic acid ester results in the effective insecticide ®Optunal (XLVII) [O-methyl O-(2-carboisopropoxyphenyl) phosphoroamidothioate], which is known under the suggested common name *isocarbophos* [876]. A related compound, O-ethyl O-(2-carbethoxyphenyl) N,N-dimethylphosphoroamidate, was synthesized, also by Schrader, in 1948 [830]. ®Optunal may therefore be regarded as a further development of this. Its synthesis follows a procedure shown in Eq. (16), according to which O-methyl phosphorodichlorothioate is reacted with salicylic acid ester in aqueous alkali:

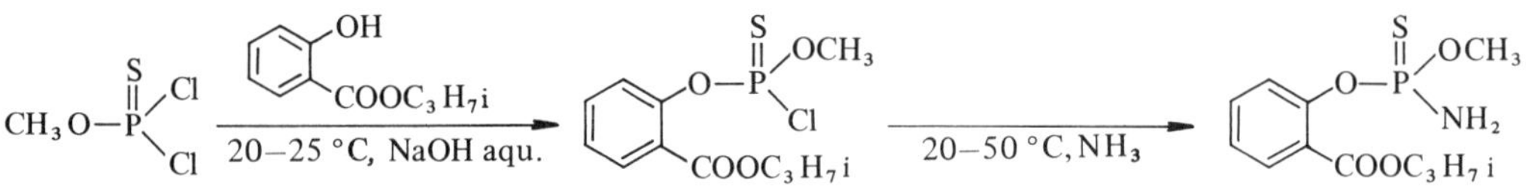

®Optunal (XLVII)

(16)

The oral LD_{50} is 50 – 100 mg/kg for the rat. The new compound is particularly successful in the control of resistant *Heliothis virescens* on cotton; it acts systemically against aphids and leafrollers and is not very persistent in the soil.
The phosphates of condensed heterocyclic compounds with the hydroxyl group on the benzene ring represent a borderline group between the esters of true phenols already discussed and heterocyclic hydroxy compounds.
Probably the oldest example of this series is ®Potasan (XLVIII) [O,O-diethyl O-(4-methyl 2-oxo-2H-1-benzopyranyl-7) phosphorothioate] which was patented by Schrader in 1948 [831, 880]. Unterstenhöfer [973] described the compound as one of the first synthetic insecticides against the Colorado beetle. (The trade name "Potasan" is derived from "potato" and the Latin "sanus".)

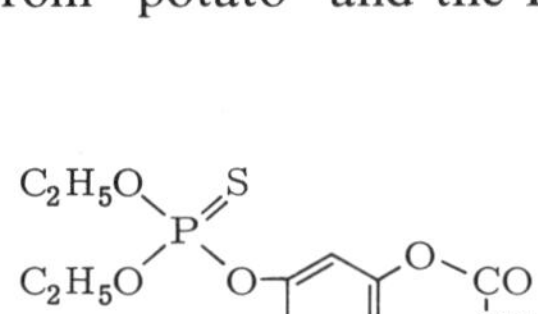

(XLVIII)

The coumarin required is prepared according to the reaction of Pechmann and Duisberg [717] from resorcin and acetoacetate in concentrated sulfuric acid, the phosphorylation following standard procedures. The oral LD_{50} for the male rat is 19 mg/kg. ®Potasan has only a slight contact insecticidal action. Hydrolysis

results in cleavage of an ethyl group giving rise to the inactive de-ethyl compound. The phosphonic acid derivatives are well known and are also good contact insecticides but have found no practical application.
A chlorine derivative of ® Potasan is *coumaphos* (XLIX) [O,O-diethyl O-(3-chloro-4-methyl 2-oxo 2H-1-benzopyranyl-7) phosphorothioate] [838]. The synthesis is analogous, α-chloroacetoacetate being used as starting material for preparation of the coumarin.

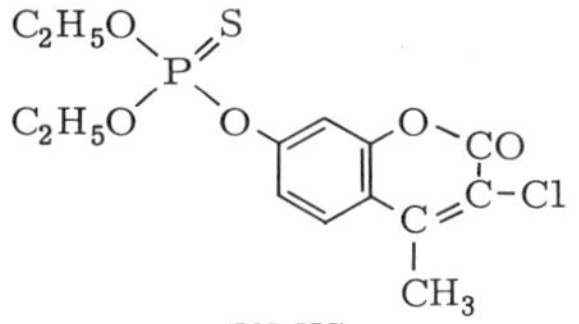

(XLIX)

This compound was also first described by SCHRADER in 1951. The toxicity is reduced by the chlorine atom to an oral LD_{50} for the male rat of 100 mg/kg.
The most surprising effect of the introduction of a single chlorine atom into the 3-position of ®Potasan is almost a thousand-fold increase in activity against mosquito larvae. The substance was therefore applied in the field of hygiene under the name ®Muscatox, later ®Resitox, as a potent larvicide. In veterinary medicine it is known as ®Asuntol (in USA ®Co-ral). Its main field of application is the control of ectoparasites such as screwworms, ticks, blowfly larvae and blowflies in cattle and sheep. On account of its stability towards hydrolysis, it can be used in dips as was suggested by BEHRENZ, FEDERMANN and BOLLE [80]. Metabolism in the mammal results in a loss of the diethyl thiophosphoric acid and opening of the lactone ring.
These metabolites are excreted mainly in the urine. Introduction of halogen atoms into the phosphoryl moiety, i.e. β-chloroethyl instead of ethyl groups results in at least a tenfold reduction in toxicity. In this way the biological activity is also frequently lost. ®Haloxon (L) [O,O-bis-2-chloroethyl O-(3-chloro-4-methyl 2-oxo-2H-1-benzopyranyl-7) phosphorothioate] is an active anthelmintic, described in 1962 by the firm of COOPER [134, 135].

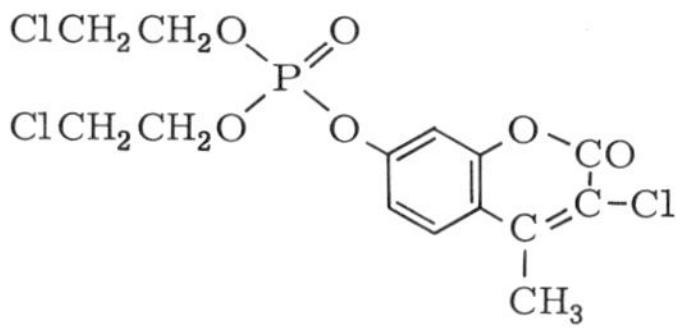

(L)

The oral LD_{50} of this compound for the rat is about 900 mg/kg. For its manufacture the synthesis of STEINBERG [921] was chosen, involving the reaction of 3-chloro-4-methyl 7-hydroxycoumarin with bis-β-chloroethyl phosphite and carbon tetrachloride in the presence of triethylamine (cf. p. 53).

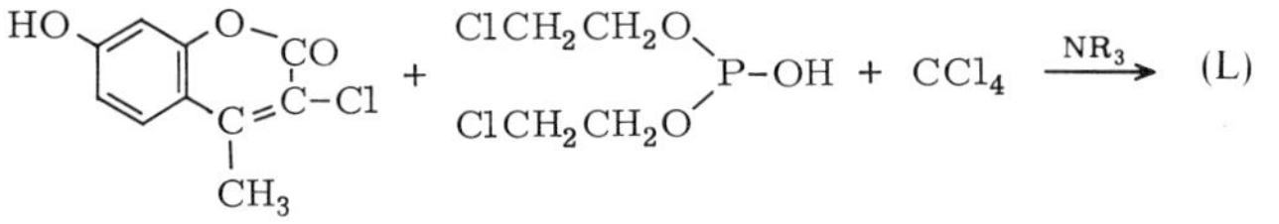

(17)

A series of highly active compounds are also derived from condensed heterocyclics containing nitrogen, such as quinoline. Phosphonic acid esters of 5-, 6-, 7-, and 8-hydroxy-quinolines can be used against ticks and resistant spider mites, maximum acaricidal activity* being found with the ester of the 6-hydroxy compound and maximum activity against ticks** with the 8-hydroxy derivative. It is surprising that these phosphonic acid esters are relatively stable to alkaline media and can, therefore, be used in a dip. An important member of this series is ®Bacdip with the common name *oxinothiophos* (LI) [O-ethyl O-quinolyl-8 benzenephosphonothioate] [279], and the derivative of 6-hydroxyquinoline having the structure (LII) [O-ethyl O-quinolyl-6 methanephosphonothioate].

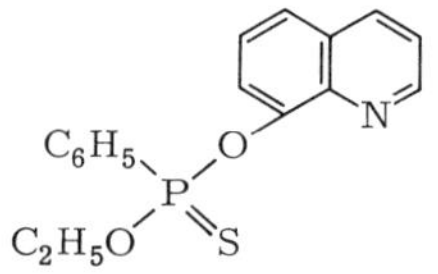

(LI)

LD_{50} for the rat ♀ per os: 150 mg/kg

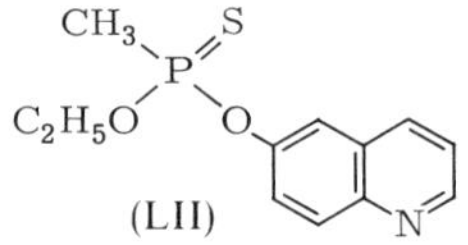

(LII)

LD_{50} for the rat per os: 5–10 mg/kg***

Since the synthesis of the requisite 6-hydroxyquinoline is rather expensive, the corresponding quinoxaline was synthesized, which is not only less expensive but has the additional advantage of possessing the structure of both the 6- and 7-hydroxyquinoline [816]:

In this way one obtains equally excellent insecticides whose acaricidal activity is maintained also against highly resistant strains, e.g. (LIII) [O,O-diethyl O-quinoxalyl-6 phosphorothioate]****.

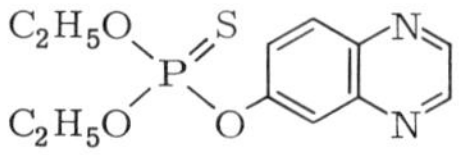

(LIII)

* The activity was discovered by I. Hammann and G. Unterstenhöfer (Farbenfabriken Bayer)

** W. Stendel (Farbenfabriken Bayer) determined the LC_{65} at $5 \times 10^{-5}\%$

*** G. Kimmerle (Farbenfabriken Bayer) determined the LD_{50}

****The acaricidal activity was determined by I. Hammann and G. Unterstenhöfer (Farbenfabriken Bayer).

The oral LD_{50} for the rat is 10 mg/kg*.
The ester series of the heterocyclics proper begins with ®Dursban (LIV) [O,O-diethyl O-(3,5,6-trichloropyridyl-2) phosphorothioate].

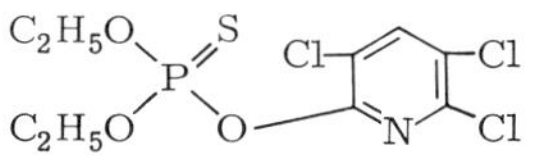

(LIV)

It is was introduced by Dow Chemical Company in 1965 as a new agent against mosquitoes [476, 772, 773]. Its larvicidal activity is some 50–100 times greater than that of many other commercial products. The LD_{50} for the rat is 135 to 163 mg/kg.
From pyrimidine is derived one of the oldest phosphoric acid esters, *diazinon* (LV) [O,O-diethyl O-(2-isopropyl 4-methylpyrimidyl-6) phosphorothioate], which was developed in 1952 by GYSIN and MARGOT [360] of Geigy.

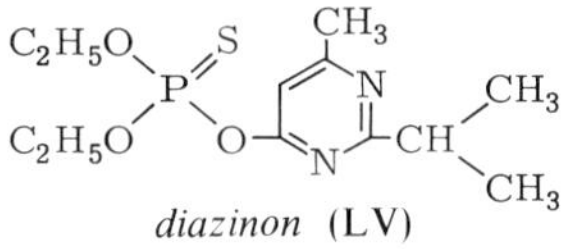

diazinon (LV)

It is a widely used, rapidly acting contact insecticide and acaricide.
The starting compound is obtained by condensation of isobutyramidine with acetoacetate.

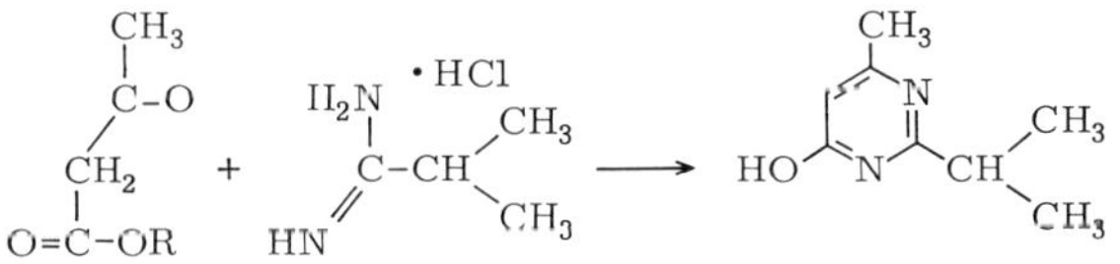

(18)

The oral LD_{50} for the male rat is 108 mg/kg. In the mammal the compound is oxidized as usual to the P=O compound (diazoxon). *Diazinon* is often applied where insects have become resistant to DDT and chlorinated hydrocarbons. It was developed alongside the analogously constructed dialkyl carbamates of enolisable heterocyclic compounds with insecticidal or repellent action of which Pyramate (LVI) is an example.

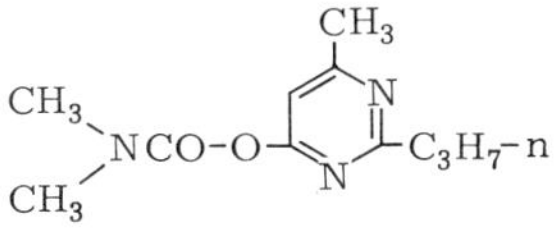

(LVI)

* The LD_{50} was determined by G. KIMMERLE (Farbenfabriken Bayer).

Closely related to *diazinon* is the ICI compound ®Diothyl (LVII) (ICI 29,661) [O,O-diethyl O-(2-dimethylamino-4-methyl pyrimidyl-6) phosphorothioate] [617]:

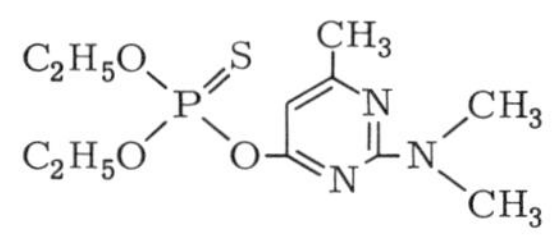

pirimiphos-ethyl, ®Diothyl (LVII)

which also possesses insecticidal, acaricidal properties and an LD_{50} of a similar order (Rat oral: 125 mg/kg).

Arrangement of the nitrogen atoms in the 1,4-position in the heterocyclic nucleus results an insecticide, nematicide and fungicide – ®Zinophos or *thionazin* (LVIII) [O,O-diethyl O-pyrazinyl-2 phosphorothioate] [246]. It is also known under the name ®Cynem and was patented in 1958 by American Cyanamid. The oral LD_{50} for the rat is 5 mg/kg.

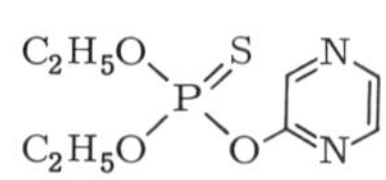

(LVIII)

For the synthesis of ®Cynem the American Cyanamid Company started with 1,2-diketones which were condensed with aminoacetamide:

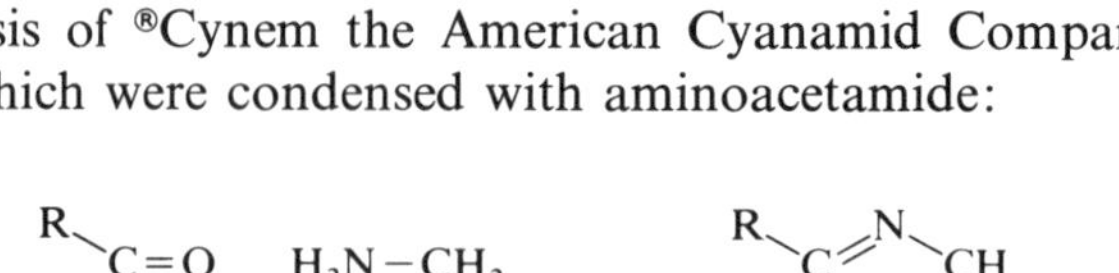

(19)

If R and R^1 together signify a benzene ring, then the process is not practicable because *o*-quinone is difficult to obtain and handle on an industrial scale. Therefore *o*-phenylenediamine is to be preferred for ring closure with 1,2-diketones.

In this way Bayer prepared the ester of 2-hydroxyquinoxaline [815]* which possesses an excellent action against biting insects, such as beetle larvae, and caterpillars, especially against resistant types of *Plutella* sp. and sucking insects such as aphids as well as against the spider mite. Substituents on both rings reduce toxicity and usually also the pesticidal activity.

®Bayrusil (LIX) [O,O-diethyl O-quinoxalyl-2 phosphorothioate] was chosen for practical application (oral LD_{50} for the rat approx. 70 mg/kg**). *Diethquinalphion* was recommended as the common name. The advantage of ®Bayrusil

* The insecticidal and acaricidal activity was found by I. Hammann and G. Unterstenhöfer (Farbenfabriken Bayer).

** G. Kimmerle (Farbenfabriken Bayer) determined the acute LD_{50} in the rat orally.

consist in an excellent initial action without notable persistence as contact and stomach insecticide.

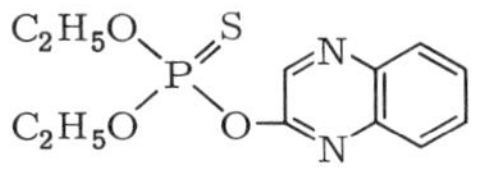

(LIX)

It is not systemic in action but is able to penetrate deeply into plants: if Bayrusil is sprayed onto the surface of the leaf, aphids on the underside of the leaf are killed within a few minutes [817].
Sandoz AG applied for a patent for this class of compounds about three months after Farbenfabriken Bayer [397].
The corresponding dimethyl ester should be mentioned because of its very favourable toxicity (LD_{50} oral for the rat >2000 mg/kg). Against diptera its activity exeeds that of Bayrusil.
Related to (LIX) is an ester of 2-hydroxy-benzo-1,2,4-triazine, [O-ethyl O-(benzo-1,2,4-triazinyl-3) ethanephosphonothioate] (LX) which possesses broad insecticidal and acaricidal activity [818]*.

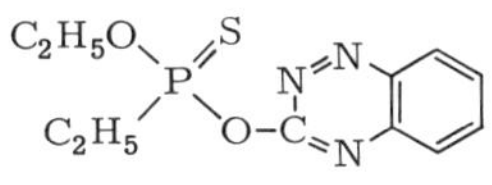

(LX)

The oral LD_{50} for the rat is 25 mg/kg**. Like ®Bayrusil (LIX), compound (LX) was developed in 1966 at Bayer. The technical synthesis of hydroxytriazine is rather simple, it proceeds as follows:

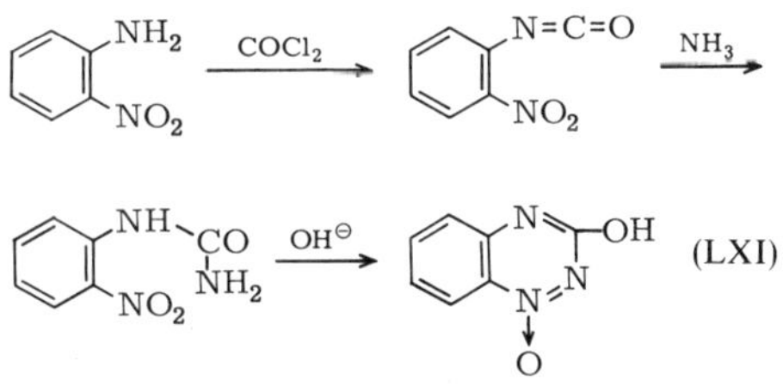

(20)

Reduction of this N-oxide (LXI), e.g. with tin in hydrochloric acid gives the dihydro-compound (LXII), which can be oxidized by various methods to triazine (LXIII) [57, 58]:

* The insecticidal and acaricidal activity was found by I. Hammann and G. Unterstenhöfer (Farbenfabriken Bayer)
** G. Kimmerle (Farbenfabriken Bayer) determined the acute LD_{50} in the rat orally

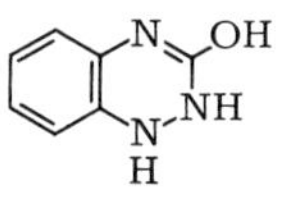

(LXII)

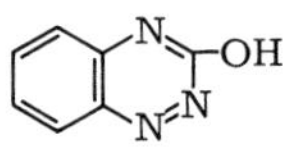

(LXIII)

On five-membered heterocyclic compounds, e.g. pyrazole are based compounds such as ®Pyrazoxon (LXIV) [O,O-diethyl O-(3-methyl pyrazolyl-5) phosphate] [360] which was filed for a patent by Geigy in 1952. ®Pyrazoxon is highly active systematically but considerably toxic to mammals (oral LD_{50} for mice: 4 mg/kg).

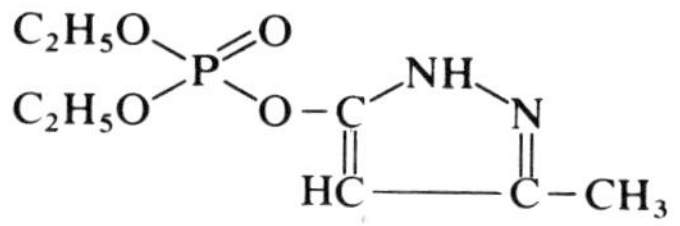

(LXIV)

The thiono-derivative ®Pyrazothion (LXV) [O,O-diethyl O-(3-methyl pyrazolyl-5) phosphorothioate] is also a systemic insecticide, but less toxic (oral LD_{50} for the rat: 36 mg/kg), with good activity against aphids and spider mites [360].

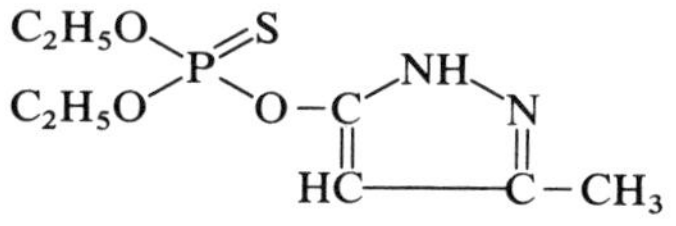

(LXV)

A true phosphoroamidate is the insecticide, acaricide and fungicide ®Wepsyn or ®Septin (LXVI) [5-amino-1-[bis-(dimethylamino)-phosphoryl] 3-phenyl 1,2,4-triazole]. This new type of product, found by the firm Philips Duphar in 1960 [506], was the first phosphoric acid derivative showing activity against powdery mildews. The oral LD_{50} for the rat is 5 mg/kg [114–118].

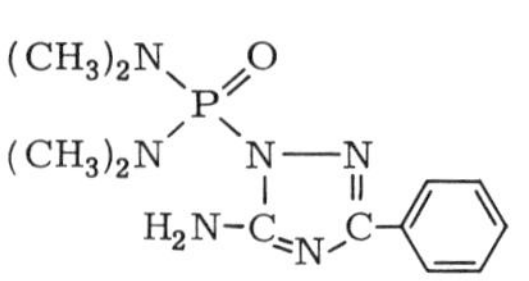

triamiphos, ®Septin (LXVI)

For its synthesis [469] (Eq. (21)), dicyandiamide is reacted with benzoyl chloride, and then with hydrazine sulfate. The urea derivative (LXVII) is hydrolyzed and the resulting amino-compound (LXVIII) acylated with bis-(dimethylamino)-phosphoric acid chloride in the presence of collidine.

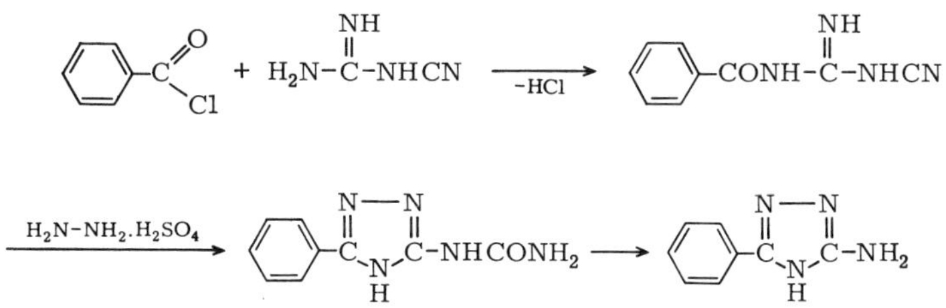

(LXVII) (LXVIII) (21)

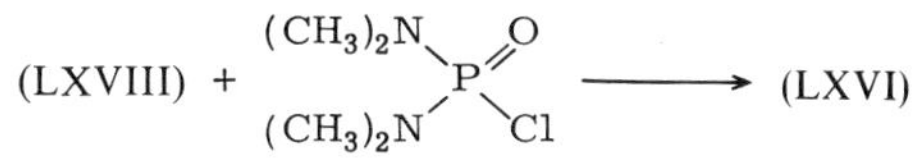

American Cyanamid patented an interesting group of esters of the formula (LXIX) with very good activity against biting and sucking insects, as well as against spider mites [3, 579].

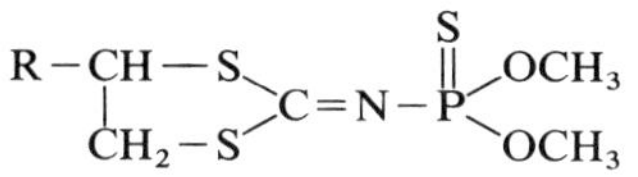

(LXIX)

4-Alkyl 2-[N-(O,O-dimethyl thionophosphoryl)]-imino-1,3-dithiolane

The necessary 2-imino-1,3-dithiolanes are obtained by reacting the corresponding 1,2-dithiols in non-polar solvents with cyanogen chloride in the presence of hydrogen chloride and catalytic quantities of alcohol. Phosphorylation is achieved by the same methods that lead to the phenol esters.
The diethyl ester was introduced under the name ®Cyolane (LXX) [2-[N-(O,O-diethyl phosphoryl]-imino-1,3-dithiolane] [2, 5]

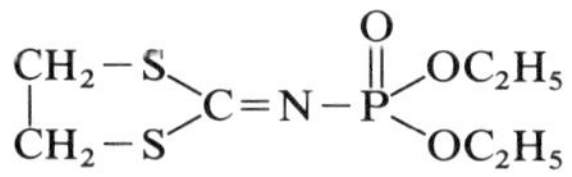

(LXX)

as a systemic insecticide for use on cotton to control a number of economically important pests. It acts both as a contact and stomach poison (oral LD_{50} for rats: 8,9 mg/kg). Besides the above-mentioned synthesis of phosphorylated 2-imino-1,3-dithiolane, ADDOR [4] described another in which potassium hydrosulfide was added to diethyl phosphoryl isothiocyanate to give potassium N-(O,O-diethyl phosphoryl) dithiocarbamate (Eq. (22)),

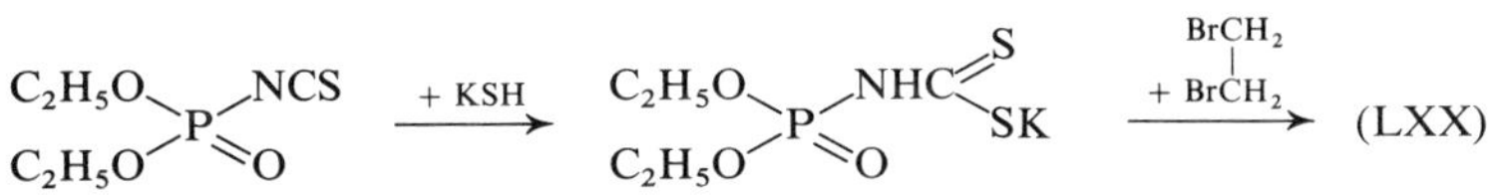

(22)

which was converted to Cyolane by reaction with ethylene bromide.

Phosphorylated hydroxamic acids have recently become known. ®Maretin (LXXI) [O,O-diethyl O-naphthaloximidophosphate] was found in 1953 at Farbenfabriken Bayer [576, 1007]; it is a very active anthelmintic, used against stomach and intestinal worms. It has also been introduced under the name ®Rametin.

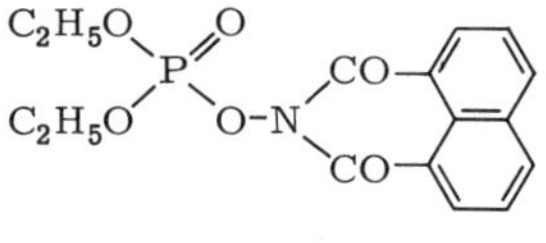

(LXXI)

For the male rat it has an oral LD_{50} of 75 mg/kg. N-hydroxynaphthalimide is synthesized from naphthalic acid anhydride and hydroxylamine. Compounds related to naphthaloxime were described in 1964 by Dow Chemical e.g. DOW 49 (LXXII) [O,O-diethyl phthalimidophosphorothioate] with an

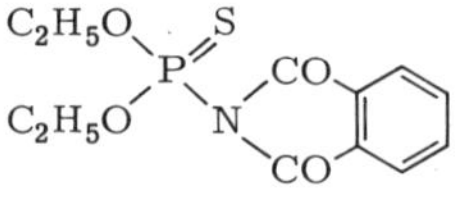

(LXXII)

oral LD_{50} of 5600 mg/kg for the male rat.
It is surprising that its spectrum also covers fungicidal activity [958].
In 1966 the same firm investigated phosphoric acid esters of dicarboximido-compounds, whose intermediates can be achieved by a Diels-Alder reaction [771]. DOW 50 (LXXIII) is O,O-diethyl O-[7-oxabicyclo-(2,2,1)-hept-5-ene 2,3-dicarboximido]-phosphorothioate. The LC_{100} for the Mexican bean beetle or two-spotted spider mite is 50 ppm.
A new group of compounds of the oximinoester type highly effective against ticks are represented by the general formula (LXXIV) [563].

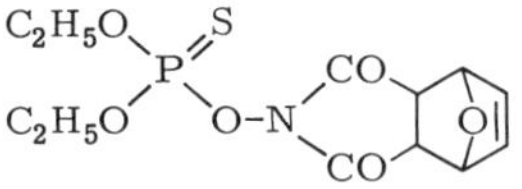

(LXXIII)

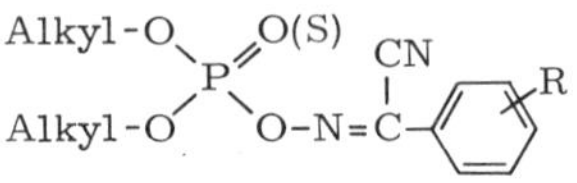

(LXXIV)

They can be obtained by the phosphorylation of α-phenyl α-hydroximinoaceto-nitriles. Some compounds of this series have interesting properties and may be

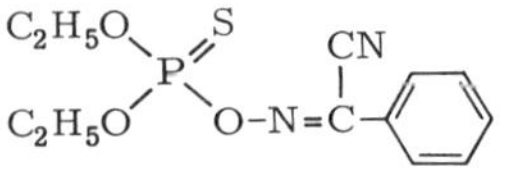

(LXXV)

regarded as very promising products for practical application, for example compound (LXXV) with the common name *phoxim* and marketed by Bayer under the names ®Valexon and ®Baythion.

The mammalian toxicity is very favourable, being of the order of 2500 mg/kg (rat oral). *Phoxim* is a powerful broad spectrum leaf insecticide, particularly suitable for the control of *Heliothis* spp. in cotton. This substance with the chemical name O-(O,O-diethyl thionophosphoryl) α-phenyl α-hydroximinoacetonitrile has the advantage of a very good initial action, is able to control advanced stages of caterpillar and is effective in rainy climates [1037].

FUKUTO, METCALF, JONES and MYERS [314] investigated *p*-substituted structural analogues of *phoxim* with the object of correlating structure and substituent properties with activity. The most active compound against the house fly is the P=O analogue of *phoxim* with a cholinesterase inhibition of $1{,}6 \times 10^{-9}$ (I_{50} molar, fly ChE).

c) Alkylation of Ester Acids

The phosphoric acid esters described in this section are obtainable by alkylation of suitable phosphoric ester acids or their salts.

The alkylating agents which are used on an industrial scale can be arranged in the following groups:

1) Methylating agents like methyl iodide and dimethyl sulfate (p. 113)
2) Alkyl-, aryl- and acyl-thiomethyl chlorides (p. 114)
3) Choline (or choline analogue) chlorides (p. 117)
4) Benzyl chloride and analogues (p. 126)
5) N-Chloromethyl compounds (p. 127)
6) α-Halogen carboxylic acid derivatives (p. 130)
7) Olefines (p. 135)

Each one of these groups has provided well-known commercial pesticides.

1) The simplest member of this group was recently developed simultaneously by Farbenfabriken Bayer and Chevron Chemical Co.:

®Monitor (trade name of Chevron Chemical Co.) (II) [586] or ®Tamaron (trade name of Farbenfabriken Bayer) [888] with the scientific name O,S-dimethyl phosphoroamidothioate is intended as an insecticide for the control of lepidopterous larvae, as well as an acaricide [365]. It is of chemical interest in that ®Monitor is one of the few phosphoric acid esters of economic importance which possesses three different substituents on the P=O group.

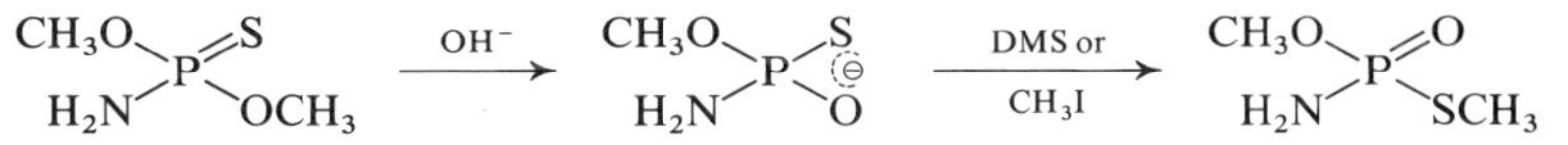

(I) ®Tamaron, ®Monitor (II) (1)

The synthesis starts with dimethyl phosphorothiochloridate and -amidate, respectively, proceeding according to Eq. (1) to the alkali salt (I) as an intermediate.

Methylation by means of methyl iodide or dimethyl sulfate results in the desired compound (II).

An investigation of QUISTAD, FUKUTO and METCALF [758] into structural analogues of Tamaron indicated that substituents on the nitrogen atom or an increase in the size of the alkyl groups on the oxygen or sulfur atom, respectively, reduce activity. Surprisingly, the insecticidal activity of Tamaron is substantially higher than might be expected from the cholinesterase inhibition data for house flies (I_{50} 3.9 × 10^{-5} molar fly ChE).

The acetylated derivative of Tamaron, ORTHO 12,420 (III) [O,S-dimethyl N-acetyl phosphoroamidothioate] was synthesized by LORENZ in 1964 and introduced by Chevron Chemical [562] at first as an experimental product and especially for the control of Lepidoptera, Hemiptera, Homoptera and Coleoptera. The oral LD_{50} ♂ is 945 mg/kg for the rat.

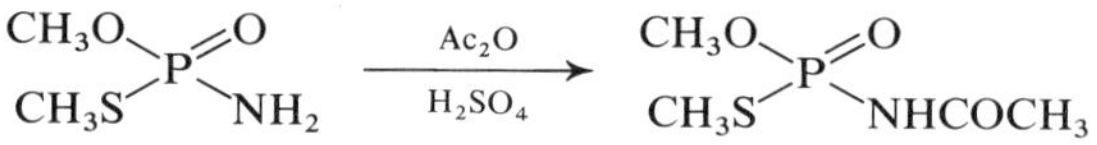

ORTHO 12,420 (III) (2)

2) Another very simple example is *phorate* or ®Thimet (IV) [O,O-diethyl S-2-ethylthiomethyl phosphorodithioate], a contact insecticide with systemic action. Processes for its manufacture have been developed by American Cyanamid (1948) (Eq. (3)) and by Farbenfabriken Bayer (1952 (Eq. (4)).

(a) Following a process by HOOK [420], diethyl phosphorodithioic acid is reacted with ethyl mercaptan and formaldehyde:

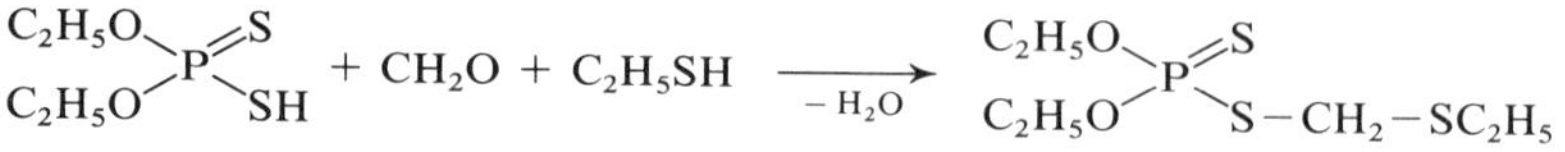

phorate (IV) (3)

(b) In another synthesis ethylthiomethyl chloride is first prepared and reacted with the salt of O,O-diethyl phosphorodithioic acid [568, 883]:

(4)

The toxicity of the compound (oral LD_{50}) is 2 mg/kg for the rat. When *phorate* is absorbed by the plant, it is converted to water-soluble metabolites; presumably oxidation takes place:

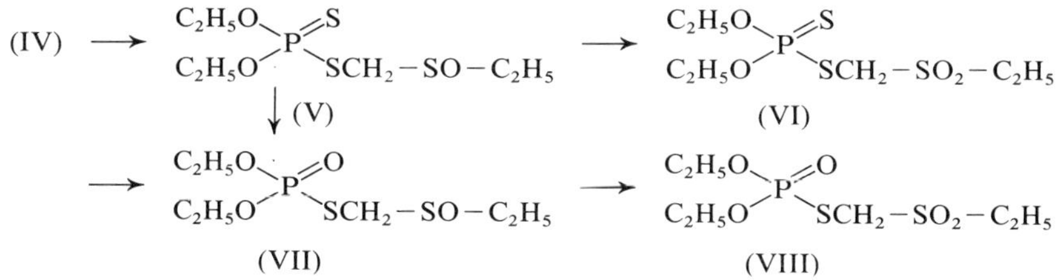

Scheme 5: *phorate* oxidation

When aliphatic groups are replaced by aromatic groups, the contact insecticidal and acaricidal properties are maintained. We would mention *carbophenothion* (®Trithion) (IX) [O,O-diethyl S-(*p*-chlorophenylthiomethyl) phosphorodithioate] and ®Methyltrithion (X) [274] [O,O-dimethyl S-(*p*-chlorophenylthiomethyl) phosphorodithioate] which were developed in 1954/55 by Stauffer Chemical Company.

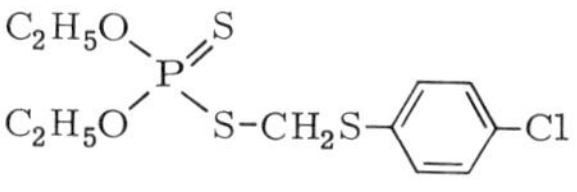

(IX)

LD$_{50}$: 28 – 100 mg/kg for the rat orally
®Trithion

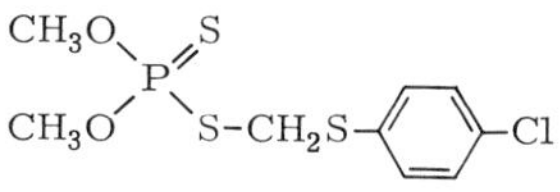

(X)

LD$_{50}$: 180 – 200 mg/kg for the rat orally
®Methyltrithion

The aromatic thioether is prepared from thiophenol with formaldehyde and hydrogen chloride [275]:

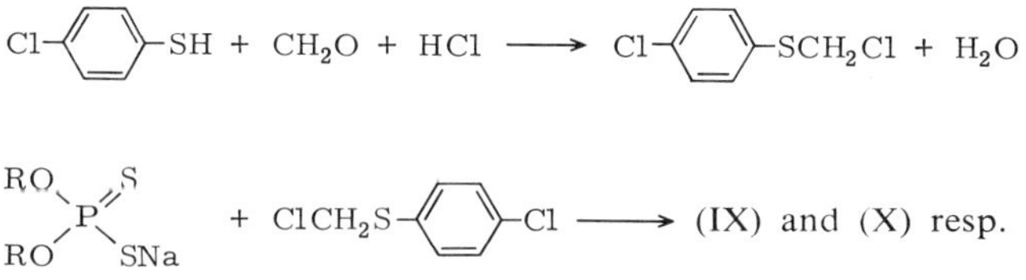

(5)

On the plant, *carbophenothion* is oxidized to the P=O compound, which is more toxic and shows an increased cholinesterase inhibiting activity. It can be prepared by means of peracetic acid.
Closely related to *carbophenothion* is *phenkapton* (XI) [O,O-diethyl S-(2,5-dichlorophenylthiomethyl) phosphorodithioate] [328], which also possesses acaricidal properties. Systemic action is lacking here. *Phenkapton* was described for the first time in 1955 by Geigy AG.

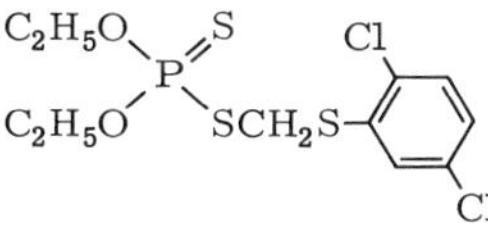

(XI)

The synthesis is analogous to that of ®Trithion. The compound has an oral LD_{50} for the rat of 200–260 mg/kg.
Joining two molecules of diethyl phosphorodithioic acid via a methylene bridge results in *ethion* (XII) [O,O,O′,O′-tetraethyl S,S′-methylene bis-phosphorodithioate]. This compound was developed independently by the Food Machinery and Chemical Corporation in 1956 and by Farbenfabriken Bayer in 1957, respectively [28, 572, 1022]. *Ethion* may also be regarded as phosphoryl-thimet.

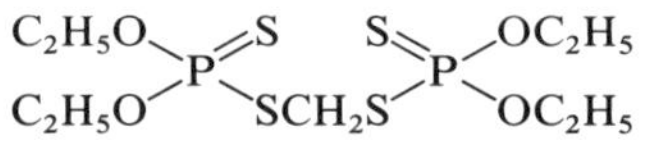
(XII)

ethion

The compound acts as a contact insecticide, acaricide and ovicide. It is manufactured from the ammonium salt of diethyl phosphorodithioic acid and chlorobromomethane or methylene bromide:

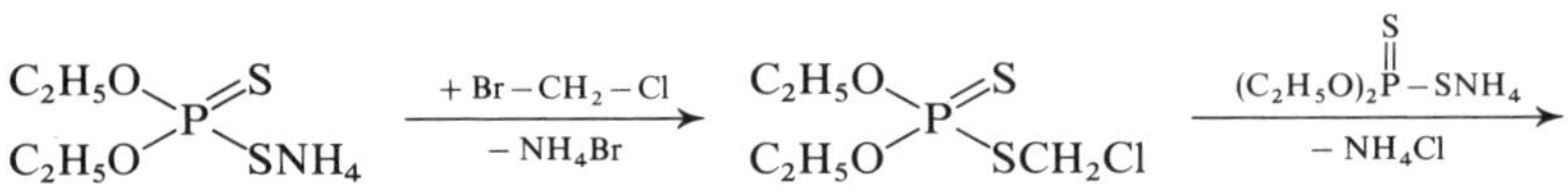
(XII) (6)

(XIII)

In another synthesis, according to CÖLLN, diethyl phosphorodithioic acid and formaldehyde are used [204]:

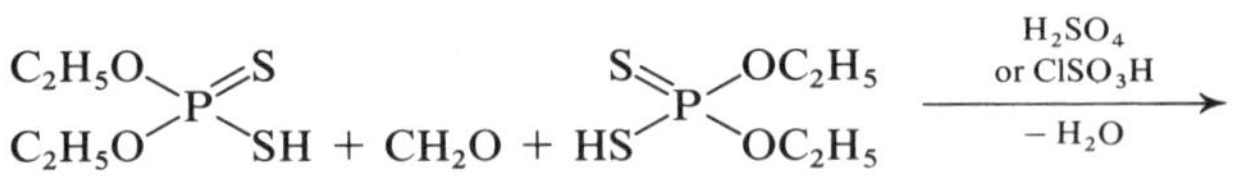
(XII) (7)

By reacting the S-chloromethyl ester (XIII) with sodium N,N-dimethyl dithiocarbamate or N,N-dimethyl thiocarbamyl chloride with the sodium salt of O,O-diethyl-S-(methylthio)-phosphorodithioic acid, ®Azothion (XIV) [O,O-diethyl S-(N,N-dimethyl dithiocarbamyl-methyl) phosphorodithioate] results. Its application as a pesticide, especially as acaricide and hygiene insecticide, was claimed by Farbwerke Hoechst in 1956 [804]:

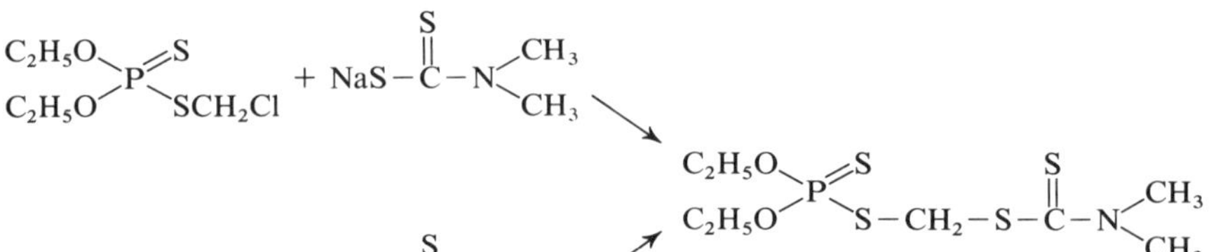
(XIV) (8)

3) One of the simplest phosphoryl choline compounds is *amiton* (XV) [O,O-diethyl S-(2-diethylaminoethyl) phosphorothioate] [106, 336]. Although *amiton* was synthesized as early as 1948, the first publications did not appear until some years later.

(XV)

It was introduced in the form of its acidic oxalate (®Tetram) by Imperial Chemical Industries as an acaricide and systemic insecticide. *Amiton* is seldom used, however, on account of its high toxicity (the oral LD_{50} for the rat is ~3 mg/kg). Established methods are used for its synthesis via the thionoderivative (Eq. (9)) [334]. By isomerization *amiton* is obtained.

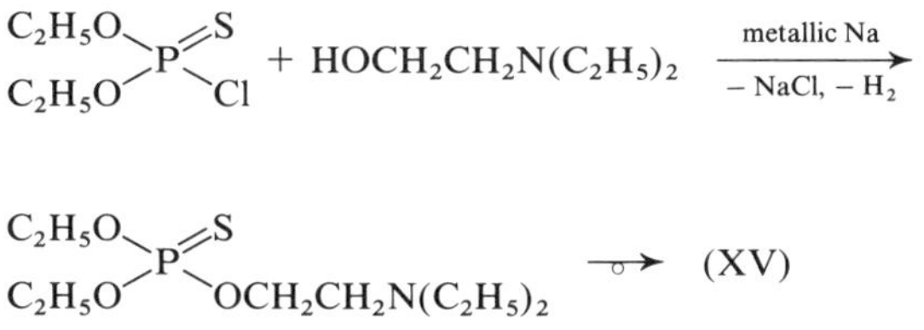

(9)

Another route for its synthesis is the reaction of sodium diethyl phosphite with the corresponding pseudo-halide [335, 850]:

(10)

With alkylating agents, quaternation occurs resulting in compounds with high cholinesterase inhibiting activity.
From a biological, biochemical and toxicological point of view, the phosphoryl analogues of acetylcholine (XVI) are of great interest. Compounds of this type, the so-called Tammelin esters have the general structure (XVII) [941, 943–945]:

Acetylcholine:

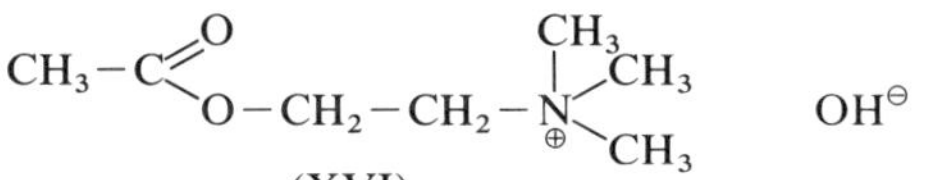

(XVI)

Tammelin esters:

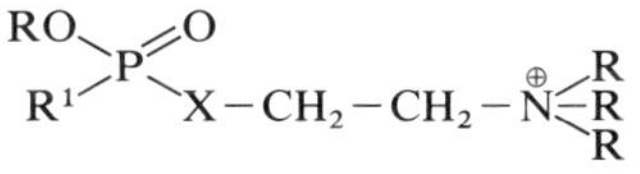

and

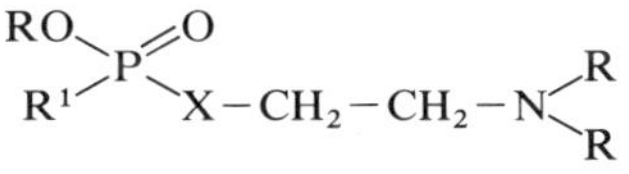

(XVII a) (XVII b)

where R^1 may be alkyl, fluorine or alkoxy, and X oxygen or sulfur. Their cholinesterase inhibiting activity is unusually high, in many compounds it surpasses even that of Sarin. For practical application as insecticides active substances with such extremely unfavourable toxicological properties are out of the question.
In order to achieve a lower mammalian toxicity, attempts were made to reduce the basicity of the choline nitrogen considered decisive for the interaction with the active sites of the enzyme. It was thus that the firm of Stauffer introduced alkoxycarbonyl [129] or alkoxy-sulfonyl groups [239]:

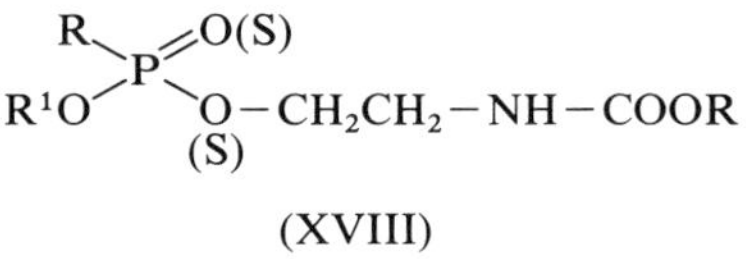

(XVIII)

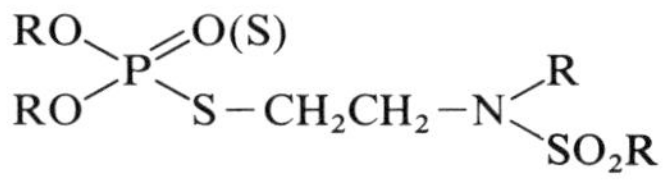

(XIX)

A second possibility consists in replacing the basic choline nitrogen by less basic atoms, such as sulfur. This results in the larger group of the ®Systox compounds.
Demeton-S-methyl (XX) was described by SCHRADER in 1950 [O,O-dimethyl S-2-(ethylthioethyl) phosphorothiolate]

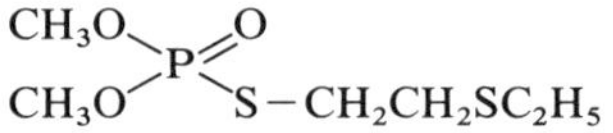

®Metasystox (i) (XX)

Under the trade name ®Metasystox (i) it is used as a systemic and contact insecticide as well as an acaricide. The substance results from alkylation of the corresponding phosphoric acid salt [836]:

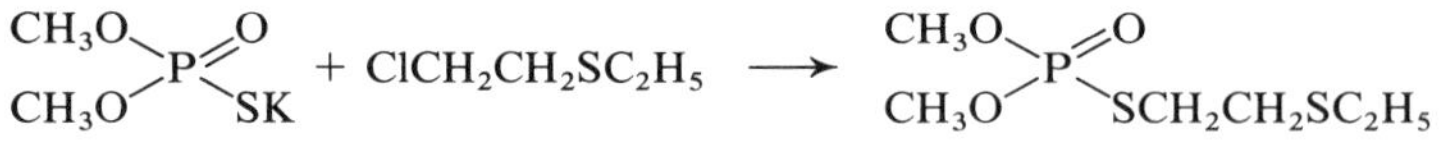 (11)

(XX)

According to PRICE and WAKEFIELD [310, 400, 750] the mechanism by which ®Metasystox (i) is formed involves a sulfonium cation of the following structure (see also p. 37):

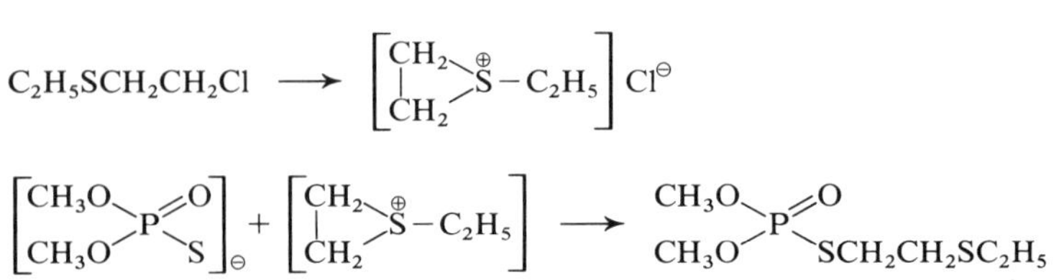 (12)

(XX)

The oral LD_{50} for the rat is 40 – 60 mg/kg.
Demeton-methyl has a lasting action, longer than might be expected from its increased rate of hydrolysis in comparison to *parathion*. A possible explanation may be its oxidation in the plant yielding metabolites less susceptible to hydrolysis. This is supported by the preparative oxidation, whereby with hydrogen peroxide [533, 666](Eq. (13)) or by means of halogen [566], the sulfinyl compound (XXI) results (Eq. (14)):

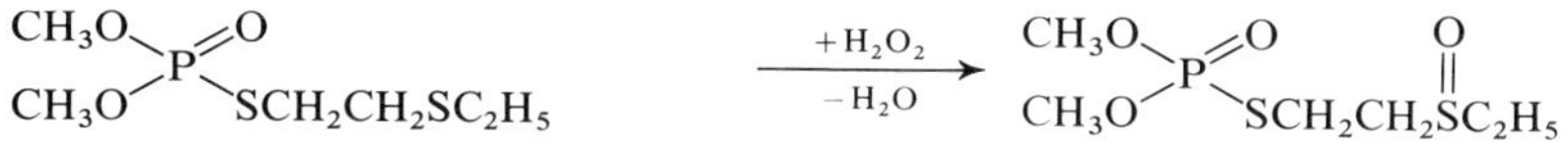

(XX) ®Metasystox R (XXI) (13)

$$\text{(XX)} \xrightarrow[-2\,HBr]{+Br_2/H_2O} \text{(XXI)} \qquad (14)$$

This ester has been manufactured as such and is available commercially under the name Metasystox R (common name: *oxydemeton-methyl*) (XXI) [O,O-dimethyl S-(2-ethylsulfinyl ethyl) phosphorothiolate] [533]. It was described by Lorenz *et al.* [565] in 1954 and 1955 as a further development of ®Metasystox (i). It can be obtained by the standard procedure:

(XXI) (15)

The oral LD_{50} for the male rat is 80 mg/kg. ®Metasystox R is compatible with all insecticides and fungicides except those with alkaline reaction. It exhibits a specific activity against aphids, spider mites, leafhoppers and similar sucking pests.
The *C*-methyl derivative of *oxydemeton methyl* is called ®Metasystox S (XXII) [O,O-dimethyl S-(1-methyl 2-ethylsulfinylethyl) phosphorothiolate], also described by Lorenz in 1955. ®Metasystox S is a less toxic (oral LD_{50} for the male rat: 105 mg/kg) insecticidal and acaricidal compound with systemic action [570].

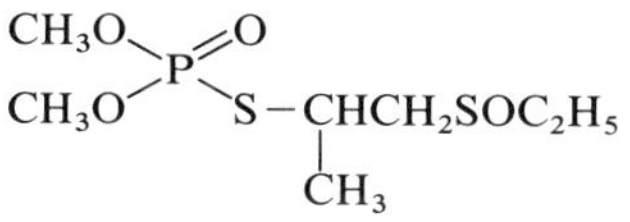

(XXII)

It is prepared by oxidizing with hydrogen peroxide the thiol ester (XXIII) obtained according to Eq. (16).

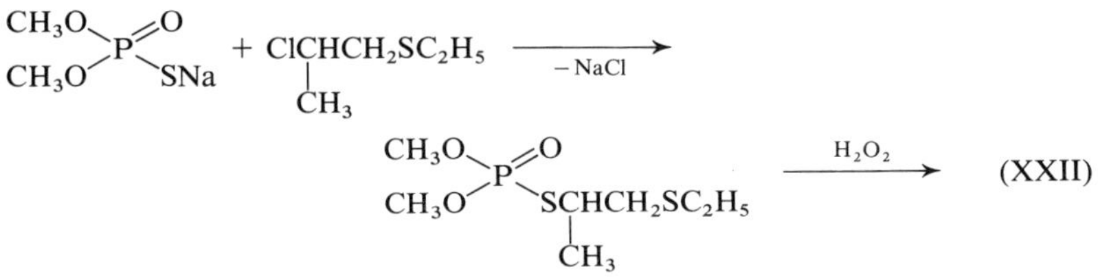

(XXIII) (16)

Like many other methyl esters, *demeton methyl* is a strong methylating agent which may itself undergo inter- or intra-molecular methylation on the sulfur atom (Scheme 6) [388]:

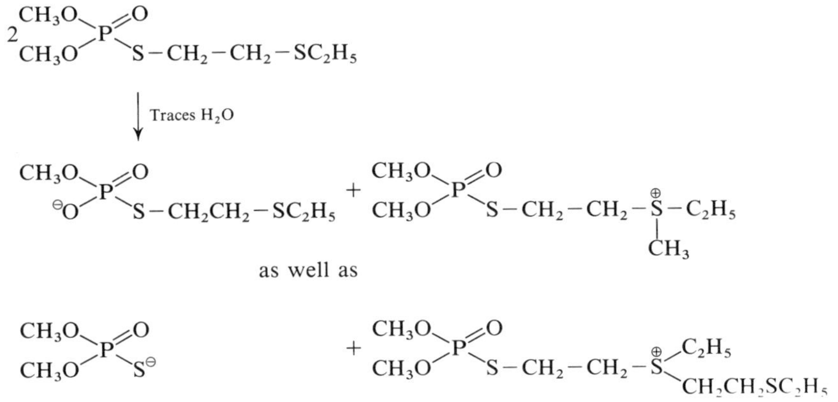

Scheme 6: Systox alkylations

The isomeric compound to *demeton-S-methyl*, i.e. ®Metasystox or *demeton-O-methyl* (XXIV) [O,O-dimethyl O-(2-ethylthioethyl) phosphorothionate] was synthesized for the first time by SCHRADER in 1950.

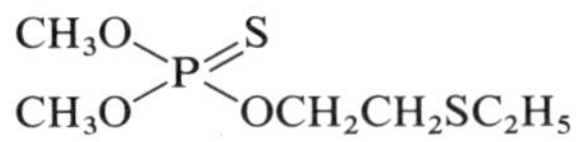

®Metasystox (XXIV)

The industrial product contains 70% of the thiono- and 30% of the thiol compound. *Demeton-methyl* operates as a systemic and contact insecticide,

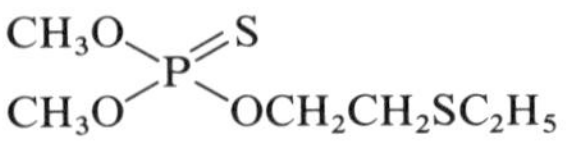

70%
demeton-O-methyl

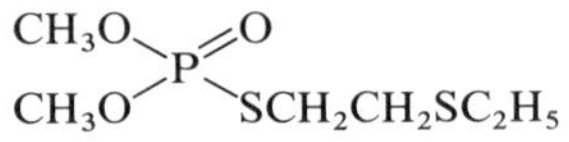

30%
demeton-S-methyl

which penetrates into the plants. There a rearrangement to the thiol form occurs [832, 836, 872] which is responsible for the systemic action. The mixture is manufactured from the acyl chloride and ethylthioethanol.

(17)

An elegant route for synthesizing *demeton* derivatives was developed by Farbenfabrik Wolfen [920]. It consists in the transesterification of dialkyl nitrophenyl phosphorothionates with monothioglycol-S-alkyl ethers to dialkyl alkylthioethyl phosphorothionates (Eq. (18)). The reaction is carried out in a two-phase system (chlorbenzene/conc. aqueous alkali) at room temperature yielding 80% thionoester free from the thiol-isomer.

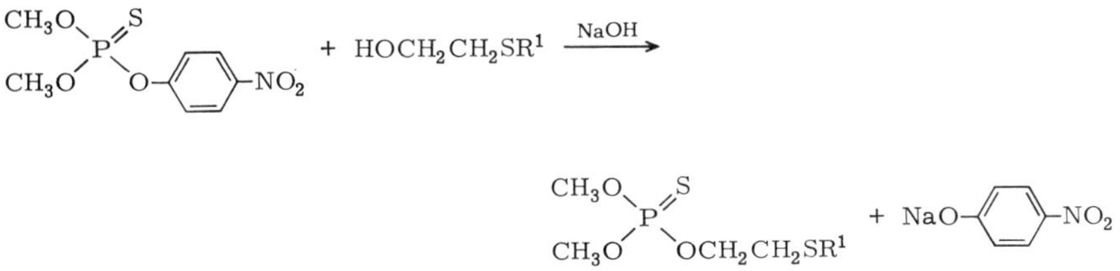

(18)

The oral LD_{50} of pure ®Metasystox for the rat is 180 mg/kg. As mentioned above, this compound easily transforms to the thiol form and exhibits the same methylating action [388] resulting in the formation of sulfonium salts. It is metabolized in the same manner as *demeton-S-methyl*.
The dithio-compound of *demeton methyl* is *thiometon* (XXV) [O,O-dimethyl S-(2-ethylthioethyl) phosphorodithioate].

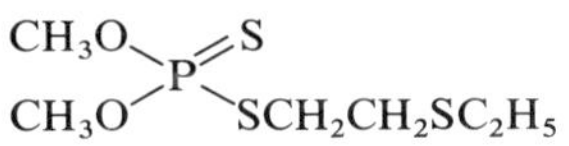

(XXV)

The compound, synthesized by Farbenfabriken Bayer in 1952 [568] and independently by Sandoz in 1953 [584] (®Ekatin), also possesses systemic and contact insecticidal activity. *Thiometon* is prepared by the usual synthesis of ®Systox compounds.

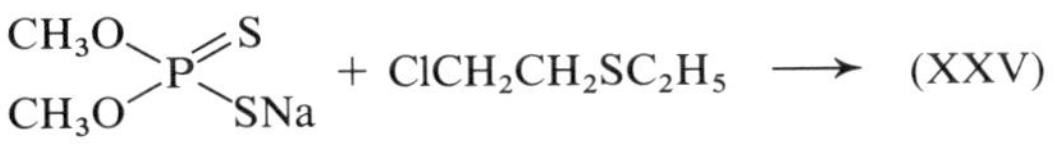

(19)

A further method was developed by Sandoz [539]:

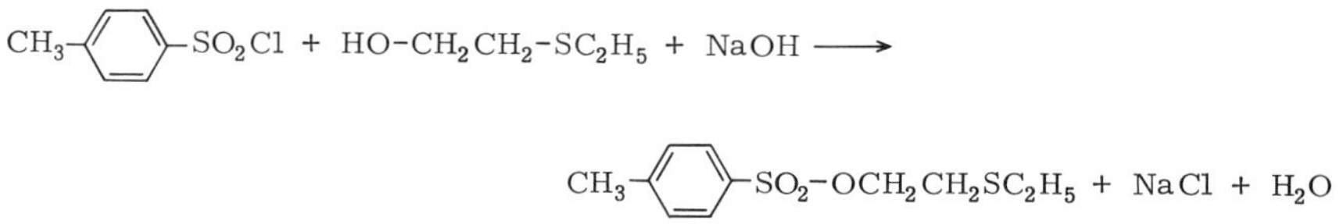

(20)

This method consists in preparing the toluolsulfonate of 2-ethylthio-ethanol and, without isolation, reacting it with the sodium salt of O,O-dimethyl phosphorodithioic acid:

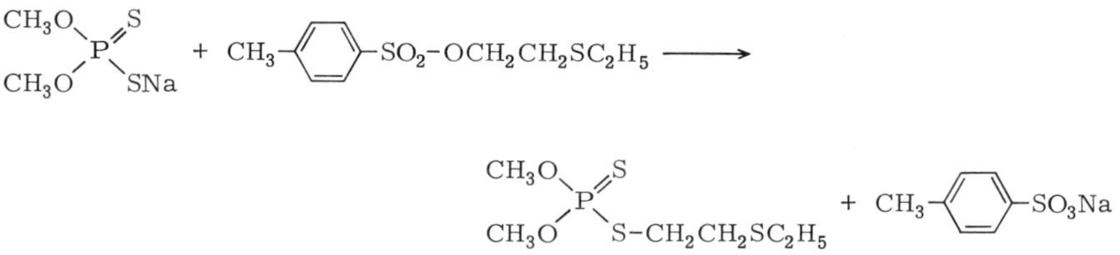

(21)

(XXV)

The oral LD_{50} for the rat is 85 mg/kg. Metabolism in the plant follows that of other ®Systox derivatives.
Not only the dimethyl phosphorothioate but also especially the diethyl phosphorothioate of this series has become a well-known commercial product under the name ®Systox or *demeton* (XXVI) [O,O-diethyl O-(2-ethylthioethyl) phosphorothioate]. It was described by SCHRADER in 1950 [310, 836, 846].
Demeton acts as a selective systemic and contact insecticide especially against aphids and spider mites. The thiol ester which penetrates rapidly into the plant is responsible for its activity [988].
The average oral LD_{50} for the mixture of the isomers for the male rat is 9–14 mg/kg, whereas the thiol ester as such is 10–20 times more toxic.
Demeton is a mixture of two isomers which appear in the ratio thiono-ester: thiol-ester of 2:1:

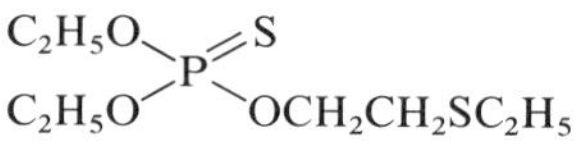

Thiono-isomer 60–70% (XXVI) *demeton-O*

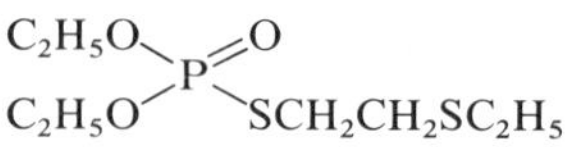

Thiol-isomer 30–40% (XXVII) *demeton-S*

There are several methods for its manufacture. On an industrial scale, O,O-diethyl phosphorothiochloridate is reacted with 2-ethylthioethanol:

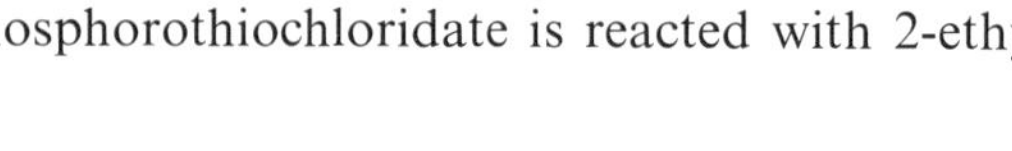

(XXVI) + (XXVII) (22)

Transesterification of triethyl phosphite with 2-ethylthioethanol [848] and thionation of the resulting ester is a straightforward route to the thiono-compound *demeton-O*:

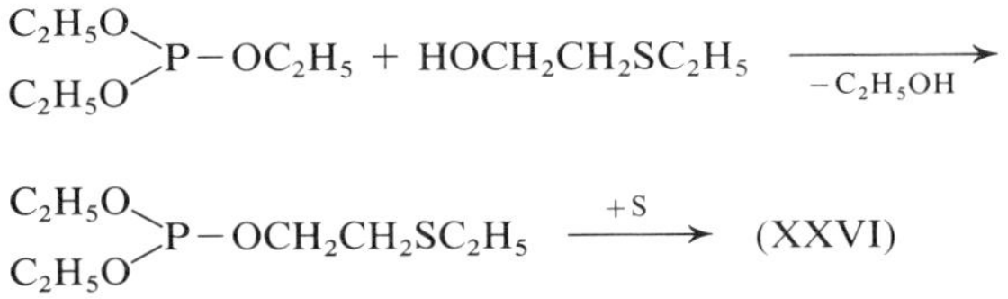

(23)

As a pseudohalide, 2-ethylthioethyl thiocyanate [832] reacts with sodium diethyl phosphite or triethyl phosphite [882] directly to the thiol ester *demeton-S*:

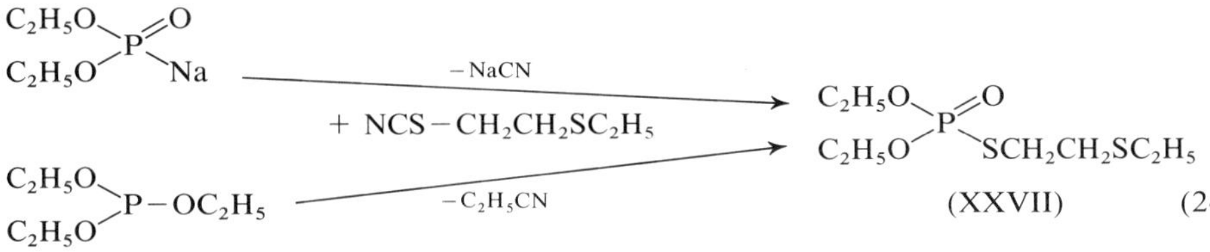

(24)

The reaction of O,O-diethyl phosphorochloridate with 2-ethylthioethyl mercaptide [834] provides a further synthesis of *demeton-S* (Eq. (25)):

(25)

also the reverse reactions [835]:

(26)

or [851]

(27)

A similar mechanism was proposed for the rearrangement of the thiono-form to the thiol form, namely, via a sulfonium ion, as for *amiton* and *demeton methyl* (see p. 118). The fact that polymerization of acryl nitrile and acryl amide is not induced by the rearrangement excludes radical mechanisms. Evidence for an ionic transition state may be provided by the fact the corresponding sulfones (XXVIII) fail to rearrange because they are unable to form anchimeric cations.

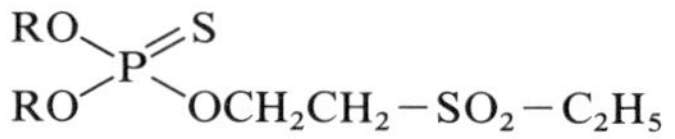
(XXVIII)

The activation energy calculated from the rearrangement of *demeton-O* is 25.4 kcal/mol and of *demeton-O-methyl* 22.8 kcal/mol. Rearrangement takes place on gentle heating and plays an important role in the plant, since the thiono-compound is almost insoluble in water, whereas the thiol compound is ten to one hundred times more soluble and thus responsible for the systemic properties.
The thiono-compound can be oxidized with bromine [357]:

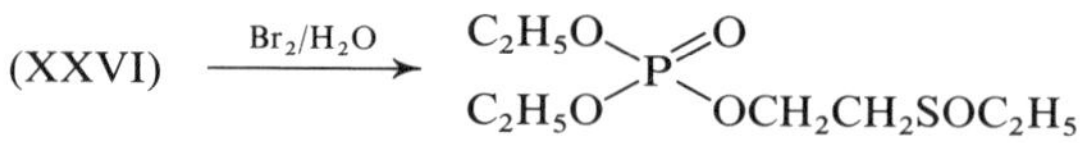
(28)

(XXIX)

Here also oxidation begins on the thioether group.
The metabolites expected in the plant are the products already cited. Oxidation of *demeton-O* to sulfoxide in the plant proceeds more rapidly than the next step, the formation of sulfone. At a suitable reaction temperature it is possible to transform also *demeton-O* into the thiol ester.
The dithio-derivative of *demeton* is *disulfoton* (*thiodemeton*, ®Disyston) (XXX) [O,O-diethyl S-(2-ethylthioethyl) phosphorodithioate], prepared by LORENZ and SCHRADER [568] in 1952.

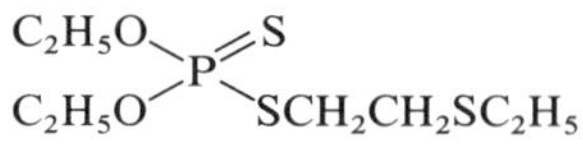
®Disyston (XXX)

The compound acts selectively as contact and systemic insecticide with remarkable stability towards hydrolysis. Because of its strongly systemic activity, ®Disyston was suggested by UNTERSTENHÖFER [979, 981] for the treatment of seed. In form of granules, the compound may also be used together with the seed [984], e.g. in potato [924] and beet growing [925]. The plants are therefore protected against aphids from the earliest stage of growth. Additionally, the active compound is continuously released over a relatively long time.
It is manufactured by conventional methods. The oral LD_{50} for the rat is 2 to 12 mg/kg.
The metabolites of *disulfoton* are the sulfoxide and sulfone. Moreover the thiono-derivative can be oxidized to the oxygen compound. In the plant, degradation to

phosphoric acid occurs [632]. The sulfoxide is known under the trade name ®Disyston-S (XXXI) [O,O-diethyl S-(2-ethylsulfinylethyl) phosphorodithioate].

oxydisulfoton, ®Disyston-S (XXXI)

It has the same spectrum of activity and scope of application as ®Disyston. The oral LD_{50} for the rat is 2.6 – 10 mg/kg.
A modification of the alkylmercapto-group of the *demeton* molecule leads to *vamidothion* (XXXII) [O,O-dimethyl *S*-[2-(1-methylcarbamoyl) ethylthio]-ethyl phosphorothioate] [637].

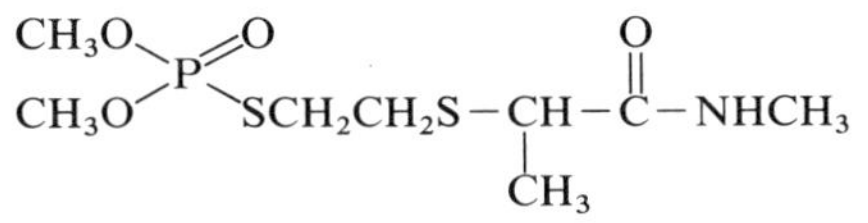

(XXXII)

This compound may be regarded both as a *demeton* and a *dimethoate* derivative, it was developed by the firm Rhône-Poulenc and synthesized similarly to *demeton*. *Vamidothion* has systemic properties and is used to control sucking insects. It has an oral LD_{50} for the rat of 64 – 100 mg/kg.

During recent decades systemic insecticides have become a basic factor in crop protection. Historically, the first member of the phosphate series to show systemic activity was OMPA [988]. This mode of action opened up a new route in controlling pests, i.e. the treatment of plants from within ("innere Therapie") [974, 976, 977], which provides several advantages: a very high expectancy of killing pests and great efficiency against such polyphages as aphids and spider mites, i.e. sucking arthropods. In principle it should become possible to control insects which live within the plant and are insensitive to insecticides applied from outside, e.g. larvae of *Pegomyia hyoscyami*. With regard to the present awareness of environmental problems, emphasis should be given to the ecological selectivity of systemic insecticides. They are non-toxic to non-phytophagous insects, especially to such beneficial insects as bees. The most important field of application seems to be indirect control of insect-borne pathogens, e.g. aphid-transmitted virus diseases [23, 409, 787]. One of the best known examples is the outbreak of yellow disease of sugar-beet in Germany. In 1951 the cultivation of sugar-beet had dropped to the economical threshold because of the extension of yellow disease, introduced shortly after the Second World War from Great Britain. As a consequence, systemic insecticides of the Systox group were developed for practical use [975] against the vector *Myzus persicae* (Green Peach Aphid). Application of systemic insecticides is of equal importance in potato growing (deterioration) [924, 1009]. Other aphid-borne virus diseases affect onions, beans, tobacco, cotton, cucumber or cereal crops [937], their transmission being effected not only by aphids but also by other arthropods [786], e.g. leafhoppers on rice plants [711], and nematodes (e.g. *Xiphinema americanum*) [753, 948, 1008]. In the Philippines, *Nephotettix impicticeps* is the vector of 'tungro', a virus disease of rice. Other virus diseases transmitted by leafhoppers are 'mentek' in Indonesia, 'penyakit merah' in Malaysia or 'suffocating' in Taiwan, which for a long time were thought to be caused by physiological factors. All such virus diseases can in principle be controlled by combatting the vector insects with systemic insecticides [27, 595, 703]. It should be mentioned that examples of vector control of great importance are found

in human medicine [1034], e.g. malaria, schistosomiasis [40, 276, 344, 354, 378], Chagas disease [132, 543, 729, 784] etc., but a discussion of vector control in hygiene would exceed the scope of this review.
Internal therapy by means of systemic insecticides involves many variables determined by the chemical substance itself, by the injurious pest and by the crop species or individual plant. It is, therefore, rather difficult to define the term 'systemic' precisely. UNTERSTENHÖFER [987] proposed three processes for systemic action:

1) Incorporation or absorption of the chemical agent into the plant
2) Translocation within the plant
3) Metabolism and detoxification within the plant

As an additional criterion of true systemic activity, UNTERSTENHÖFER requires compounds to have a lasting action within the plant of at least seven days.
The most important absorption organs of the living plant are the roots, although the whole surface is able to absorb the active compound. A positive correlation seems to exist between the solubility of a substance in water and its systemic activity via the roots.
Penetration through the leaves has the advantage that contact and systemic action are combined. The result is a high initial activity in the control of virus vectors, particularly in the series of *demeton* derivatives.
Ectodesmata, as the main connection between environment and the interior of plants, may also be expected to be the main route for absorption of systemics.
The mechanism of translocation depends on many factors such as chemical structure, chemical reactivity, plant species, absorption organs and environmental conditions. The mode of translocation is, therefore, relatively unknown. Members of the *demeton* series [987] are translocated quite rapidly acropetally into parts of the plant above ground. The route can be assumed to be mainly via the xylem, i.e. translocation takes place with the transpiration flow.
The metabolism and detoxification of systemic phosphates is similar to that of other organophosphates, as is discussed on page 234f. in detail.

4) Phosphoric acid alkyl esters also play a role as rice fungicides, for example ®Kitazin (XXXIII) [O,O-diethyl S-benzyl phosphorothioate].
In 1955 Farbenfabriken Bayer applied for a patent for the chlorinated derivatives [884]. ®Kitazin, which was then already known, was described in 1964 by the firm of Ihara (Japan) as a fungicide against rice blast [467].

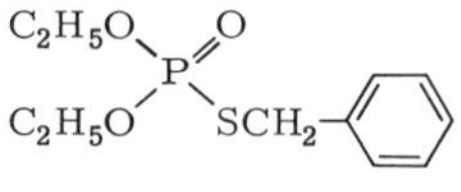

(XXXIII)

A further development by Ihara of ®Kitazin is the tetrachlorophenyl derivative [467].
The synthesis follows from the structural formula. The dioxymethylene derivative has also become known (XXXIV).

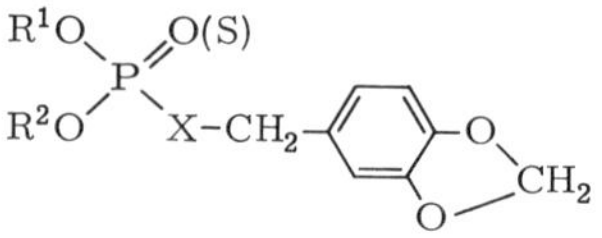

(XXXIV)

An alkyl ester of dimethylphosphorothioic acid is *endothion* (XXXV) [O,O-dimethyl S-[(5-methoxy-4-oxo-4H-pyranyl-2) methyl] phosphorothioate] [636] which was synthesized in 1955 by MÉTIVIER.

It can be prepared by familiar methods:

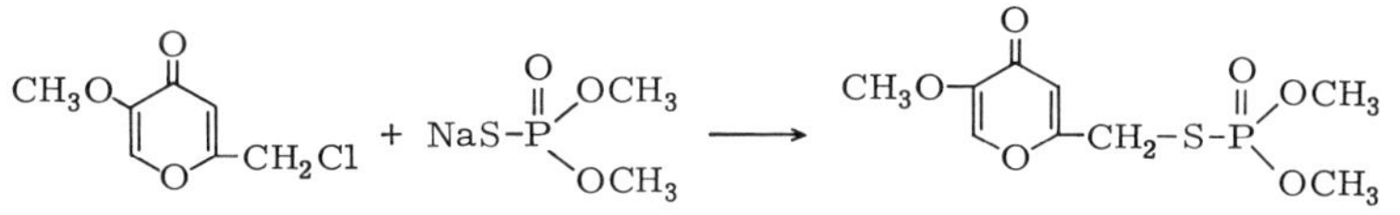

(XXXV) (29)

Endothion exerts a marked systemic action; its oral LD_{50} for the rat lies between 30 – 50 mg/kg.
Menazon (XXXVI) [O,O-dimethyl S-(4,6-diamino-1,3,5-triazinyl-2) methyl phosphorodithioate] is a less toxic, selective insecticide, a phosphoric acid ester which is heterocyclic substituted in the alkyl group. The substance was developed in 1957 by I.C.I. Ltd. [38, 156].

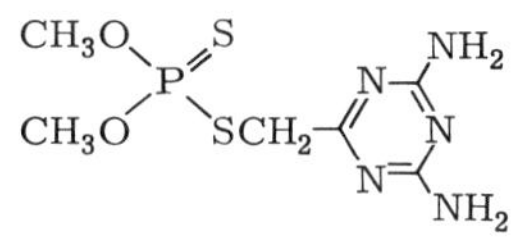

(XXXVI)

The synthesis consists in alkylating the alkali salts of dimethyl phosphorodithioic acid with the heterocyclic chloromethyl compound. The oral LD_{50} for the rat is 900 – 2000 mg/kg. *Menazon* also functions as a systemic insecticide and is used preferably to combat aphids and, in the veterinary sector, against *Hypoderma lineatum* [348].
5) A group of substances which has been thoroughly studied comprises the N-chloromethyl derivatives of heterocyclic compounds such as phthalimide [558], benzotriazole [885], indazole [248], quinazolone [560], which were reacted with salts of phosphoric acids. One of the most active of these compounds is derived from benzazimide and is called *azinphos-methyl* (®Guthion) (XXXVII) [O,O-dimethyl S-4-oxo-1,2,3-benzotriazin-3(4H)-ylmethyl phosphorodithioate] which was synthesized for the first time in 1953 by LORENZ [557]. The insecticidal properties of ®Guthion were described by UNTERSTENHÖFER [980]. The oral LD_{50} for the male rat is 15 – 20 mg/kg.

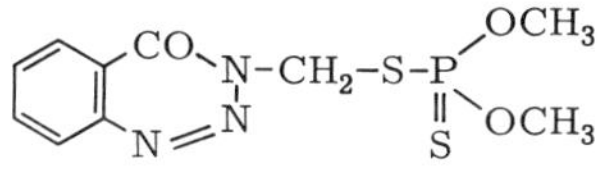

azinphos-methyl (XXXVII)

The diethyl ester is in use as a pesticide. Both compounds are also available commercially as a mixture. *Azinphos-ethyl* has an oral LD_{50} for the rat of 17.5 mg/kg and, in contrast to the methyl derivative, is effective against resistant spider mites. *Azinphos-methyl* shows a residual activity significantly higher than that of most

of the other non-systemic phosphorus insecticides. It finds particular use in fruit-growing against aphids and spider mites, as well as against cotton pests – acting therefore as contact and stomach insecticide and acaricide.
It is prepared by alkylating dimethyl phosphorodithioic acid with N-halomethyl benzazimide:

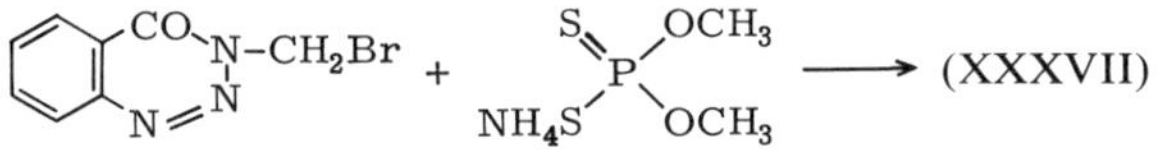

(30)

or the benzazimide is reacted directly with formaldehyde and the dithioacid in the presence of hydrogen chloride [206]:

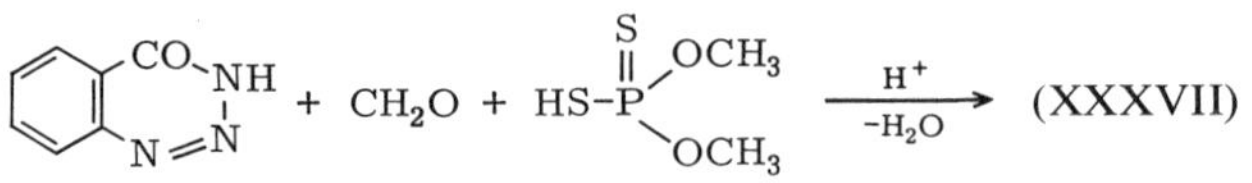

(31)

The synthesis of N-chloromethyl benzazimide starts with anthranilic acid and proceeds in the following manner (Scheme 7) [1, 280, 398, 1005, 1043]:

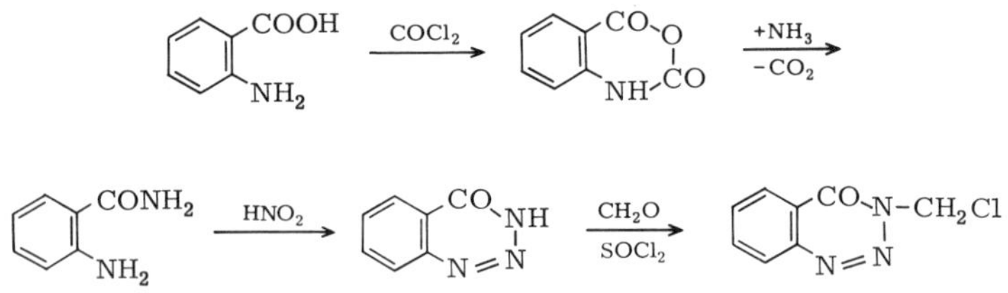

Scheme 7: Technical synthesis of N-chloromethyl benzazimide

On oxidation ®Guthion is converted into the considerably more toxic P=O compound, though this conversion does not take place on the plant under the influence of sunlight.
The use of phthalimide as the nitrogen-containing component, results in ®Imidan (XXXVIII) [O,O-dimethyl S-(N-phthalimidomethyl) phosphorodithioate] [273, 558]. It was introduced by the Stauffer Chemical Company.
®Imidan can be obtained in an analogous way to *azinphos methyl*

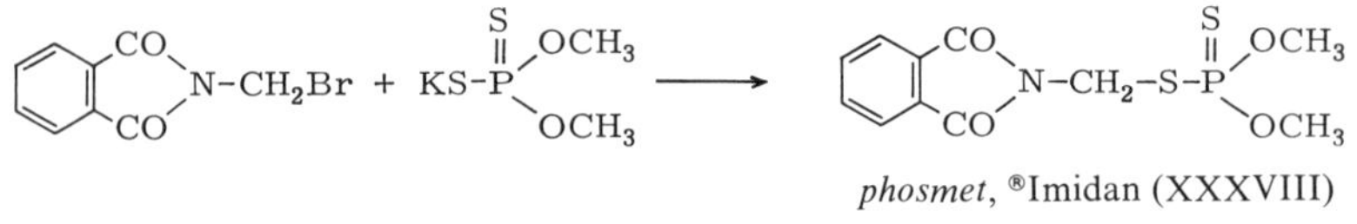

(32)

phosmet, ®Imidan (XXXVIII)

The oral LD_{50} for the rat is 147 – 216 mg/kg. The N-ethyl derivatives, for example (XXXIX), are inferior in action to the N-methyl compounds [569].

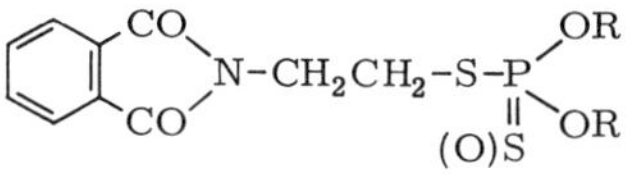

(XXXIX)

Benzoxazolone as an heterocycle leads to the compound *phosalone* or ®Zolone (XL) [O,O-diethyl S-(2-oxo-6-chloro-2H-benzoxazolinyl-3) methyl phosphorodithioate] [130, 247, 638].

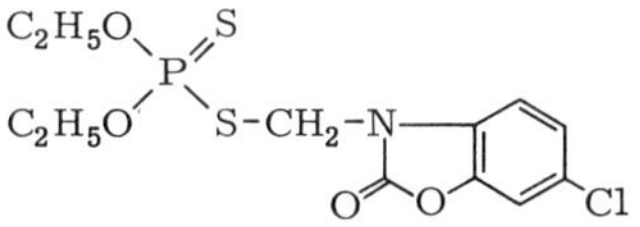

(XL)

It was introduced on the market by Rhône-Poulenc. N-chloromethyl 5-chlorobenzoxazolone serves as the starting material. *Phosalone* exerts a broad activity against spider mites, Hemiptera, Coleoptera, Diptera, having an LD_{50} of 135 mg/kg when administered orally to the rat.

Recently ®Supracide or ®Ultracide (*methidathion*) (XLI) [O,O-dimethyl S-((2-methoxy-5-oxo-Δ^2-1,3,4-thiadiazolinyl-4) methyl) phosphorodithioate] was developed by Geigy as an insecticide, especially acaricide [37].

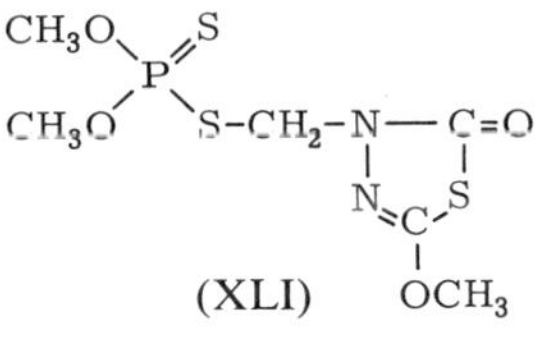

(XLI)

The heterocyclic intermediate is synthesized as follows:

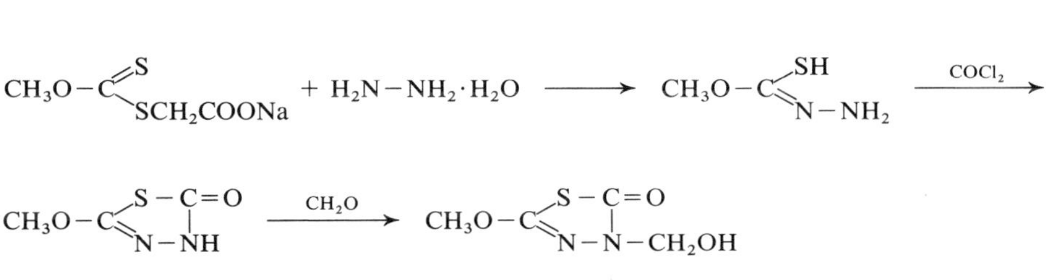

Scheme 8: Synthesis of 2-methoxy 4-hydroxymethyl 1,3,4-thiadiazolinone

The oral LD_{50} for the rat is 25 – 48 mg/kg. ®Supracide acts against biting, sucking insects, resistant spider mites and is suitable for application in fruit-growing and arable farming.
The ethoxythiadiazole derivative is also an effective insecticide and acaricide with an oral LD_{50} for the rat of 268 – 443 mg/kg.
In this connection the following heterocyclic compound, which was introduced by American Cyanamid, must be mentioned (XLII) [O,O-dimethyl S-(5-amino-1,2,4-oxadiazolyl-3) methyl phosphorodithioate] [433].

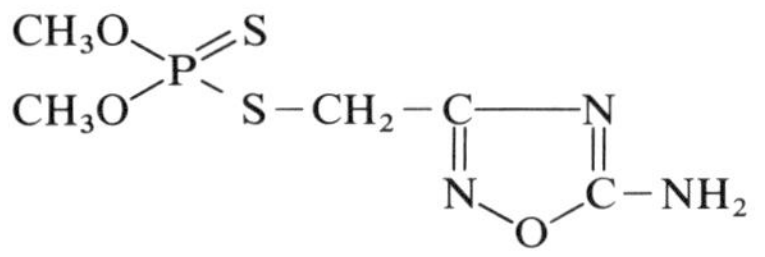

(XLII)

The heterocyclic starting material results from ortho-ester and sodium cyanamide via N-cyanoimidate and ring closure with hydroxylamine (Eq. (33)).

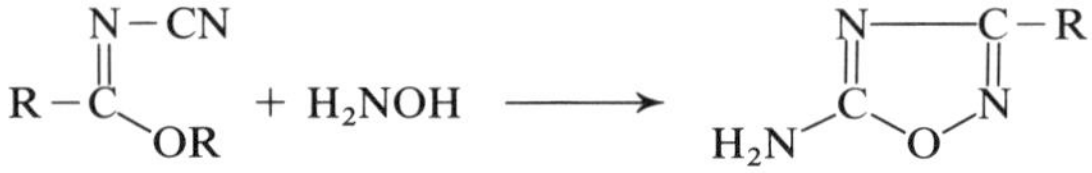

R = $-CH_2Cl$ (33)

Subsequent reaction with dimethyl phosphorodithioate yields the desired product.
6) When salts of dialkyl phosphoric acid are alkylated by derivatives of α-halogen-carboxylic acids, a series of well-known and important compounds result which have become valuable insecticides.
A basic member of this series is *acethion* (XLIII) [O,O-diethyl S-carbethoxymethyl phosphorodithioate], whose synthesis was achieved by the firm of Boehringer in 1955 [25].

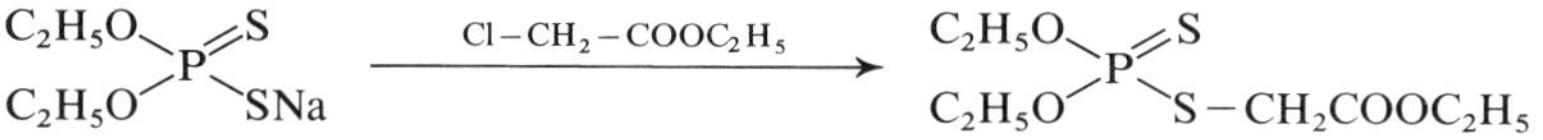

(XLIII) (34)

The oral LD_{50} for the rat is 1050 – 1100 mg/kg. In *Musca domestica acethion* is metabolized primarily by oxidation of the thiono-group [863]. (For further degradation mechanisms see p. 234 f.).
α-Phenyl α-halogen-acetic ester yields ®Cidial (*phenthoate*) (XLIV) [O,O-dimethyl S-(ω-carbethoxy) benzyl phosphorodithioate], which has been prepared by

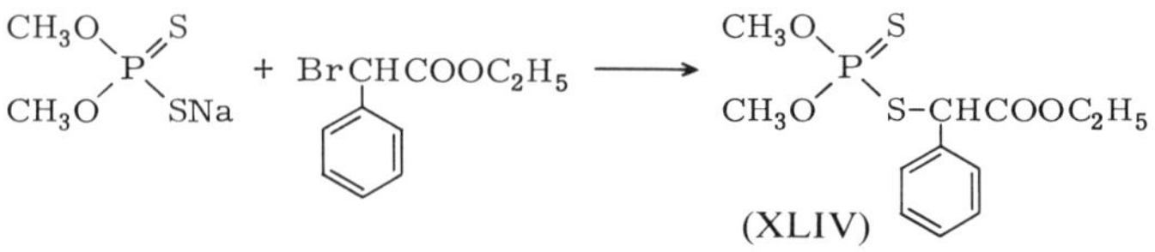

(XLIV) (35)

numerous firms. In 1955 SCHRADER (Farbenfabriken Bayer) [853] synthesized Cidial by this route given in patent D.A.S. 1.011.416.
In 1964 FUSCO *et al.* [320] (Montecatini) described a process in which esterification is performed subsequently:

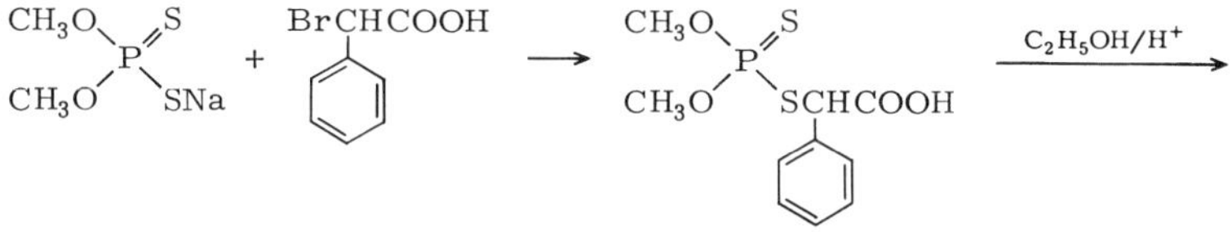
(XLIV) (36)

This compound is used especially in Japan under the name ®Erusan as a contact insecticide and special agent against scale insects. The oral LD_{50} for the rat is 250 mg/kg.
If the acetic acid molecule is varied in the carboxyl group, a series of well-known compounds result, of which *dimethoate* or ®Rogor (XLV) [O,O-dimethyl S-(N-methylcarbamoyl) methyl phosphorodithioate] is best known.

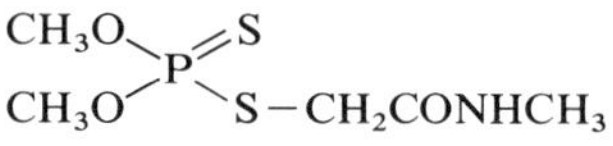

dimethoate, ®Rogor (XLV)

There are numerous syntheses available, which indicate the interest of industry in this potent product. In 1948 American Cyanamid [168] claimed a process involving the reaction of alkali salts of O,O-dialkyl phosphorodithioic acids with chloroacetic acid amides.

(XLV) (37)

In 1955 Montecatini [719] and Boehringer [25] described the following process in different patents:

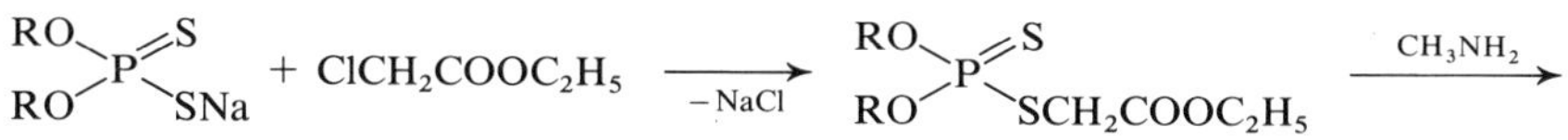
(XLV) (38)

In 1958 Boehringer filed a new application for a further process [894]:

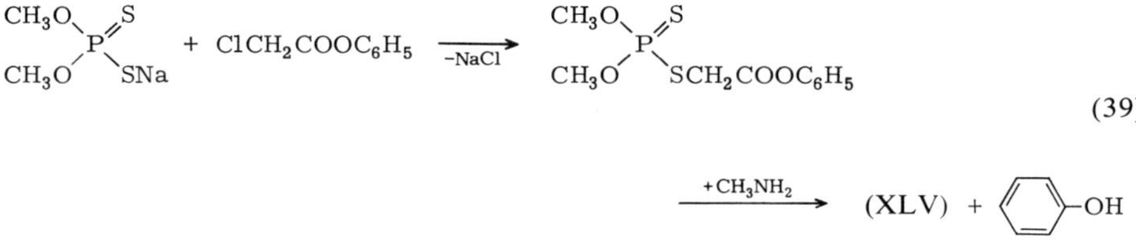

(39)

In the same way that O,O-dialkyl phosphorodithioic acids react with thiosulfuric acid mono-esters to give disulfides, the salts of O,O-dialkyl phosphites and thiol phosphites [573] react with the thiosulfuric acid mono-esters to give *dimethoate*:

(40)

In 1959 Boehringer applied for another patent [26]:

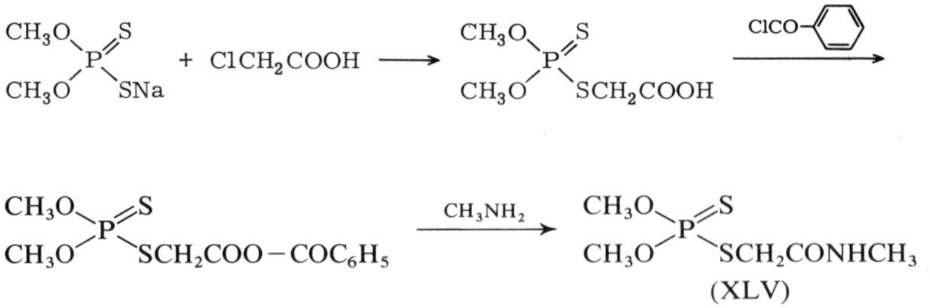

(41)

Also in 1959, American Cyanamid developed a synthesis with a different acid anhydride as intermediate [1039]:

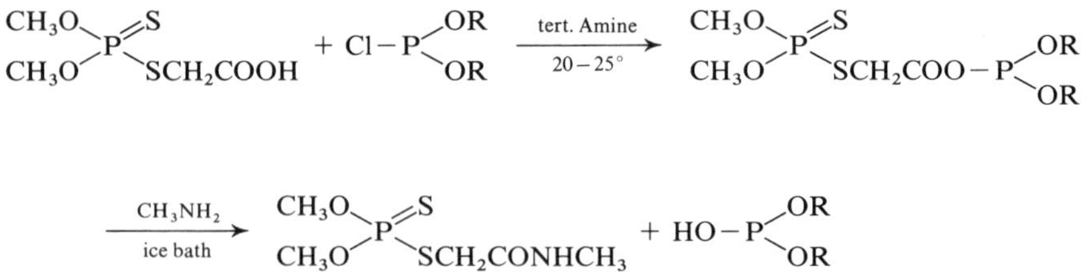

(42)

In 1962 the following reaction was investigated by Farbenfabriken Bayer [692]:

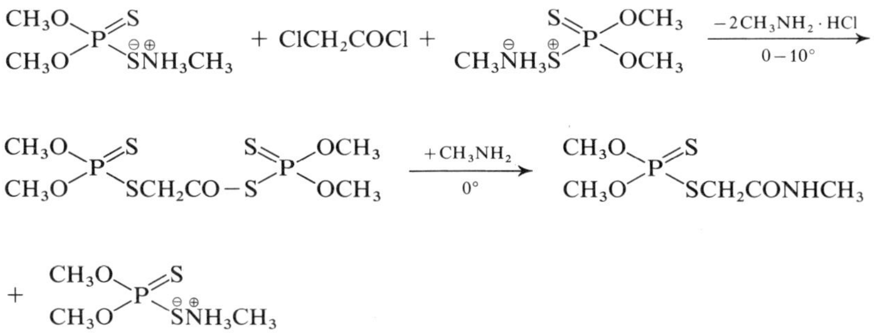

Scheme 9: *dimethoate* synthesis

In G. D. R. patent 49.605 an improved synthesis of *dimethoate* is described [503]. In principle, it concerns alkylation of dimethyl phosphorodithioic acid by chloroacetic acid monomethylamide in a two-phase system at pH 1 – 4 i.e. in an acid medium. The oral LD_{50} for the rat of the pure compound is 215 – 267 mg/kg. ®Rogor is systemic in action and effective particularly against sucking insects, Diptera and susceptible spider mites. This compound is used in particular to control the olive fly (*Dacus olae*). Unlike most phosphoric acid esters, *dimethoate* is not taken up by the oily phase and therefore has very good residue properties [733].
Rearrangement of *dimethoate* as a thiono-thiol compound into the dithiol derivative is feasible because of the alkylating properties of *dimethoate* (see p. 36) [253].

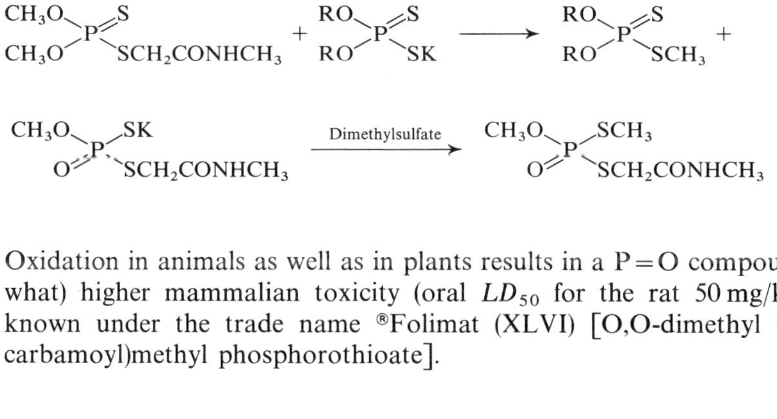

(13)

Oxidation in animals as well as in plants results in a P=O compound of (somewhat) higher mammalian toxicity (oral LD_{50} for the rat 50 mg/kg), which is known under the trade name ®Folimat (XLVI) [O,O-dimethyl S-(N-methylcarbamoyl)methyl phosphorothioate].

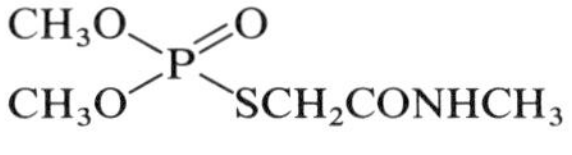

omethoate, ®Folimat (XLVI)

In 1961 Sumitomo [48] found a synthesis for *dimethoate* and in 1962 Farbenfabriken Bayer another for ®Folimat, both using dimethyl thiophosphoryl mercaptoacetic acid and methyl isocyanate [693]:

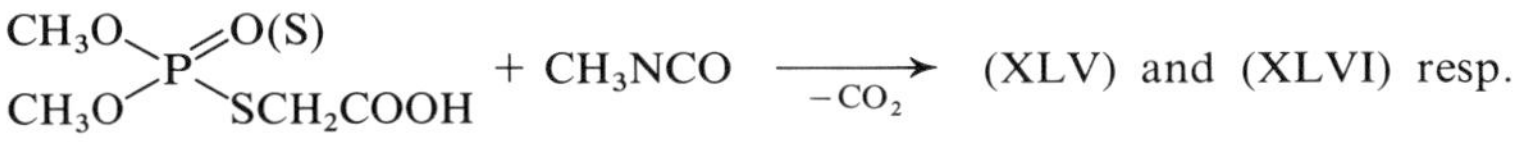 (44)

®Folimat is recommended by Farbenfabriken Bayer as a systemic insecticide in agriculture and also as an effective acaricide against resistant spider mites. On degradation in the plant, demethylation, oxidation and hydrolysis take place. The use of ®Folimat as a systemic agent to control ecto- and endoparasites has been patented by American Cyanamid [404].
An example of a derivative of *dimethoate* with another amide group is ®Ekatin F or *morphothion* (XLVII) [O,O-dimethyl S-(morpholinocarbonyl)methyl phosphorodithioate], which was described by Sandoz AG in 1956 [44].

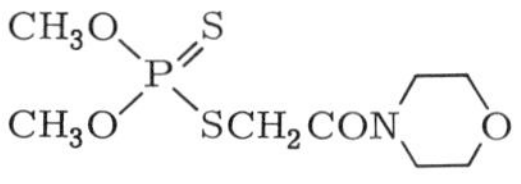

(XLVII)

The compound shows contact insecticidal and systemic activity.
An analogous substance, ®Fac 20 (*prothoate*) (XLVIII) [O,O-diethyl S-(N-isopropyl carbamoylmethyl) phosphorodithioate] may be mentioned, which was prepared in 1948 by American Cyanamid [168], in 1957 by Montecatini [720] and in 1958 by Boehringer [894]. It is manufactured by the same process as that described for ®Rogor. Its toxicity is considerably higher having an oral LD_{50} for the rat of 8 mg/kg. Its insecticidal properties are similar to those of *dimethoate.*

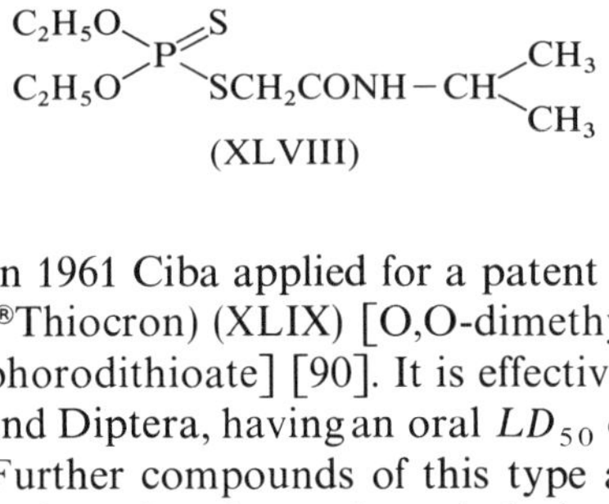

(XLVIII)

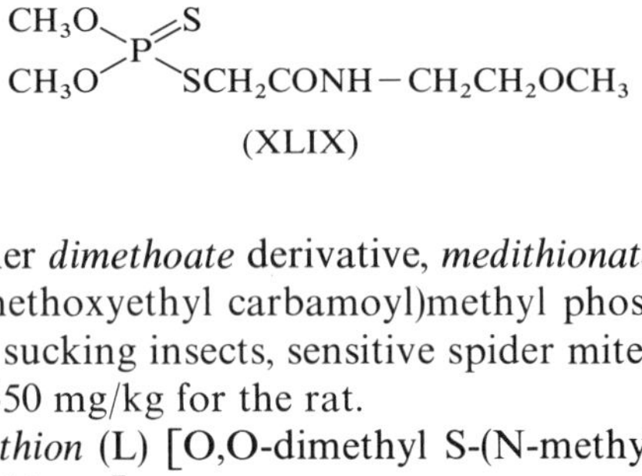

(XLIX)

In 1961 Ciba applied for a patent for another *dimethoate* derivative, *medithionate* (®Thiocron) (XLIX) [O,O-dimethyl S-(N-methoxyethyl carbamoyl)methyl phosphorodithioate] [90]. It is effective against sucking insects, sensitive spider mites and Diptera, having an oral LD_{50} of 420–650 mg/kg for the rat.
Further compounds of this type are *formothion* (L) [O,O-dimethyl S-(N-methyl N-formyl carbamoyl) methyl phosphorodithioate]

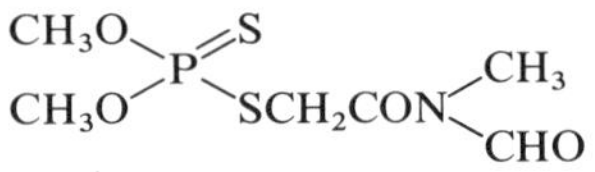 (L)

of Sandoz (1960) [583] with an oral LD_{50} for the rat of 353 mg/kg, and *cyanthoate* (LI) [O,O-diethyl S-[N-(1-cyano-1-methyl) ethyl carbamoyl] methyl phosphorodithioate],

(LI)

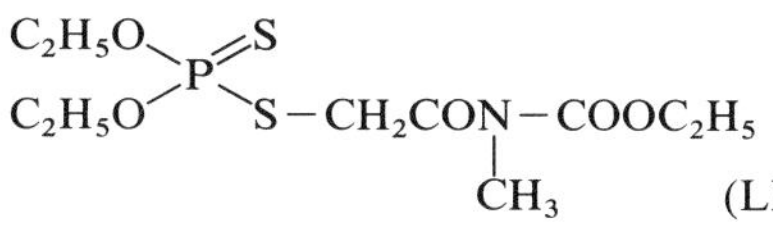
(LII)

proposed by Montecatini [577] as an insecticide and acaricide, and finally *mecarbam* (LII) [O,O-diethyl S-[(N-methyl N-carbethoxy) carbamoyl] methyl phosphorodithioate] developed in 1961 by Murphy Chemical Company. *Mecarbam* is an aphicide and acaricide with ovicidal action, having an oral LD_{50} for the rat of 31 – 35 mg/kg [731, 732].

7) One of the oldest methods for the synthesis of substituted alkyl esters of dialkyl phosphorodithioic acids consists in the addition of dialkyl phosphorodithioic acids to unsaturated compounds. In this way *malathion* (LIII) [O,O-dimethyl S-(1,2-dicarbethoxy) ethyl phosphorodithioate] resulted [167]. The model reaction was the addition of sodium hydrogen sulfite to unsaturated dicarbonic acid esters, in this case to sulfonosuccinic acid esters such as Aerosol OT, the bis-(2-ethyl hexyl) succinate. Compounds of this type were used as detergents [601]. *Malathion* is a product of American Cyanamid (1950).

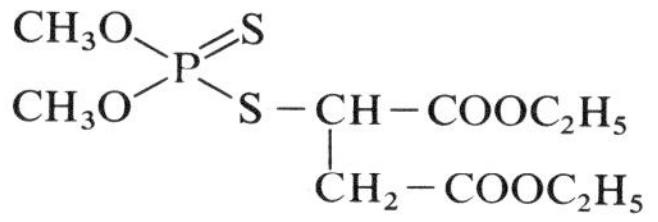
malathion (LIII)

The synthesis is a one-step process in which dimethyl phosphorodithioic acid is first prepared from phosphorus pentasulfide and methanol and then added directly to maleic acid diethyl ester under the influence of catalytic quantities of alkali. The oral LD_{50} for the male rat is 1200 mg/kg. In this connection it is worth mentioning that diethyl phosphorodithioic acid does not undergo an analogous addition reaction, for example, with dimethyl propynyl phosphonate, but is methylated almost quantitatively on the thiol group [892]:

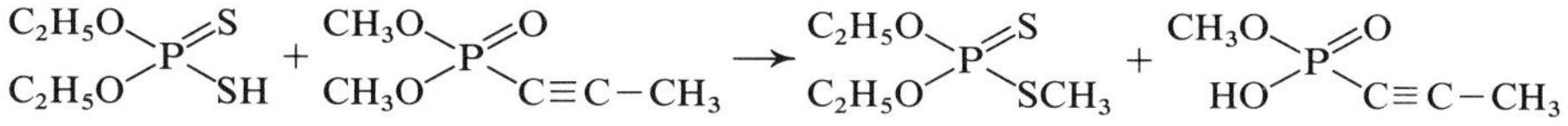
(45)

Oxidizing agents such as nitric acid convert *malathion* into malaoxon (LIV) [O,O-dimethyl S-(1,2-dicarbethoxy) ethyl phosphorothioate] [450].

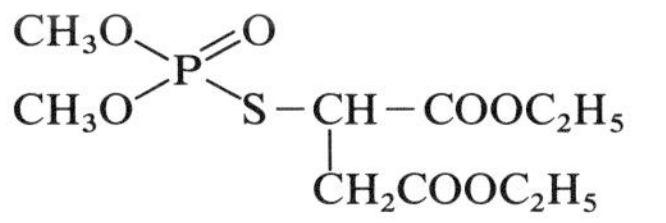
(LIV)

Yet another synthesis for malaoxon is from α-halosuccinic acid [849]:

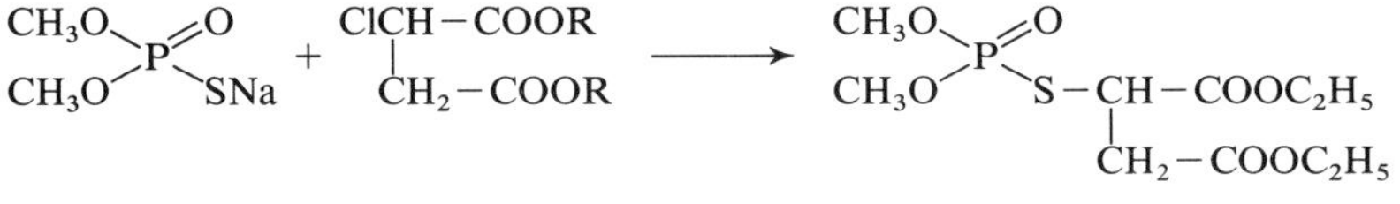

(LIV) (46)

The oxygen compound is considerably more toxic, having an oral LD_{50} of 87 to 90 mg/kg for the rat. Both compounds are effective against biting, sucking insects and susceptible spider mites.

As indicated by its name, *malathion* has been applied to control the vectors of malaria. Considerable quantities are employed to protect stored products. On heating, isomerization to the methyl thiolester (LV) takes place.

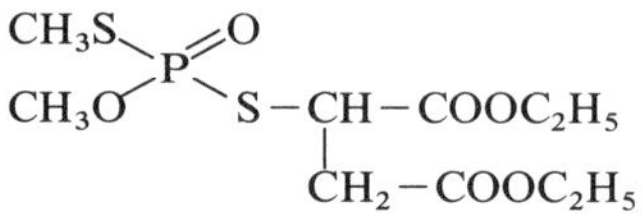

(LV)

Malathion is metabolized differently in mammals and insects. In insects predominantly malaoxon results, in mammals the monocarboxylic acid ("malathionic acid"). Recently some resistance to *malathion* has been observed in insects.

A related synthesis, the addition of bis-(O,O-diethyl thionophosphoryl)-disulfide to *p*-dioxen [245, 521] leads to *dioxathion* (®Delnav) (LVI) [1,4-dioxane-2,3-(bis-O,O-diethyl thionophosphoryl)-dithiol],

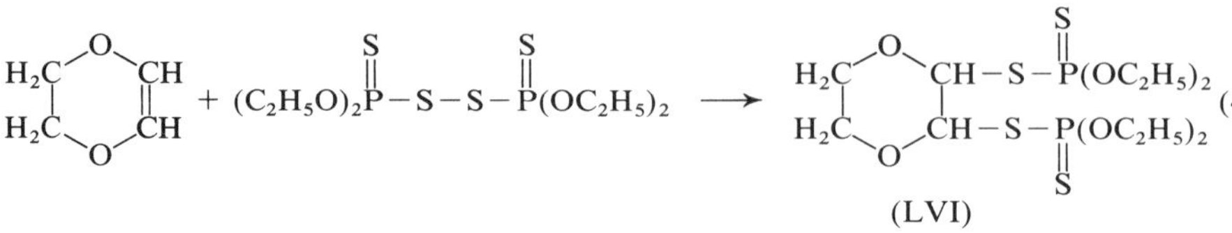

(LVI) (47)

which was described in 1954 by Hercules Powder Company. Another possibility for synthesis consists in reaction of diethyl phosphorodithioic acid with 2,3-dichloro-*p*-dioxan [74, 242, 243, 244, 381]:

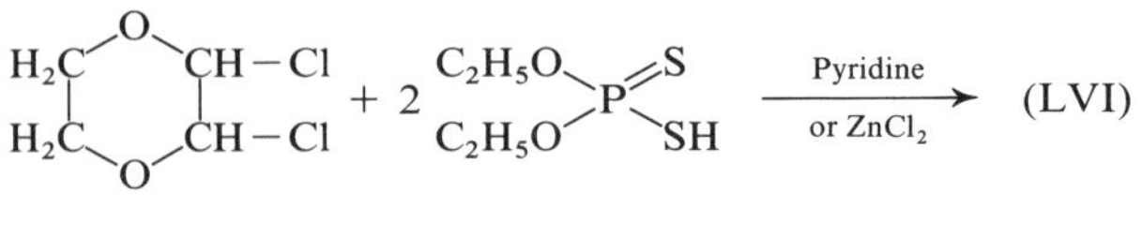

(48)

This compound exists in the cis and trans forms with different toxicities. The subcutaneous LD_{50} for the rat is 65 mg/kg for the cis-compound, and 240 mg/kg for the trans-compound. ®Delnav is used as contact insecticide and acaricide against ticks.

d) Phosphite Reactions

Commercial importance has been attached to derivatives of the DDVP and Phosdrin group as insecticides, particularly as they are generally easy to prepare.
Addition of dimethyl phosphite to chloral [72, 556] (Eq. (1)) leads to a stable product with the structure (I)

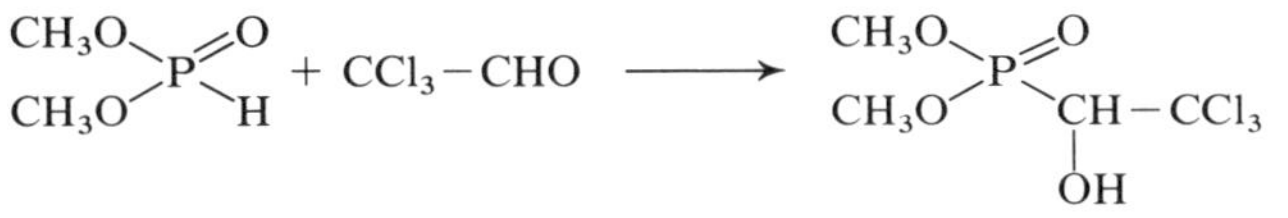
(1)

trichlorfon (®Dipterex) (I)

which itself has developed into one of the most important insecticides and become known under the name *trichlorfon* (®Dipterex) (I) [O,O-dimethyl (1-hydroxy-2,2,2-trichloro)-ethanephosphonate]. A second synthesis [299] involves a single-step process:

$$PCl_3 + CCl_3CHO + 3\,CH_3OH \xrightarrow{-3\,HCl} (I) \qquad (2)$$

The resulting hydrogen chloride is removed by blowing air through the mixture. *Trichlorfon* was developed in 1952 by Farbenfabriken Bayer. It has a low mammalian toxicity (oral LD_{50} 625 mg/kg for the male rat). The main application of ®Dipterex, which was reported for the first time by UNTERSTENHÖFER [978], is as a stomach insecticide against Lepidoptera such as the Egyptian cotton worm (*Prodenia* sp.), also against Diptera and Heteroptera in vegetables, rice, corn, sugar cane and fruit culture. This substance is used in the field of hygiene to control flies and is known under the name ®Tugon. BEHRENZ, FEDERMANN and BOLLE proposed the use of this compound in dips against ectoparasites in sheep [80]. On account of its favourable toxicity it has become established in the veterinary sector under the name of ®Neguvon for application against ecto- and endoparasites. *Trichlorfon* is

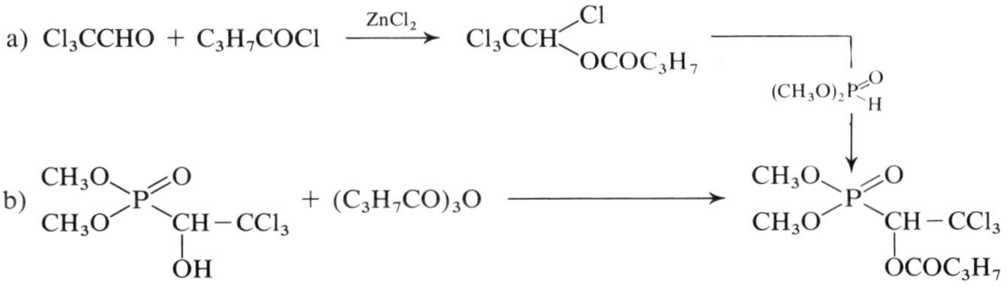
(3)

butonate (II)

relatively harmless to bees. *Butonate* (II) may be regarded as an acyl (butyryl) derivative of *trichlorfon* [O,O-dimethyl (2,2,2-trichloro-1-*n*-butyryloxy)-ethane-phosphonate].
It was developed by the Wisconsin Alumni Research Foundation (WARF) as an agent to control house pests. WARF has given two syntheses [161].
If the hydroxyphosphonate *trichlorfon* is reacted with alkali, rearrangement [71] to the phosphate *dichlorvos* (III) [O,O-dimethyl O-(2,2-dichlorovinyl) phosphate] occurs as was found by LORENZ in 1954 [559, 564].

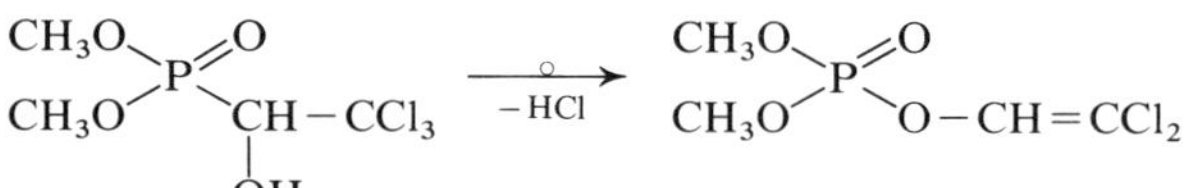

dichlorvos, DDVP

(I) (III) (4)

A direct method yielding *dichlorvos* is the reaction of trimethyl phosphite with chloral [306, 789]:

$$(CH_3O)_3P + CCl_3CHO \xrightarrow[-CH_3Cl]{} (III) \qquad (5)$$

The toxicity of *dichlorvos* is significantly higher than that of *trichlorfon* (oral LD_{50} for the rat: 62 mg/kg). In contrast to *trichlorfon*, *dichlorvos* is remarkably volatile and serves as contact, stomach and respiratory insecticide. In the field of hygiene its use is mainly against flies (®Oko, ®Mafu, ®Vapona [293] etc.); in the agricultural sector it is used against sucking, biting pests and leaf miners.
In the plant, *dichlorvos* is very rapidly hydrolyzed to dimethyl phosphoric acid and dichloroacetaldehyde, it can, therefore, be applied shortly before harvest.
Condensation of triethyl phosphite with trichloroacetic acid, instead of chloral, leads to Forstenon (IV) [O,O-diethyl O-[1-(2-chloroethoxy) 2,2-dichlorovinyl] phosphate]:

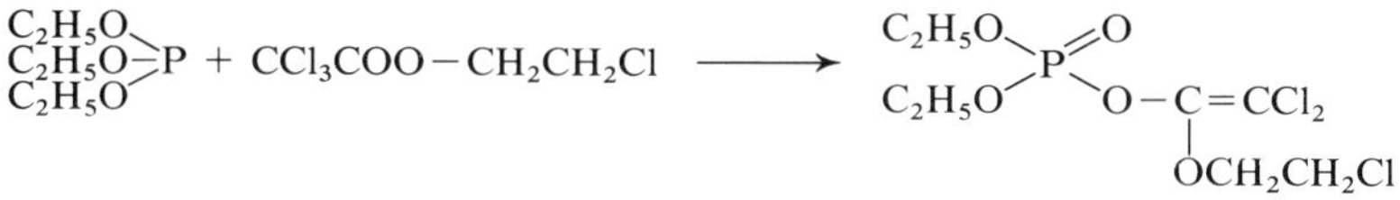

(IV) (6)

The oral LD_{50} for the rat is 6.8 – 9.7 mg/kg. Thus the compound is about ten times more toxic than *dichlorvos*. It is used as a contact insecticide [479, 788].
A compound which may be regarded as a phenyl derivative of *dichlorvos* is *chlorfenvinphos* (®Dermaton or ®Birlane) (V) [O,O-diethyl O-[1-(2,4-dichlorophenyl) 2-chloro-vinyl] phosphate].

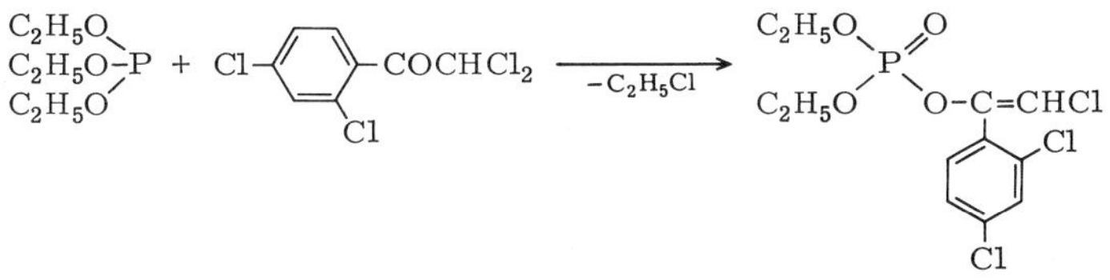
(7)

chlorfenvinphos (V)

It is synthesized from the appropriately substituted dichloroacetophenone and triethyl phosphite [1014, 1016]. Under the code number G.C. 4072 (Gen. Chem. Div. of Allied Chem. and Dye Corp.) ®Dermaton was described by DRUMMOND as a systemic insecticide for veterinary medicine [251, 350].
Another compound introduced by Shell Oil Co. and in 1964 patented in the USA [999] is ®Akton (VI) [O,O-diethyl O-1-(2',5-dichlorophenyl) 2-chloro-vinyl phosphorothioate]. It is noteworthy that this is a thiono-compound. The use of phosphorothioite in this synthesis is too expensive on an industrial scale and therefore another route is chosen using a phosphorothiochloridate (Eq. (8)):

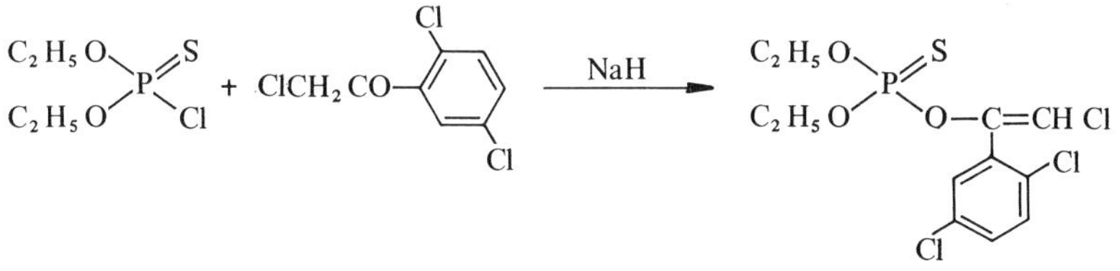
(8)

®Akton (VI)

The oral LD_{50} is 146 mg/kg for the rat. The new compound was proposed mainly for the control of onion maggot and mosquito larvae.
Chlorfenvinphos, like the trichloro derivative *tetrachlorvinphos* (®Gardona) (VII) [O,O-dimethyl O-1-(2,4,5-trichlorophenyl) 2-chloro-vinyl phosphate], was patented by Shell in 1952 [730].

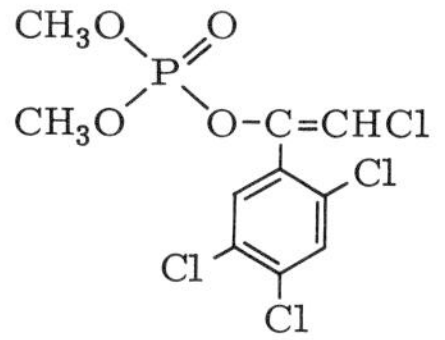
(VII)

Gardona acts as a contact insecticide with an extremely powerful stomach poisoning action. It is highly active against lepidopterous and dipterous pests and also controls certain Coleoptera, but is not very persistent in the soil and does not show activity against the major soil pests. The oral LD_{50} for the rat is within the range 4000–5000 mg/kg.

The olefinic double bonds of the vinyl phosphates add halogens. *Naled* (VIII) [O,O-dimethyl O-(1,2-dibromo-2,2-dichloroethyl) phosphate] results from the addition of bromine to *dichlorvos* [702, 930].

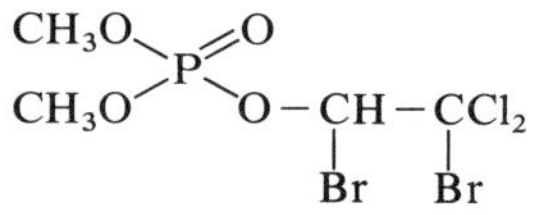

(VIII)

As a pesticide of low toxicity (oral LD_{50} for the rat is 450 mg/kg) it is marketed in the hygiene sector. There have also been reports of numerous analogous compounds with perhalogenated ethyl or vinyl-oxy groups [608].
A very important product of this series is *mevinphos* (®Phosdrin) (IX) [O,O-dimethyl O-(1-methyl 2-carbomethoxy) vinyl phosphate], which was developed in 1957 by Shell. There are two possible routes for its synthesis. O,O-dimethyl phosphorochloridate is reacted with the sodium salt of methyl acetoacetate [844].

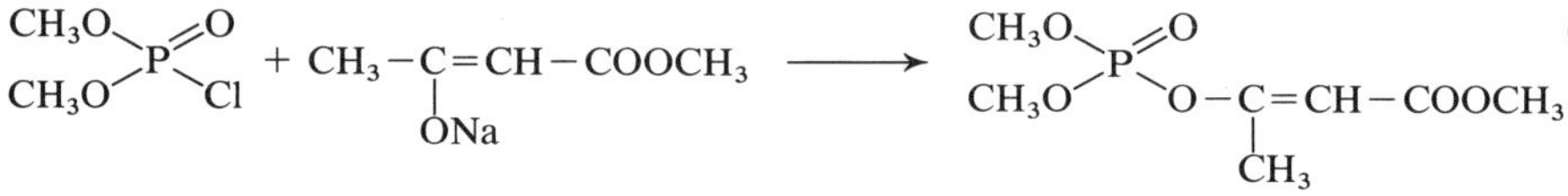

(9)

mevinphos (IX)

The technically applied synthesis [926] uses trimethyl phosphite and chloroacetic acid ester:

$$(CH_3O)_3P + CH_3COCH(Cl)COOCH_3 \xrightarrow[-CH_3Cl]{} (IX) \qquad (10)$$

Mevinphos is a very toxic compound, having an oral LD_{50} for the male rat of 3.7 mg/kg. It exists in both a cis and trans form. The cis form is important for its insecticidal properties, being about a hundred times more active than the trans isomer.

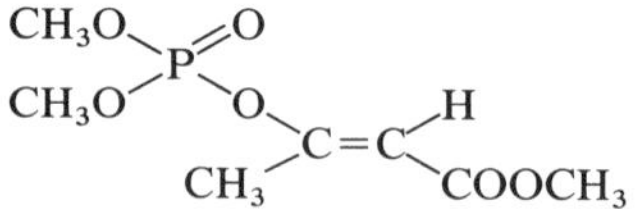

cis-mevinphos

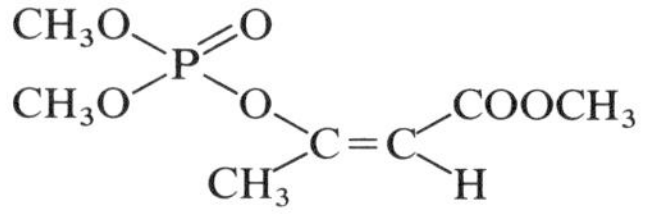

trans-mevinphos

Mevinphos is a contact insecticide, acaricide and exhibits systemic activity. However, due to its increased susceptibility towards hydrolysis insecticidal activity rapidly decreases. In the plant, ®Phosdrin is degraded in the following manner [18]:

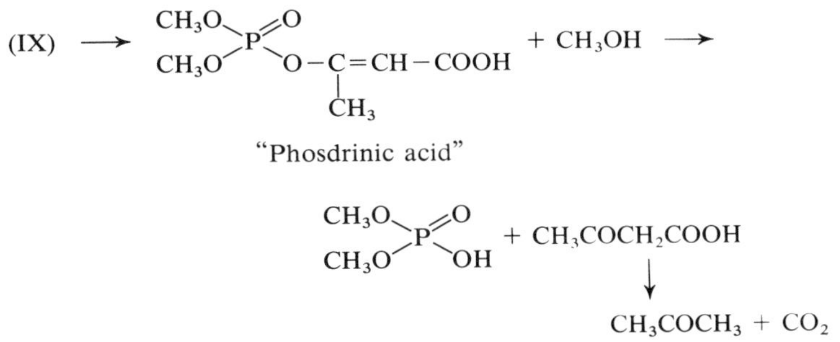

"Phosdrinic acid" (11)

A compound closely related to *mevinphos* is the methylbenzyl ester *crotoxyphos* (X) [O,O-dimethyl O-[1-methyl 2-carbo-(α-methyl) benzyloxy] vinyl phosphate] [45, 953, 1015].

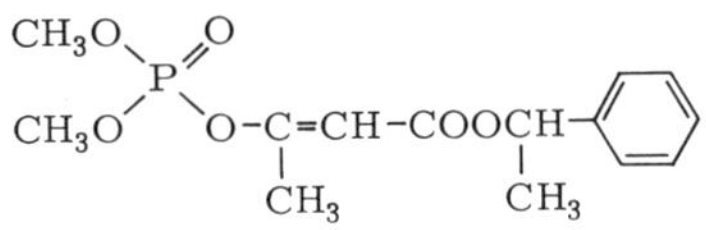

crotoxyphos, ®Ciodrin (X)

®Ciodrin is considerably less toxic than *mevinphos* but shares the latter's property of being rapidly detoxicated in the plant. It is also effective as a systemic and contact insecticide having an oral LD_{50} for the rat of 140–200 mg/kg.
The dimethylamide of *mevinphos* is called *dicrotophos* (®Bidrin) (XI) [213, 953] [O,O-dimethyl O-[1-methyl 2-(N,N-dimethyl)carbamoyl] vinyl phosphate].

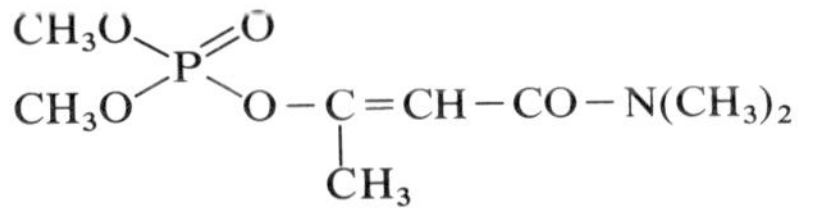

(XI)

Its uses are similar to those of ®Ciodrin. *Monocrotophos* (®Azodrin) is the cis isomer of the analogous N-monomethyl compound [O,O-dimethyl O-(1-methyl 2-N-methylcarbamoyl) vinyl phosphate] with an acute toxicity of 20 mg/kg for the rat.

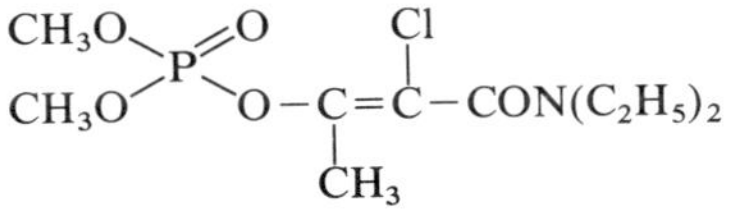

phosphamidon (XII)

Another important member of this series is *phosphamidon* (XII) [O,O-dimethyl O-[1-methyl 2-chloro-2-(N,N-diethylcarbamoyl)] vinyl phosphate] for which Ciba in 1955 applied for a patent [92].
The synthesis follows a standard procedure employing trimethyl phosphite and α,α-dichloroacetic acid diethylamide. In its toxicity *phosphamidon* resembles ®Phosdrin. It has an oral LD_{50} for the rat of 10 mg/kg.
Chlorination of acetone-dicarboxylic acid dimethyl ester and subsequent reaction with trimethyl phosphite yields O,O-dimethyl O-1,3-di(carbomethoxy)-1-propen-2-yl phosphate (XIII). In 1959 Allied Chemical Corporation offered this compound

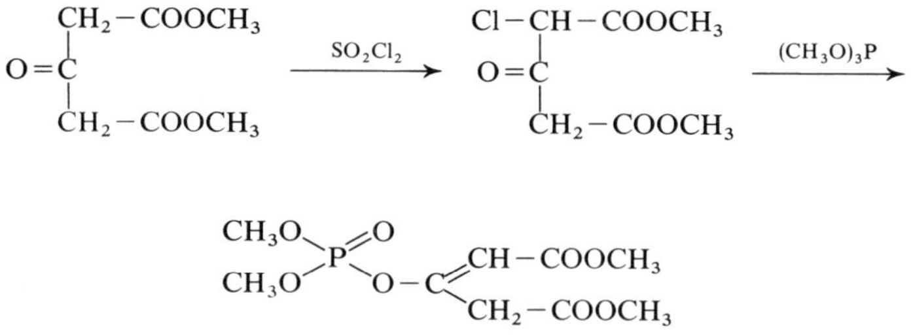

®Bomyl (XIII)

under the trade name ®Bomyl as a contact insecticide and acaricide [337]. It has an oral LD_{50} for the rat of 32 mg/kg.

e) Miscellaneous Reactions

Compounds of the type

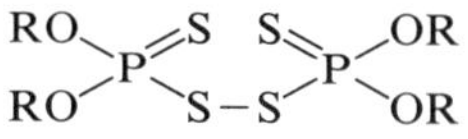

which belong to this section have, in practice, acquired no importance as insecticides. The only commercial product of this series that has become known is ®Phostex (I) [Bis-(O,O-dialkyl thionophosphoryl) disulfides] a compound developed by Food Machinery and Chemical Corporation in 1954 [28, 1021, 1023].

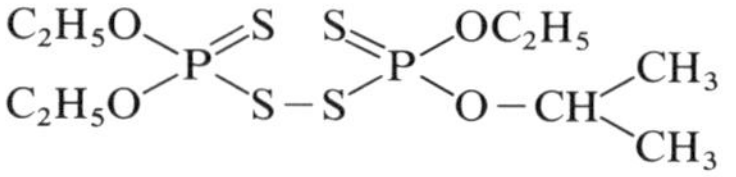 (I)

®Phostex is not represented by a definite structural formula, but contains isopropyl and ethyl groups in the ratio 1 : 3. Synthesis according to Eq. (1):

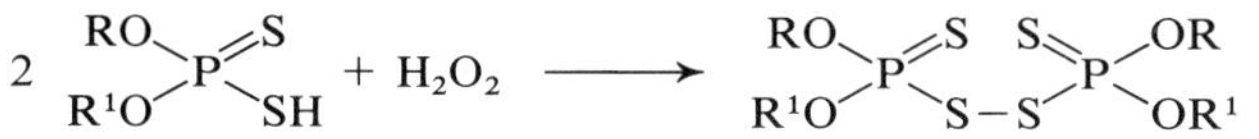
(1)

If the phosphorodithioic acid is prepared not with one alcohol but with a mixture of two, then the disulfide contains different alkyl groups in varying combinations. The toxicity of ®Phostex is 1000 mg/kg (oral LD_{50} for the rat). It has poor contact insecticidal and very good ovicidal action.
It is the chemical properties rather than the insecticidal properties which make the disulfides such interesting substances, for example, they can be desulfurized with potassium cyanide [667], and cleaved [867]:

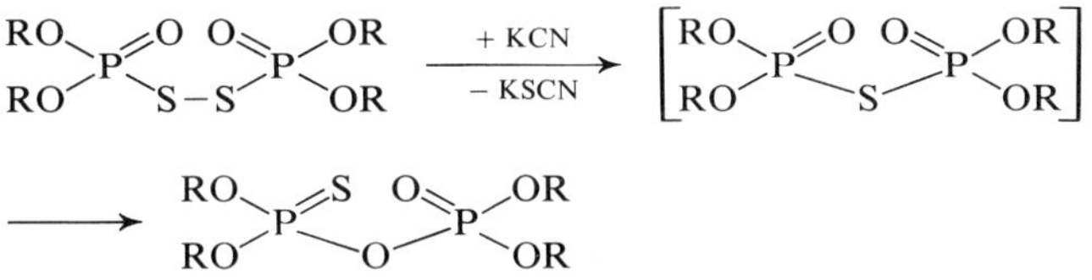
(2)

(3)

The de-sulfurization with O,O-dialkyl sodium (thiol) phosphite [574] proceeds as follows:

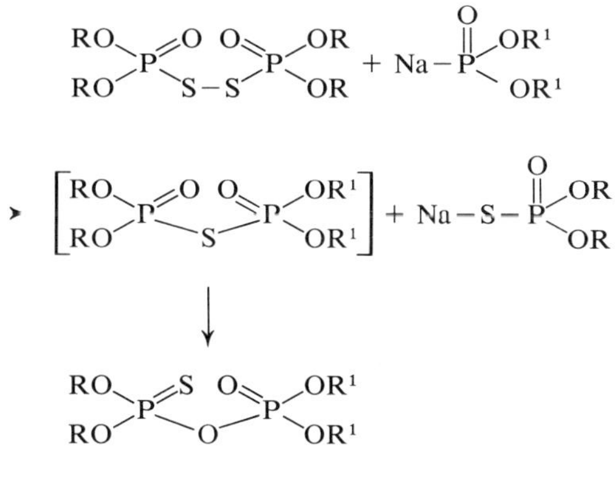
(4)

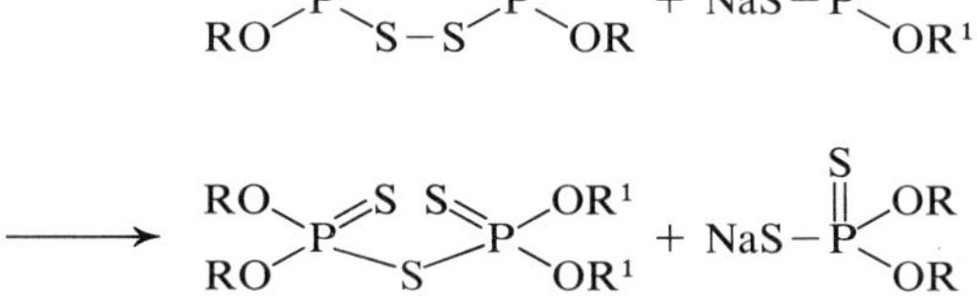
(5)

Trialkyl phosphites react in a different way [622]:

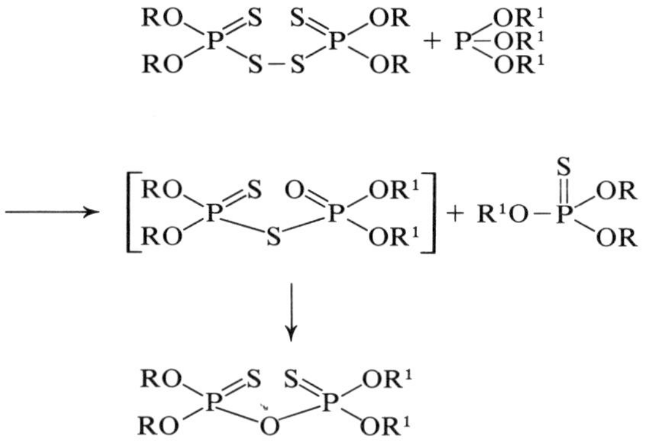

(6)

A particularly interesting reaction is the cleavage of disulfides with alkines, alkenes and Grignard compounds [651],

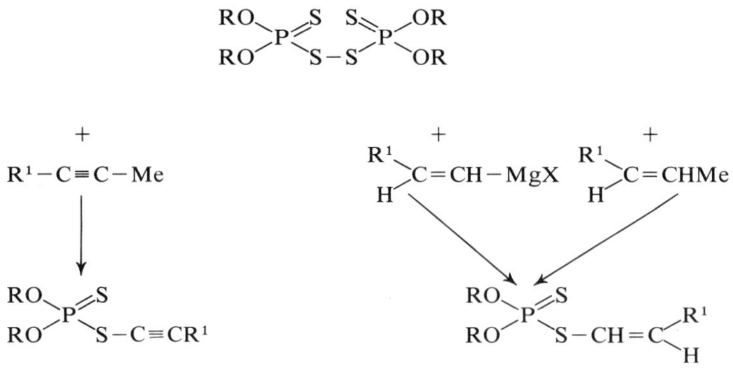

Scheme 10: Cleavage of phosphoryl sulfides

which have, to date, been the only means of synthesizing thiol derivatives of the thiono-*dichlorvos* type.

3.3. Other Compounds

a) Fungicides

The spectrum of activity of the above-mentioned types of organophosphates not only comprises insects and other arthropods but may extend beyond the plants to the fungi. For this reason the phosphoric acid esters as fungicides and

herbicides are treated separately in the following paragraphs. They are also arranged according to their method of preparation.
In the early days of crop protection, mainly inorganic compounds were used to control fungus diseases; although to some extent this is still true today, the quantities applied are decreasing.
In viticulture, in particular, elemental sulfur has proved effective for the control of powdery mildew. Cereal seed was protected from pathogenic fungi with preparations of copper sulfate, this has the disadvantage of impairing the germinating power of the seed.
Copper compounds are still frequently used as leaf fungicides, in wine-growing for example. Mercurial compounds, e.g. corrosive sublimate, also exhibit fungicidal properties in the disinfection of seed. Because of the toxicity, seed and germ-damaging action of the inorganic mercurials, organo-mercurial compounds (e.g. chlorophenol-mercury, ®Semesan, ®Uspulun, ®Ceresan etc.) were developed.
Dithiocarbamates and thiuram derivatives (Ferbam, Zineb etc.) are a class of fungicides whose action depends upon the formation of thioureide ions, which probably react with the mercapto-groups of the co-enzymes and thereby block them [602].
There is a general tendency today for the heavy-metal fungicides to be replaced by organic compounds which are more readily metabolized. The trend is being encouraged by the authorities of many countries on account of the toxicology of the residues.
In comparison to the heavy-metal compounds, the phosphoric acid esters are particularly favourable as regards the residue problem. They have recently been gaining importance in the control of pathogenic fungi. There are still no clearly defined structure-activity relationships in the fungicidal action of the organophosphates for work done in this field is barely a decade old. This is reflected in the rather small number of commercial products available as fungicides in comparison to the insecticides.
The simplest organophosphate is trimethyl phosphorothioate (I). It is an effective, selective soil fungicide used to control *Pythium* sp. [542] without harming the crop. This substance can be prepared by sulfurization of trimethyl phosphite

$$(CH_3O)_3P + S \longrightarrow \underset{\text{(I)}}{(CH_3O)_3P{=}S} \qquad (1)$$

or from O,O-dimethyl phosphorothiochloridate and sodium methylate.

$$(CH_3O)_2\overset{\overset{\displaystyle S}{\|}}{P}-Cl + CH_3ONa \longrightarrow \text{(I)} \qquad (2)$$

O,O-diethyl S-methyl phosphorodithioate (II) [874] is a very effective agent recommended against *Rhizoctonia solani*.

$$(C_2H_5O)_2P(=S)-SCH_3 \qquad \text{(II)}$$

The oral LD_{50} for the male rat is 156 mg/kg.
In this class of substances the S,S,S-trimethyl phosphorotetrathioate (III) [893] is even more active.

$$(CH_3S)_3P{=}S \qquad \text{(III)}$$

The toxicity of the trialkyl tetrathiophosphates to fungi decreases with increasing length of the alkyl chain. Compound (III) can be obtained from phosphorus pentasulfide and an excess of methyl mercaptan, or from thiophosphoryl chloride with mercaptan in the presence of anhydrous zinc chloride [105].
The dithiol ester (IV) appeared as a trial insecticide, but found application as a rice fungicide and is now marketed in Japan [889].

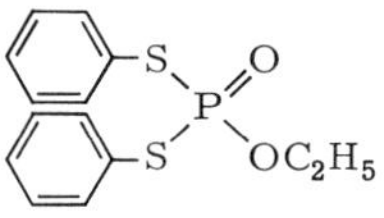

®Hinosan (IV)

O-ethyl S,S-diphenyl phosphorodithioate (IV) was developed by Farbenfabriken Bayer in 1965–66 and is known under the name ®Hinosan. The oral LD_{50} for the male rat is 212 mg/kg. It is synthesized from phosphoryl chloride which is converted to the dichloride with alcohol and then reacted with thiophenol in the presence of an acid-binding agent to give the dithiol ester:

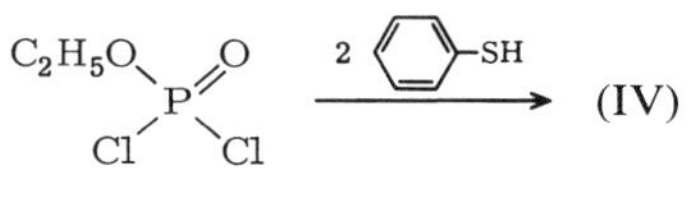

(3)

Another method of preparation consists in reacting O,O-diethyl S-phenyl phosphite with phenylsulfenyl chlorides which may be substituted as required [891];

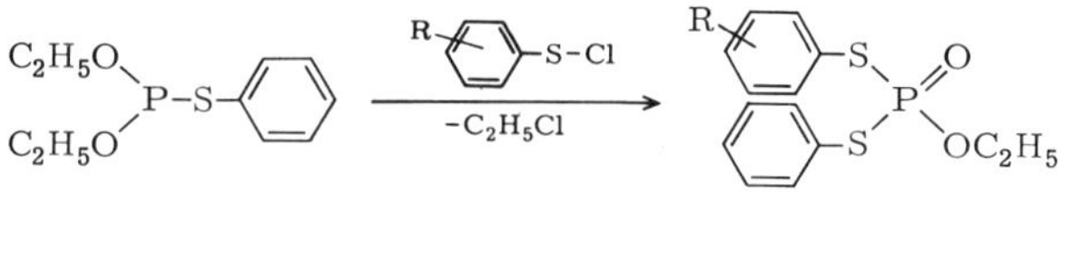

(4)

®Hinosan is effective against *Piricularia oryzae* without causing crop damage [801].
An important rice fungicide developed by the Sumitomo Kagaku is ®Conen (V) [O-*n*-butyl S-ethyl S-benzyl phosphorodithioate] [49].

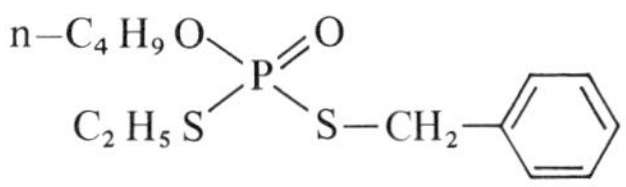

®Conen (V)

(Oral LD_{50} for mice is 118 mg/kg).
This new ester is prepared from the sodium salt of O-*n*-butyl S-benzyl phosphorodithioic acid and bromoethane:

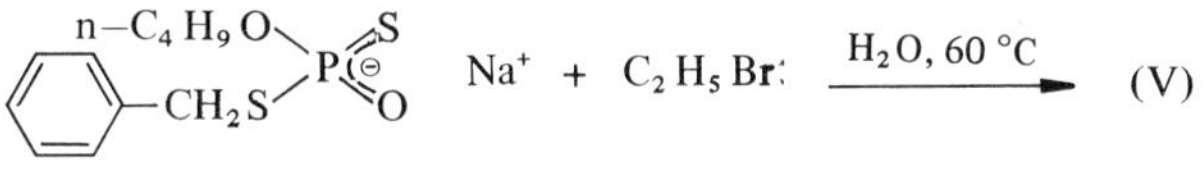 (5)

Conen exerts a curative rather than a protective action.
Alkylation of phosphonic ester acids results in a compound developed by the firm of Nissan Kagaku Kogyo K.K. called ®Inejin F 254 (VI) [O-methyl S-benzyl benzene-phosphonothioate] [681, 856].

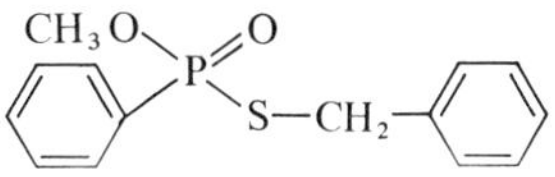

®Inejin F 254 (VI)

For its synthesis O-methyl benzenephosphonothiochloridate is first hydrolyzed to the ester acid and subsequently alkylated with benzyl chloride.

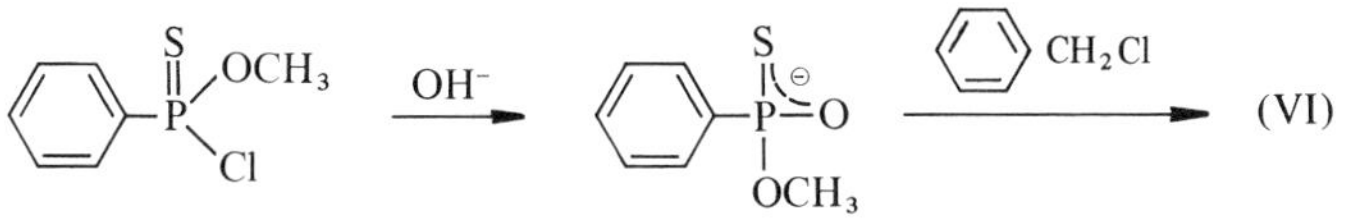 (6)

®Inejin is used as an agent against *Piricularia* sp. The O-ethyl ester is designated ESBP and also marketed as an agent against *Piricularia* sp.
A related monothiol ester is ®Cerezin (VII) [O-methyl O-cyclohexyl S-(4-chlorophenyl) phosphorothioate] [801, 802, 862] (oral LD_{50} 160 mg/kg for the rat).

It is prepared by reacting O-methyl O-cyclohexyl phosphite in carbon tetrachloride with *p*-chlorophenylsulfenyl chloride [865].

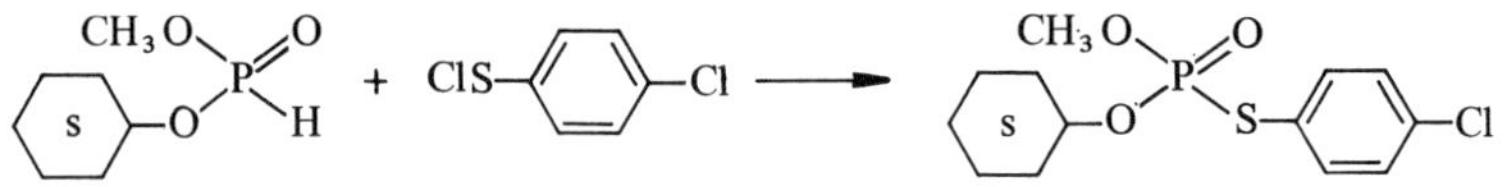

(7)

®Cerezin (VII)

Cerezin is also effective against *Piricularia oryzae* and shows a good residual action. On account of its curative action it can be used after infection has already become established.

The fungicide TH 184-F (VIII) [N,N,N',N'-tetramethyl O-pentachlorophenyl phosphorodiamidate] [821] is said to be very effective against powdery mildew.

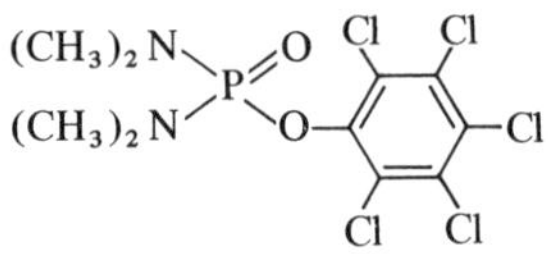

TH 184-F (VIII)

It is prepared by established methods.

In the course of his work with thionophosphoryl amides, Tolkmith [956] described a new class of compounds comprising phosphoric acid ester amides of NH-acidic compounds:

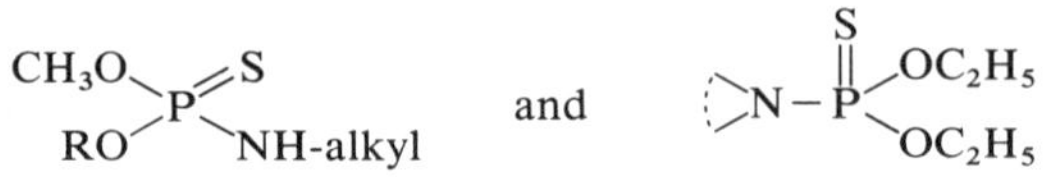

The insecticidal action of compounds containing the groups

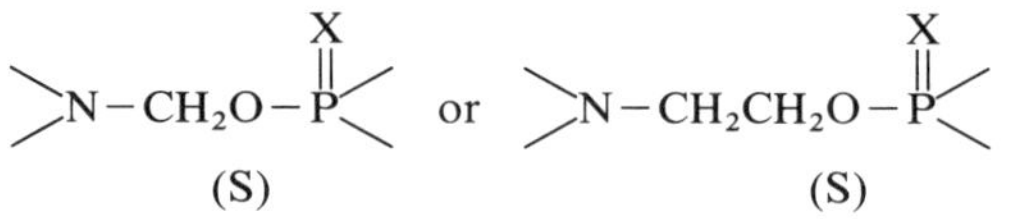

was already known, as well as the fungicidal activity of ®Phaltan, for example [46]:

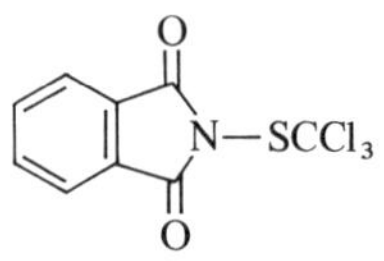

The fungitoxicity of peracylated amides, the best-known representative being DOWCO 199 (or 49) (IX) [956], was predicted and was confirmed by experiment (see page 112).

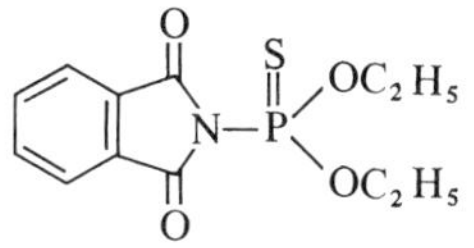

DOWCO 199 (49 resp.) (IX)

Its extremely low toxicity (oral LD_{50} 4930 mg/kg for the female rat) is attributed to the weak cholinesterase inhibiting action of the compound. The biological activity is explained by the ready hydrolytic and nucleophilic opening of the imido-ring. During its metabolism it is possible that amidases are the first to attack before the esterases. This compound does not participate in thionothiol isomerization. There are several interesting methods of preparing (IX) [698, 960]. The simplest would appear to be the reaction of potassium phthalimide [958, 959] with O,O-diethyl phosphorothiochloridate in N-methyl-pyrrolidone-(2),

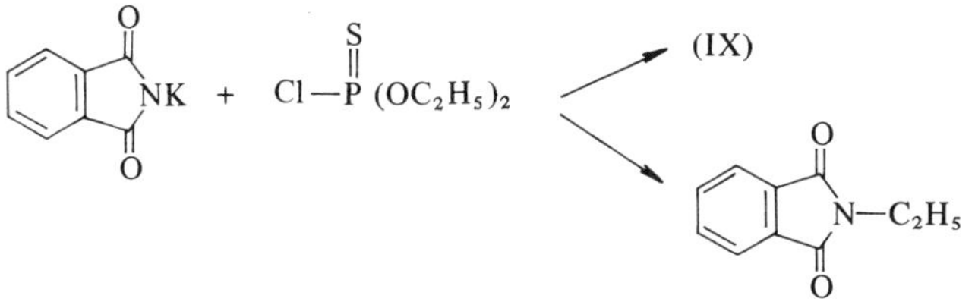

(8)

where ethylation may occur as a side reaction.

Phthalic acid anhydride and O,O-diethyl phosphorothioamidate react in the presence of sodium hydride [8] by the following equation:

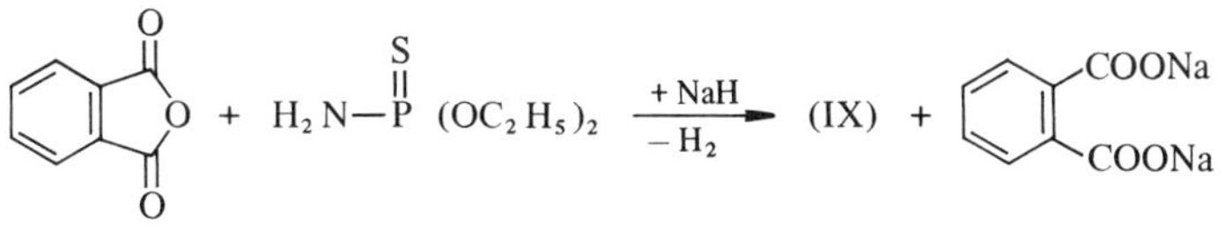

(9)

This reaction is not successful with other than 5-membered anhydride rings. In place of phthalic acid anhydride, phthalyl chloride may also be used, triethylamine being used in place of sodium hydride. N-chlorophthalamide reacts with triethyl phosphite, but not with triethyl thiol phosphite, in a Michaelis-Arbusov reaction, so that in this way the oxygen analogue may be obtained [698]. Dowco 199 (or 49) is extremely effective against powdery mildews at a concentration of 0.00001 % and has both a curative and protective action. The following table illustrates its biological activity [960]:

Table 6. Activity of O,O-diethyl phthalimidophosphorothioate

Crop	Disease	Spray conc. of IA (% by wt.)	Rating of disease control
Apple	Apple scab *(Venturia inaequalis)*	0.03 to 0.06	Good
	Powdery mildew *(Podosphaera leucotricha)*	0.023 to 0.015	Excellent
	Frogeye leaf spot *(Physalospora malorum)*	0.045	Excellent
Red cherry	Leaf spot *(Coccomyces hiemalis)*	0.03 to 0.06	Good
	Powdery mildew *(Podosphaera oxycanthae)*	0.023 to 0.045	Excellent
Peach	Brown rot *(Monilinia fructicola)*	0.03 to 0.045	Good
	Rhizopus nigricans	0.03 to 0.045	Excellent
Cucumber squash	Powdery mildew *(Erysipphe cichoracearum)*	0.015 to 0.03	Excellent
Rose	Black spot *(Diplocarpon rosae)*	0.03 to 0.06	Good
	Powdery mildew *(Sphaerotheca humuli)*	0.015 to 0.03	Excellent
Turf	Powdery mildew *(Erysiphe graminis)*	0.023 to 0.045	Good

Substitution of the thionogroup by oxygen, the introduction of methylene groups, or of an oxygen atom (oxim) between the nitrogen and phosphorus atom [957, 960, 801] enhance mammalian toxicity and lead to a marked decrease in fungicidal activity. In the same way unsubstituted phthalamides are more effective than those substituted in the ring.

N,N-diethyl N′-(2-methyl imidazolyl-1) benzenephosphonoamidothioate (X) [140, 957] belongs to the same group of NH-acidic compounds and like many imidazole derivatives it shows fungitoxic activity.

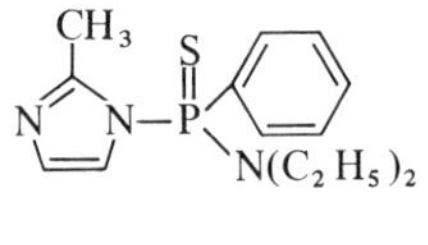

(X)

This imidazole derivative, which is particularly toxic for *Phytophthora infestans*, can be obtained from benzenephosphonodichlorothioate, diethyl amine and 2-methyl-imidazole.

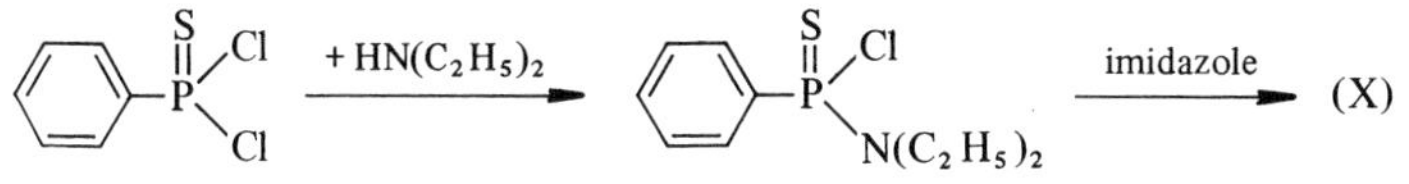

Its activity can be seen from the following Table 7:

Table 7. Optical and biological activity of N,N-diethyl N′-(2-methyl imidazolyl-1) benzenephosphonoamidothioate (X)

Optical form	Racemate	*d*-Isomer	*l*-Isomer
M.p. (°C)	77.5–79	80–83	82–84
$[\alpha]_D^{25}$ in $CHCl_3$		+7.9°	−9.3°
Acute oral toxicity (LD_{50} in mg/kg for white mice)	147	316	147
Min. concentration (p.p.m.) for complete control of *Phytophthora infestans*	~200	~200	~200

®Wepsyn (XI) represents a phosphoryl trisamide (see p. 110).

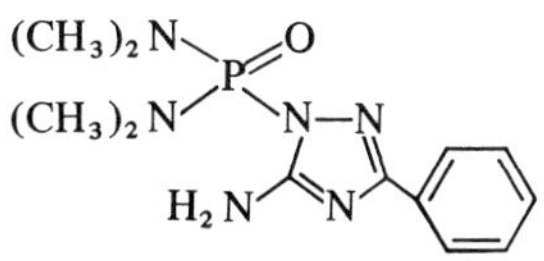

triamiphos, ®Wepsyn (XI)

It was discovered and developed in Holland before the aforementioned phosphoryl amides. Wepsyn is applied chiefly for the control of powdery mildew in roses and in apple culture [801].
This compound has a surprisingly broad biocidal spectrum, possessing at the same time a relatively high mammalian toxicity (oral $LD_{50} \sim 5$ mg/kg for the rat), insecticidal, acaricidal and fungicidal activity, together with systemic properties.
®Kitazin (XIIa) [803] (see p. 126) belongs to the class of thiol esters and is commercially available for the control of *Piricularia oryzae*.

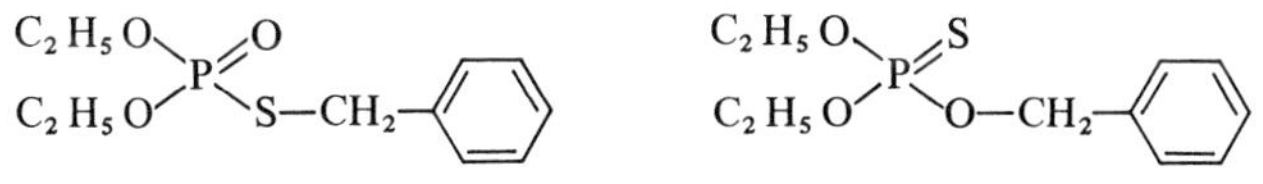

®Kitazin (XIIa) (XIIb)

The isomeric O-benzyl ester (XIIb) has virtually no fungicidal properties. A variant of ®Kitazin is the diisopropyl ester ®Kitazin P (XIII) [O,O-diisopropyl S-benzyl phosphorothioate] of the Ihara Chemical Company [466].

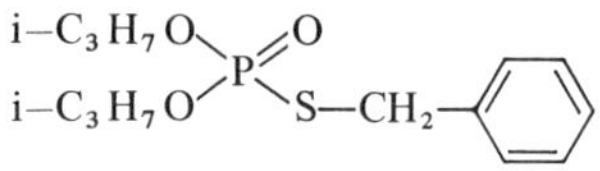

®Kitazin P (XIII)

®Kitazin P is a systemic rice fungicide applied specially to control *Piricularia oryzae*, and, like ®Kitazin, it inhibits mycelial growth in tissue.
For both preventive and curative control of powdery mildew, Farbwerke Hoechst AG have recently introduced the product '*Hoe* 2873' (XIV) [O,O-diethyl O-(5-methyl 6-carbethoxy pyrazolo-(1,5a)-pyrimidyl-2) phosphorothioate] [33].

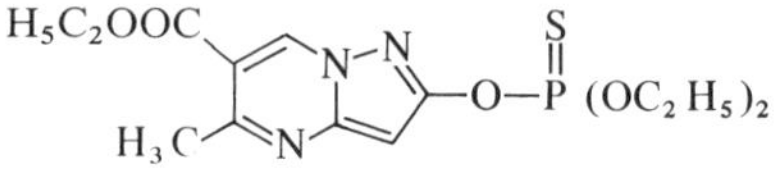

®Afugan, HOE 2873 (XIV)

Oral LD_{50} 140 mg/kg for the rat.
With the exception of ®Wepsyn, no other systemically effective compounds have been found among the fungicidal phosphoric acid derivatives; on the other hand, the curative activity has repeatedly been mentioned which presupposes penetration into the leaf. Protectively-acting phosphoric acid esters are not dependent upon this property.
This explains the experience that very good protective fungicides seldom exhibit marked curative properties, while those phosphoric acid derivatives which have very good curative properties are usually less protective in action or less persistent. It is desirable that one molecule should possess both types of action, but this ideal combination has not yet been found in the phosphoric acid ester series.
Phosphonomycin (XV) is an antibiotic recently discovered by Merck & Co. Inc. It was isolated from fermentation broths on which *Streptomyces fradiae* was grown. Its structure was demonstrated by synthesis [183, 184, 373, 399].
The new antibiotic has a low mammalian toxicity (LD_{50}: 4000 mg/kg for mice intraperitoneally), a broad spectrum of activity and inhibits irreversibly pyruvate-uridine diphospho-N-acetylglucosamine transferase in extracts of gram-positive and gram-negative micro-organisms. It compares favourably with tetracycline and chloramphenicol. The calcium salt of phosphonomycin is absorbed from the gastrointestinal tract in man and appears in the serum within two hours. It is excreted unchanged by the kidneys.

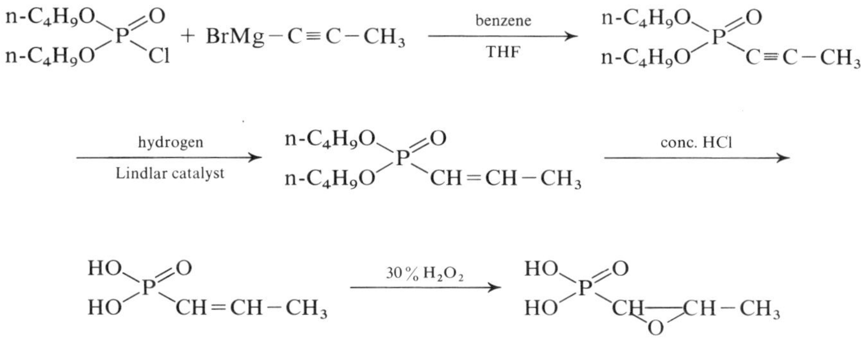

Scheme 11: Synthesis of Phosphonomycin

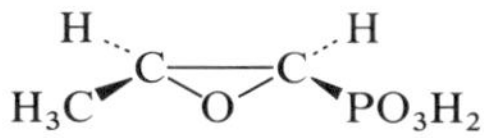

Phosphonomycin configuration

Phosphonomycin is not the only naturally occurring phosphonate. How the living organism is able to synthesize the phosphorus-carbon bond is completely unknown. All preparative and technical processes require a high expenditure of energy and also need, as starting material, trivalent phosphorus compounds which are too reactive to exist *in vivo*. There is no evidence that living organisms are able to concentrate phosphonates in any quantity from their environment. Therefore knowledge of this synthetic pathway *in vivo* would be of great theoretical and practical importance. Reference will, therefore, be made to some naturally occurring compounds, although they are of no practical significance and their metabolic function is unknown. These products can hardly be considered to be present by accident, since they constitute a considerable proportion of the total phosphate content (5 – 13 %) in the organism.

Examples are the N-methyl derivatives of 2-amino-ethanephosphonic acid isolated by KITTREDGE, ISBELL and HUGHES [489] in crystalline form from ethanolic extracts of the sea anemone *Anthopleura xanthogrammica* (XVI, XVII, XVIII):

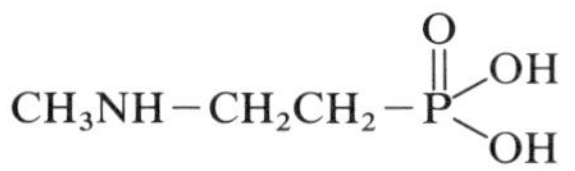

(XVI)

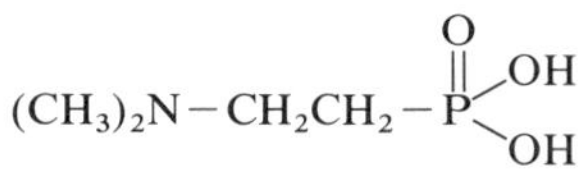

(XVII)

(XVIII)

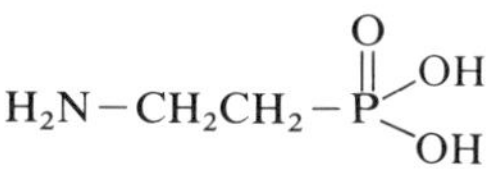

Ciliatine (XIX)

The trimethyl derivative is present in the form of its inner salt (XVIII). The phosphonic acid analogue of β-alanine itself, Ciliatine (XIX), was demonstrated by QUIN [757] e.g. in *Metridium dianthus.*

b) Herbicides

The range of herbicides which can be derived from the phosphoric acid esters is very large indeed. We find here phosphites, disulfides and phosphorus-containing ring compounds etc. It is difficult to correlate structure and activity. For this reason promising trial compounds will be mentioned here, because for the commercial product activity is only one factor among many, such as ease of preparation, stability and, of course, cost.
Systematically we must begin with phosphorous acid esters like ®Falone (I) [tris-(2,4-dichlorophenoxy-ethyl) phosphite] [374].

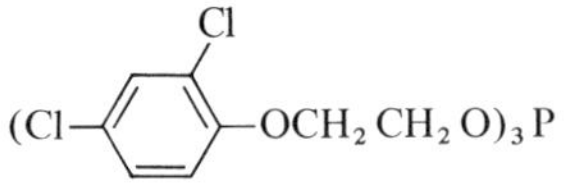

®Falone (I)

which was introduced by Uniroyal Inc. as herbicide and plant-growth regulator. Another interesting compound is the spiro-phosphite of pentaerythritol with the structure (II) [765].

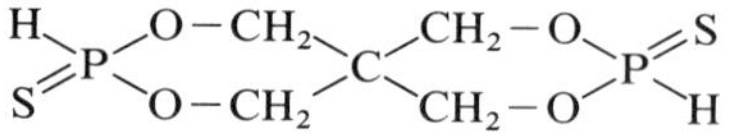 (II)

Using pentaerythritol as the starting material, ring closure occurs with phosphorus trichloride.

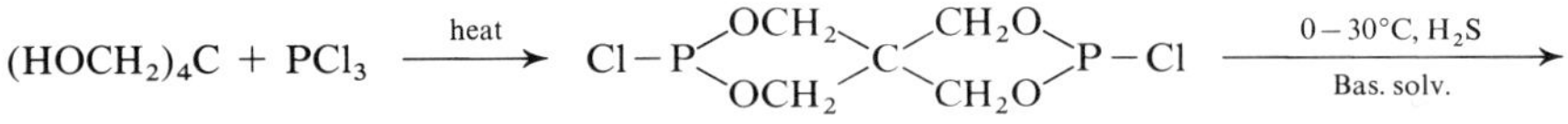 (II)

(1)

The compound is a herbicide for pre-emergence application and is particularly recommended for weed mustard.
In cotton growing, defoliants play an important role, since they facilitate mechanical harvesting. A well-known compound is ®Folex (III) [S,S,S-tri-*n*-butyl phosphorotrithioite] [1035] which is synthesized from finely dispersed white phosphorus and di-*n*-butyl disulfide in dimethyl sulfoxide under nitrogen.

$$P_{white} + n\text{-}C_4H_9S-SC_4H_9\text{-}n \longrightarrow (n\text{-}C_4H_9S)_3P \qquad (2)$$

®Folex (III)

Also the manufacture from phosphorus trichloride and sodium butyl mercaptide is possible on a technical scale:

$$PCl_3 + 3\,NaSC_4H_9\text{-n} \longrightarrow (III) \qquad (3)$$

One of the simplest pentavalent phosphorus compounds, a derivative of phosphoryl fluoride, is the phosphoryl isocyanate (IV) [694].
The product is manufactured from phosphoryl isocyanate dichloride by fluorination with antimony trifluoride:

$$Cl_2\overset{O}{\overset{\|}{P}}-N=C=O \xrightarrow[-SbCl_3]{+SbF_3} F_2\overset{O}{\overset{\|}{P}}-N=C=O \qquad (4)$$

(IV)

It is recommended as a growth inhibitor for grasses.
An important defoliating agent is ®DEF (V) [S,S,S-tri-*n*-butyl phosphorotrithioate] [697, 966],

$$(n\text{-}C_4H_9S)_3P=O$$

®DEF (V)

(oral LD_{50} for the rat is 177 mg/kg) starting materials for its synthesis being *n*-butyl mercaptan and phosphoryl chloride.

$$3\,n\text{-}C_4H_9SH + POCl_3 \xrightarrow{\text{Na, emuls.}} (V) \qquad (5)$$

®DEF is also readily obtained by oxidation of ®Folex. It defoliates plants already in the growth phase and suppresses the development of new growth.
An example of a phenyl phosphate is ®Zytron (VI) [O-methyl O-(2,4-dichlorophenyl) N-isopropyl phosphoroamidothioate] (see p. 93) [538].

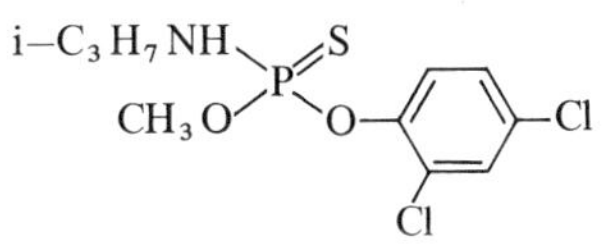

®Zytron, DMPA (VI)

®Zytron suppresses the growth of germinating seeds and undesirable species of plant. Closely related compounds are the corresponding O-nitrophenyl esters (VIIa and VIIb) [873]

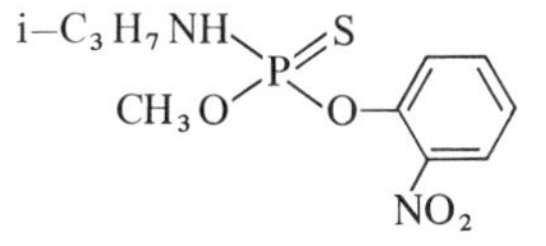

(VIIa)

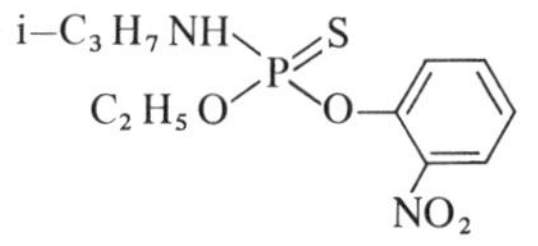

(VIIb)

the synthesis of which is evident from the formula. These esters are recommended as germination inhibitors and for the destruction of weeds.

TEPA (VIII) [tris-(1-aziridinyl) phosphine oxide] is described in a patent filed by the Dow Chemical Company [457] as a growth-inhibiting compound; besides which the substance

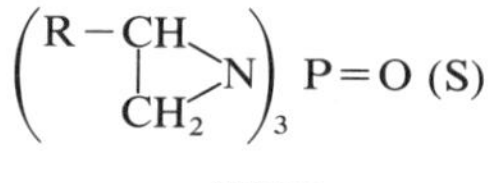

(VIII)

is said to increase the sugar content of corn, sugar beet and sugar cane. An indication of the mechanism of action is offered by the alkylating property of these azaridine compounds which, as miotic poisons, (inhibitors of cell division) have been described in the section on chemosterilization.

Active phosphoric acid amide esters are also derived from the *s*-triazines (IX), a group well known for their herbicidal activity [31]:

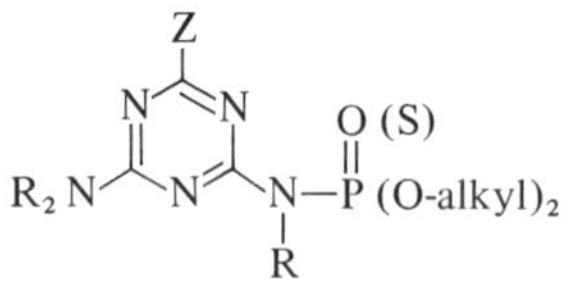

(IX)

where R may represent saturated or unsaturated alkyl or alkoxyalkyl groups, Z halogen atoms, cyano-, alkoxy or mercapto-groups. These compounds are obtained by reaction of (thio) phosphoryl chloride with aminotriazines:

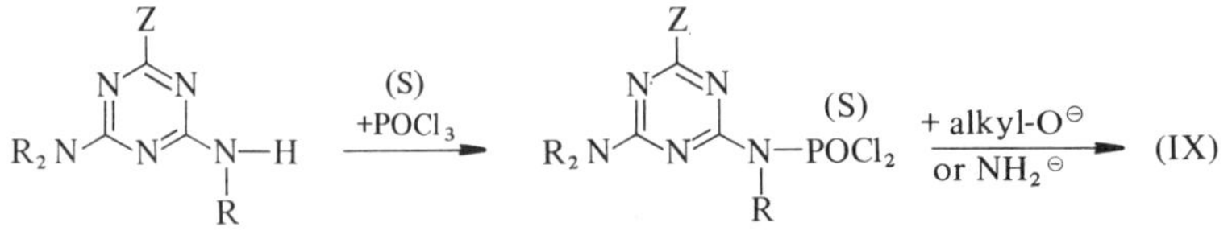 (6)

Alkylation of an ester acid results in *bensulide* (X) [O,O-diisopropyl S-[2-(N-benzenesulfonyl) aminoethyl] phosphorodithioate] [239]:

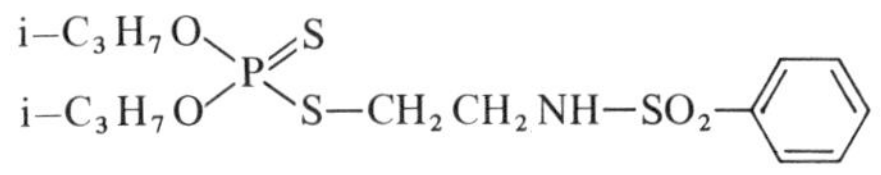

bensulide, ®Prefar, ®Betasan (X)

N-(β-chloroethyl) benzenesulfonamide and a slight excess of O,O-diisopropyl phosphorodithioic acid ammonium salt leads to the compound (IX) which may be considered as an "acyl amiton" (c.f. p. 118):

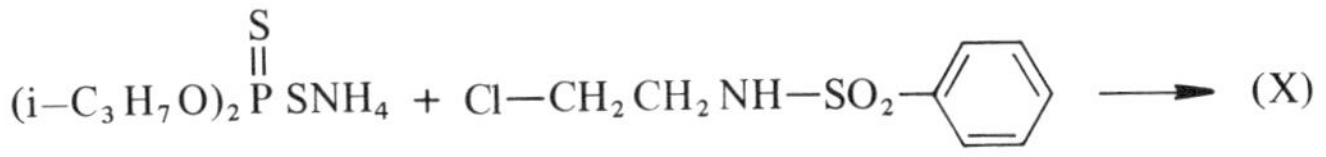

(7)

®Betasan (lawn) or ®Prefar (grain) is a post and pre-emergence herbicide with a duration of action of 4–12 months (oral LD_{50} for the male rat is 340 mg/kg).
Varying the molecule of dimethoate results in herbicidal compounds which are described in a series of Monsanto patents [188, 189, 914, 915, 916, 917, 918]. The general structure of such dimethoate derivatives is (see p. 131):

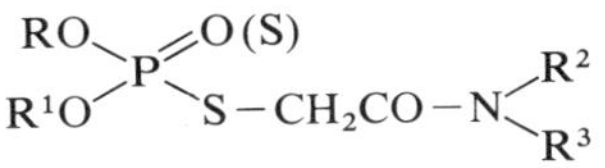

(XI)

If R, R^1, R^2, R^3 represent alkyl of alkoxyalkyl radicals, these compounds inhibit germination and are, therefore, effective as pre-emergence herbicides when applied at the rate of 0.5 to 5 pound per acre.
R^2 and R^3 may also represent olefine groups, for example, an allyl group. The herbicidal activity is also maintained in the case of the monoalkyl derivative or when R^2 is an alkyl group and R^3 an aryl radical, furthermore, by R^2 may be represented hydrogen or alkyl and by R^3 a hydrocarbon ring. The nitrogen atom can also belong to a heterocyclic ring (e.g. oxazine, thiazine).
An interesting type of very active herbicide, especially against wild oats, includes the asymmetrical disulfides (XII) [752]

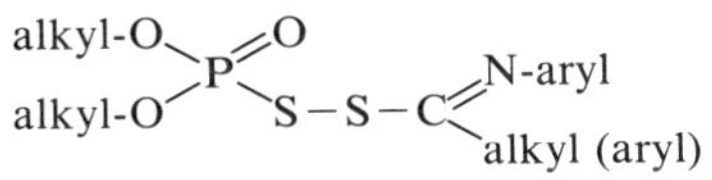

(XII)

which can be obtained from O,O-dialkyl phosphorylsulfenyl chloride and thioamides:

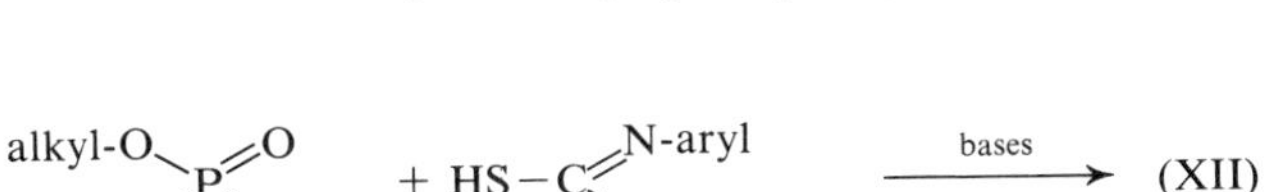

(8)

Phosphorus-containing ring compounds, of which to date no member has appeared notable as insecticide, have proven to be good herbicides.
Exchange of an halogen atom in trimeric phosphonitrile chloride [42, 43] also yields herbicides (XIII).

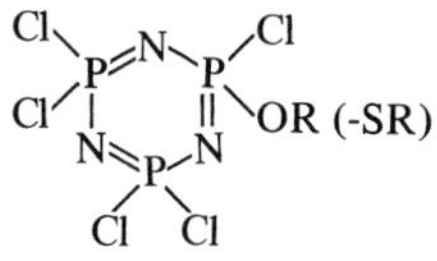

(XIII)

The compounds are obtained from triphosphatriazinyl chloride by reaction with the corresponding amount of alcohol or phenol, in the presence of an acid binding agent.
Ethrel [2-chloroethanephosphonic acid] (XIV) manufactured by American Chemical Products Inc. is a plant-growth regulator which is able to form ethylene:

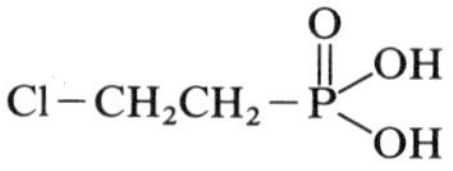

Ethrel (XIV)

Ethrel increases protein content, accelerates the ripening of fruit and stimulates seed germination [20].
CASELEY [159] found that Ethrel, applied to foliage or soil, released dormant buds from apical dominance in aerial shoots and rhizomes of *Agropyron repens.* Plants treated with Ethrel showed dwarfing, and had twice the peroxidase activity of the controls. In this way, the competitive capacity of this weed is reduced, and it becomes more susceptible to subsequent chemical and mechanical treatments.

c) Chemosterilants

In addition to the direct insecticidal control of pest populations, other indirect methods are applicable: for example, the use of sterile insects for their self-destruction [908]. KNIPLING has discussed various ways in which this goal may be achieved [498, 499, 500]:

1) Liberation of male insects that have been sterilized by gamma-radiation, X-rays or other means (sterile-male method) [734].
2) Use of chemical substances which cause sterility in both sexes in a natural population.
3) Rearing and liberation of mutants of a species which show damaged or lethal genetic characteristics.
4) Liberation of insects which are infected with pathogens (viruses, bacteria, fungi) and serve as vectors.

Using the first method, BAUMOVER, BUSHLAND *et al.*, during the years 1955 and 1957, succeeded in eradicating the screwworm *(Callitroga hominivorax)*, one of the most important domestic animal pests [75, 151, 152, 153], from Curaçao and later, around 1957–1959, from the southern part of the USA. In less than 18 months, reproduction of the pest was arrested [497, 501]. Since 1962 there have been similar campaigns in Texas and New Mexico. For the USA program alone, more than a thousand million *Callitroga* males were reared, sterilized and liberated on an assembly-line basis [906]. In a similar way *Dacus curcubitae* was exterminated from the Pacific island of Rota and *Dacus dorasalis* from Guam. The "sterile-male" method is being used on a trial basis on the Californian-Mexican border to combat *Anastrepha ludens*, in Egypt, Israel and Central America to control *Ceratitis capita* and in Africa against *Glossina morsitans* (Tsetse-fly), the carrier of sleeping sickness [19, 657].
All these examples show, however, that not every population can be depressed in this manner for, where there is possibility of immigration into the controlled area, a fertile population can once again be built up. For these reasons, geographically or ecologically isolated islands must be chosen. In these circumstances the method can be successful but it requires considerable technical, biological (i.e. mass-rearing) and financial resources and a high degree of organization.
It was, therefore, of great interest that the discovery of chemical substances with sterilizing properties led to the second method of control; these substances were already known as inhibitors of cell division (mitotic poisons) [108, 358, 907]. From a point of view of economy and potency, the best compounds include the particularly important phosphorus compounds of ethylene imine, of 2-methyl and 2,3-dimethyl ethylene imine. In 1961 LABRECQUE [527] was the first to describe their suitability for sterilization of house flies *(Musca domestica)*. The first example of a compound of this type, synthesized in 1954 by RÄTZ and GRUNDMANN [768], was 2,2,4,4,6,6-hexakis-(1-aziridinyl)-2,4,6-triphospha-1,3,5-triazine or 2,2,4,4,6,6-hexakis-(1-aziridinyl)-1,3,5,2,4,6-triazaphosphorin with the common name *apholate*, which was tested successfully by CHAMBERLAIN [173] against *Callitroga hominivorax*.
It is synthesized by reacting phosphorus pentachloride with ammonium chloride to give the triphosphatriazinyl chloride and by replacing the halogen atoms by ethylene imine or its substituted derivative in the presence of equivalent quantities of tertiary amine [704, 706, 766, 768, 769]:

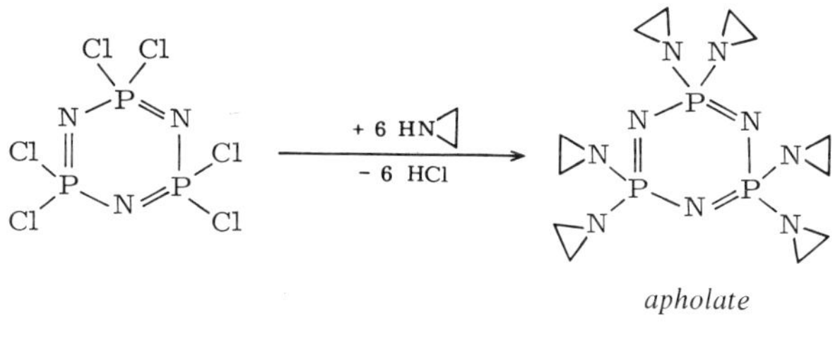

apholate

(1)

The second example, phosphoric acid tris-ethyleneimide or tris-(1-aziridinyl)-phosphineoxide with the common name TEPA is synthesized in an analogous manner. The third example, described by CHANCE is N,N′-ethylene-bis-(P,P-bis-(1-aziridinyl)-N-methyl) phosphoroamidate or *aphomide* [175]:

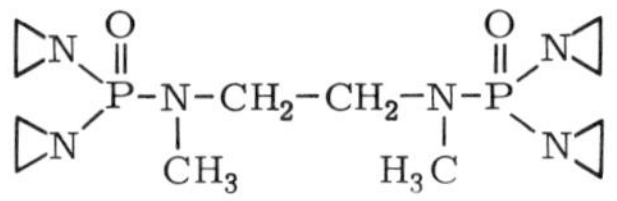

aphomide

Since 1960 numerous papers have appeared [cf. 95, 112, 178, 225, 661, 709] describing the sterilizing action of analogues of the parent compounds *apholate* and TEPA. The use of chemical contact sterilizing agents is certainly attractive but the intrinsic problem is that the sterilizing action of these alkylating and mitotic poisons is not restricted to insects, for they can exert carcinogenic and teratogenic actions on mammals [68, 322, 382]. The type of compounds most frequently investigated are as follows [6, 59, 94, 109, 111, 113, 158, 176, 224, 528, 710, 736].

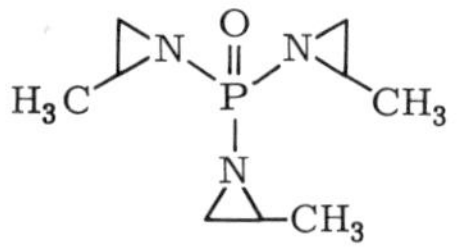

METEPA (= Methaphoxide)

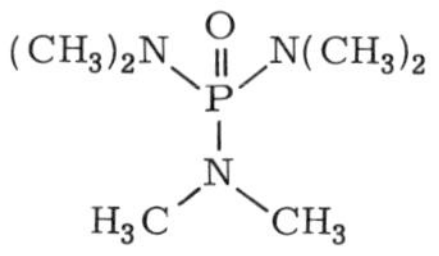

HEMPA

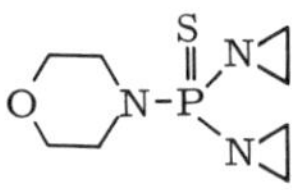

MORZID (=OPSPA, MSPA)

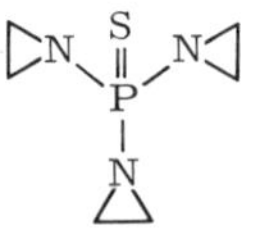

THIOTEPA

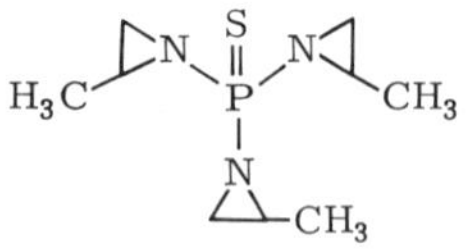

METHIOTEPA

An N-phosphoryl urea with the following structure deserves particular attention because of its synthesis:

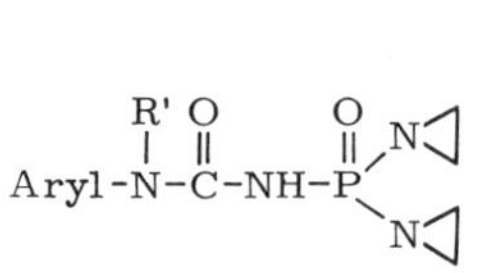

(I)

RÄTZ and GRUBER [767] obtained dichlorophosphoryl isocyanate (II) from PCl_5 and ethyl carbamate which, in situ, reacts with aromatic amines to form the N-arylureido-phosphoryl dichloride (III). The chlorine atoms may be exchanged in the usual manner.

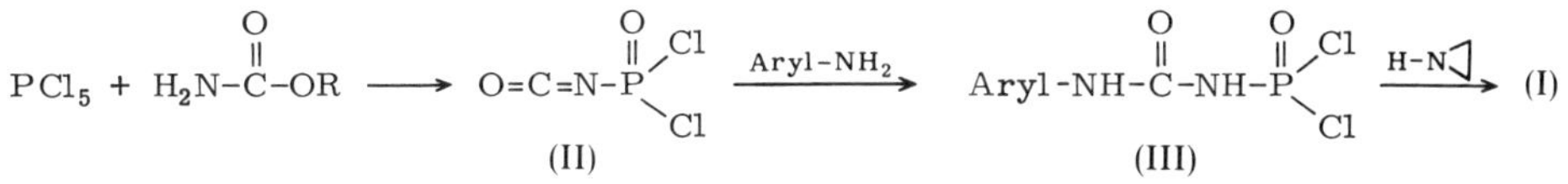

Spiro-compounds of the formula (IV) are of structural interest. RÄTZ [764] synthesized these types from penta-erythritol, phosphoryl chloride and ethylene imine. The most active compound was bis-3,9-(1-aziridinyl)-2,4,8,10-tetraoxa-3,9-diphosphaspiro (5,5)-undecane-3,9-dioxide.

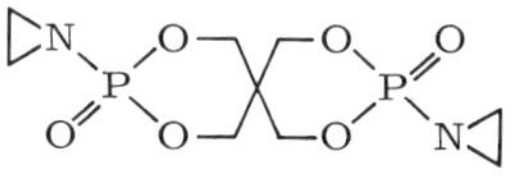

(IV)

Typical information on structure and activity is provided by an investigation of RISTICH, RATCLIFFE and PERLMAN, as well as by CASTLE and RISTICH [170, 705, 776], on the sterilizing action, cytostatic properties and oral toxicity in mice for the *apholate* series. Substituents chosen were ethylene imino-, subst. ethylene imino-, amino-, alkylamino-, alkoxy groups and chlorine atoms. The most important result was the demonstration of a positive correlation between the sterilizing properties of a compound and the number of ethylene imino-groups in the molecule (alkylating action), as well as its solubility in water, which, together with the sterilizing action, falls with increasing number of chlorine atoms. In the presence of at least four ethylene imino-groups, radicals like methoxy, amino- or methylamino-groups do not substantially influence the alkylating properties, while the hydrazino-group, substitution of the ethylene imine itself, or the formation of metal complexes weaken the sterilizing action. The first members of a given series, each with five ethylene imino-groups and another substituent e.g. halogen atoms, exhibit increased activity. Compounds analogous to *apholate* without ethylene imino-groups are inactive. The tetrameric analogue of *apholate* with eight ethylene imino-groups is no more active than *apholate* itself. CHANG and BOŘKOVEC [177] found similar relationships between structure and activity also in the TEPA series. They replaced an ethylene imino-group in TEPA by monoalkylamino-groups and found a decreasing ability to sterilize. Isopropylamine was, however, found to be an exception. CHANG and BOŘKOVEC determined both the sterilizing activity and the toxicity in *Musca domestica.*

If LD_0 represents the maximal dose at which no mortality results, and ED_{100} the minimal dose at which complete sterilization occurs, then a "safety factor" *(SF)* can be calculated from both values. Using probit analysis (see p. 201), the $LD_{0.01}$ and $ED_{99.99}$ (lethal dose and effective dose) are determined and the safety factor obtained:

$$SF = \frac{LD_{0.01} - ED_{99.99}}{ED_{99.99}}\,.$$

If % mortality of % sterilization in probit units are plotted against the logarithm of the dose (μg per male injected) then for each compound two curves are obtained, as shown in Figs. 14 to 16. Fig. 14 illustrates the conditions pertaining to a negative safety factor. Sterilizing action is revealed only at a dose rate at which the substance is lethal. The borderline case 0 is shown in Fig. 15. Here the $ED_{99.99}$ and the $LD_{0.01}$ coincide, CHANG and BOŘKOVEC call this "exact dose". Fig. 16 shows a positive safety factor, i.e. the effective dose is substantially smaller than the lethal dose, complete sterility is achieved without entering the mortality dose range.
At the present time the main targets for chemosterilants of the TEPA and *apholate* type are economically important cotton pests such as the boll weevil *(Anthonomus grandis)* [230, 383, 391, 392, 393, 552], the pink boll worm *(Pectinophora gossypiella)* [351, 707] and the cotton leaf worm *(Prodenia litura)* [962].
Contact sterilizing agents for the extermination or even the suppression of a harmful insect population would offer considerable advantages over physical sterilization, because the mass rearing and liberation of sterile males would no longer be necessary.

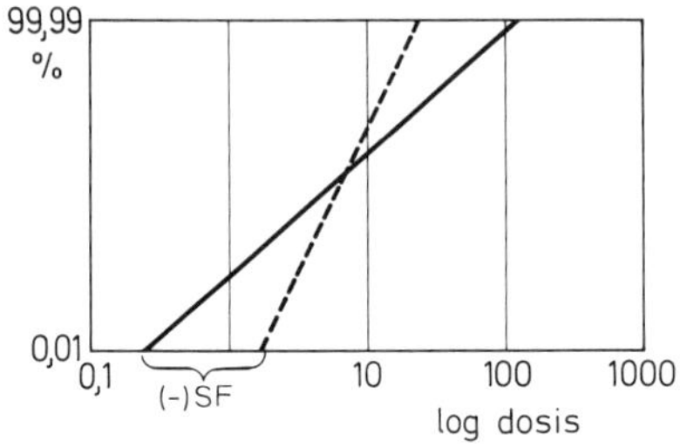

Fig. 14. Mortality after sterilization

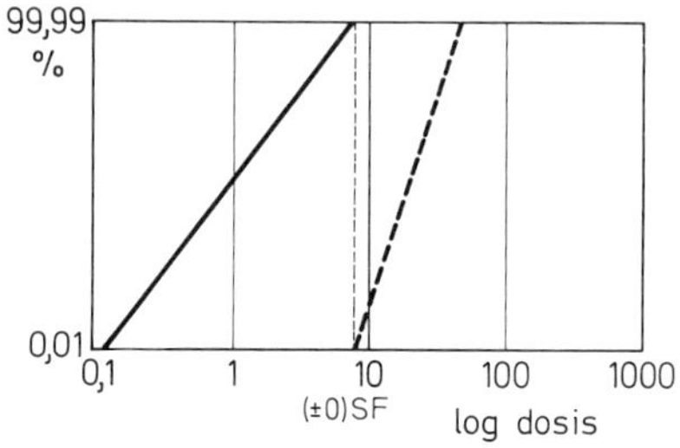

Fig. 15. "Exact dose"

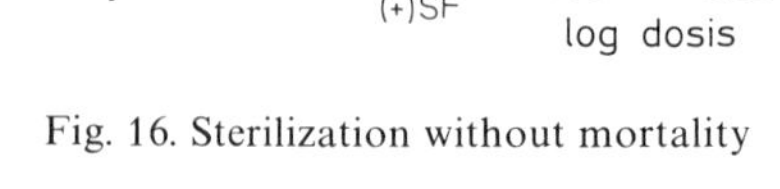

Fig. 16. Sterilization without mortality

Furthermore, as KNIPLING [496, 498] propounded, these compounds are theoretically superior in comparison to direct-acting insecticides. If it can be assumed that the mating behaviour of the insects is not disturbed by sterilization, that the population from generation to generation increases five-fold and that, in case A,

90% of the population is killed by an insecticide, while in case B, 90% would become infertile due to a sterilizing agent, then in case A, 10% fertile insects would remain which were capable of building up a population and under circumstances of developing resistance. In case B after sterilization, 90% sterile insects would meet 10% fertile insects and reduce their biotic potential by 90% down to a total of 1%. Moreover, sterile insects are effective over a long period of time and over a considerable area against the building up of a new population. In the case of insecticide control, these factors are not involved. Starting with one million insects, insecticides require about 20 generations to reduce the population to one individual, whereas with sterilizing agents the population is reduced to six in only four generations. KNIPLING suggested furthermore that the classic methods of control using insecticides, parasites and predators might be combined with general cultivation methods, the "sterile male" method or contact sterilization. This integrated control which KNIPLING calls "Total Insect Population Suppression" (TIPS) [501] appears, however, to be economically possible only with important pests and in large areas of monoculture.

BOŘKOVEC [110] has recently published a review of the various classes of chemical substances in which chemosterilizing agents are to be found, also their pharmacology and practical application.

4. Biochemistry

4.1. Mechanism of Action

a) Mechanism of Action in Mammals

The organophosphates exert their biological action in mammals and arthropods by attacking the system of neural transmission and thus interfering with the function of the target organs. The scope of this publication permits reference to be made only to the most important conceptions and interrelationships necessary for a general understanding of the mode of action of organophosphates. Detailed discussions of the hypotheses and experimental results will be found in the monographs published by HEATH [387], by O'BRIEN [686], more recently by TRIGGLE [968], as well as in a review of the literature on acetylcholinesterases by ENGELHARD, PRCHAL and NENNER [260].

The basic structural unit of the nervous system is the neuron, essentially a cell having a nucleus and numerous fibres (dendrites) leading away from it. One of these fibres, recognized by its considerably greater length, is responsible for the transmission of impulses from the cell body to other neurons or to the receptors. The simplest neural action in vertebrates involves the participation of several neurons which form a common functional path. This functional conjunction is established in the synapses of the ganglia. They consist of a presynaptic membrane (end of a dendrite A) and a postsynaptic membrane (beginning of a dendrite B or target organ) which are 100 – 200 Å apart (synaptic gap). The action potential (presynaptic impulse) is built up by selective Na^+ and K^+ concentration changes ("ion pump"), and arriving at the synapse induces a corresponding postsynaptic reaction. The impulse is transmitted in the synapse by a chemical mechanism. On the presynaptic side, acetylcholine or noradrenaline is liberated and absorbed by receptors on the postsynaptic side, thus altering the permeability of the membrane to ions.

This results in the build-up of a postsynaptic potential; ECCLES discussed the mechanism of postsynaptic inhibition in his Nobel laureate lecture [257]. In a second Nobel laureate lecture, HODKIN referred to the ion movements as the basis of nervous transmission [411]. The quantitative analysis of nerve stimulation and neural transmission constituted the subject of a third Nobel laureate lecture by HUXLEY [434]. Further details on these papers cannot, however, be given within the scope of this publication. According to O'BRIEN [686] the whole system can be well-illustrated by the following scheme (Fig. 17).

In this connexion, the anatomic description of the nervous system is of less interest than the fact that a chemical classification is possible. The synapses and myoneural junctions of the motor and the parasympathetic system as well as the ganglia of the sympathetic system transmit the presynaptic impulse by means of acetylcholine to the postsynaptic side. The neuromuscular junctions of the sympathetic system are stimulated by adrenaline or noradrenaline. It is, however, possible that the adrenergic mechanism, too, is set in motion primarily by acetylcholine [619]. In order to bring the acetylcholine-induced action potential back to the resting potential, in other words to facilitate the transmission of a new impulse, the stimulant in the synapse must be degraded. Degradation is effected by the enzyme acetylcholinesterase (which, according to its source, is also known as erythrocyte cholinesterase, true cholinesterase, or in the classification of enzymes suggested by the Enzyme Commission [290], acetylcholine acetylhydrolase [3.1.1.7.]). This enzyme hydrolyzes acetylcholine to acetic acid and inactive choline. A second enzyme, choline-O-acetyltransferase [2.3.1.6] [290], is capable of esterifying both compounds to acetylcholine again. ATP and CoA are required for this reaction.

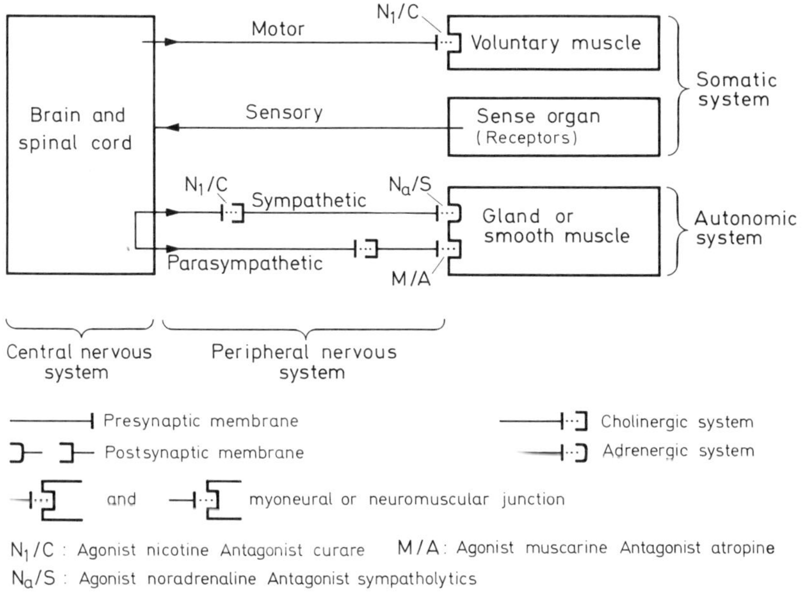

Fig. 17. Scheme of nervous system

The active organophosphates function by blocking acetylcholinesterase. This inhibition results in an accumulation of acetylcholine at the postsynaptic membrane which is then unable to return to its original (resting) state. Depending upon the part of the nervous system in which the synapses are thus kept in a state of permanent stimulation, symptoms of predominantly nicotine, muscarine and sometimes CNS poisoning are noted. With some organophosphates, the accent on action may lie more in the one or the other direction. A description of the physical and kinetic properties of acetylcholinesterase is given by ENGELHARD, PRCHAL and NENNER [260].

KRYSAN and CHADWICK [520] found the molecular weight of house fly head cholinesterase to be about 160,000. However, several aggregation states or molecular sizes were encountered, which made the isolation of pure insect AChE very difficult (as is also true for the mammalian cholinesterases).
LEUZINGER and his colleagues have recently succeeded in purifying acetylcholine acetylhydrolase [3.1.1.7] in large quantities and in crystallizing it [544, 545]. They determined the molecular weight by sedimentation equilibrium, finding a value of 260,000 $\pm$ 10,000 [546]. The native molecule of AChE consists of four subunits of average molecular weight 64,000 $\pm$ 4,000. The enzyme is a dimer, each protomer consisting of two non-identical chains. LEUZINGER suggested that the molecule may be a dimeric hybrid. The α-chains may be said to contain the active site, the function of the β-chains remains unknown at present; or the $\alpha + \beta$-chains together may form an active site. If the latter is true, it becomes possible to set up simpler models for the mechanisms of action and for the inhibition of acetylcholinesterase than if the α-chain alone is involved.
In addition to acetylcholinesterase, a second cholinester-hydrolyzing enzyme is found in the serum of mammals as well as in insects, viz. acylcholine-acylhydrolase-[3.1.1.8] (E.C. trivial name: cholinesterase) which formerly was also referred to as pseudocholinesterase or serum-ChE. Its synthesis in vertebrates takes place in the liver and requires ATP, CoA-SH, acetate or citrate, choline and the enzymes "acetylkinase" or choline-acetyltransferase. CLITHEROW, MITCHARD and HARPER [196] assume that in the degradation of fatty acids containing an even number of carbon atoms, butyryl co-enzyme A is formed, which presumably is also involved in the cholinester synthesis. The biological function of acylcholine-acylhydrolase might then be preferentially to hydrolyze butyrylcholine (which causes strong nicotinic effects) at the site of formation, for acylcholine-acylhydrolase hydrolyzes butyrylcholine at a considerably faster rate than does acetylcholine-acetylhydrolase.
JAMIESON [437], on the other hand, is of the opinion that acetylcholine-acetylhydrolase normally degrades neurogenic and non-neurogenic acetylcholine in the ileum of vertebrates (rats, guinea pigs), while acylcholine-acylhydrolase is capable only of hydrolyzing non-neurogenic acetylcholine in the intestine.
Both enzymes belong to a group of hydrolases [3.1.1] which, at their site of action, contain similar amino acid sequences with participation of serine. They are, therefore, also referred to as "serine enzymes". Other hydrolases of the serine group are, for example, trypsin [3.4.4.4] and α-chymotrypsin [3.4.4.5]. They are all inhibited by DFP.
A third complex of aliphatic ester hydrolyzing enzymes are the carboxyl esterases [3.1.1.1] ('ali-esterases'). MYERS, TOE and DEJONGE [672] have suggested that carboxyl esterases may play a role in protein metabolism because of their ability to hydrolyze esters and amides of amino-acids. The carboxyl esterases are certainly of great significance in the detoxification of foreign substances with ester bonds; their physiological function in normal metabolism is, however, unknown. They are likewise inhibited by DFP and partly by other phosphoric acid esters; for example, chymotrypsin, trypsin and C'Ia (activated form of the first component of haemolytic complement) are inhibited by phosphonic acid esters of the formula (I)

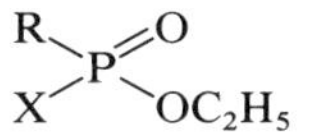

(I)

where R represents an alkyl or aralkyl group, which in the aliphatic part is substituted by alkoxy or acetoxy radicals. X may be nitro- or fluorophenyl [78].

DuBois, Kinoshita and Frawley [252] developed a method of determining quantitatively *in vivo* the inhibition of ali-esterase, acyl amidase and cholinesterase by *dioxathion* and EPN.

Certain organophosphates, depending upon their structure, may also inhibit other systems in addition to hydrolases, e.g. aminoparathion inhibits peroxidases in milk [481, 482]. The inhibition of xanthine dehydrase in milk would appear to be limited strictly to derivatives of the *parathion* series as was found by Wildbrett, Kiermeier and Lettenmayer [1019]. Thiono-esters such as *methyl, ethyl* and isopropyl *parathion*, methyl and ethyl ®Chlorthion may inhibit xanthine dehydrases *in vitro*, but not *in vivo* (concentrations of 10^{-3} mol.). Thiono-esters of other series, e.g. *diazinon, demeton, malathion* or *azinphos*, are completely inactive. In contrast, the oxon compounds of the *parathion* series caused marked inhibition also *in vivo*. For organophosphate inhibitors of xanthine dehydrase in milk it is possible to draw up the following general formula:

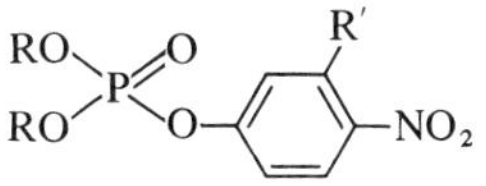

R : CH_3-, C_2H_5-

R′: H −, Cl−

where R must be lower alkyl. Larger alkyl groups reduce the activity. The nitro-group is also a prerequisite, for aminoparathion is completely inactive. Other P=O esters, such as *trichlorvos*, OMPA, or malaoxon, show no inhibitory effect [483]. By dialysis of the inhibited enzyme the inhibition can be abolished. Wildbrett *et al.* suggest, therefore, as mechanism a reversible adsorption of the inhibitor onto the protein molecule; this is plausible in that only the relatively polar oxon esters are active.

There are various possibilities for the inhibition of the normal reaction of an enzyme with its specific substrate, e.g. an inhibitor may react with one or more sites of the enzyme, the substrate or the enzyme-substrate complex. Webb [547] presents a classification of the sites of inhibition based on the component primarily involved in the inhibition:

(1) Reaction of inhibitor with apoenzyme

A. Chemical reaction with specific protein groups:
These groups, such as sulhydryl, amino- or phenolic groups, may react irrespective of their position relative to the active center.

B. Specific reaction with sites on the apoenzyme:
The specificity of interaction here resides in the spatial pattern of matter and charge over the site rather than in a simple chemical group.

a) Substrate site
b) Coenzyme site
c) Activator site

C. Generalized adsorption onto the protein surface:
A relatively nonspecific and weak interaction by substances, frequently nonpolar or amphotropic compounds, that associate with the protein side-chains and may interfere with binding of any component.

D. Denaturation of the protein:
An alteration of the basic protein structure, usually by substances reacting with those protein groups responsible for the bonds holding the polypeptide fabric in a specific orientation.

E. Hydrolysis of apoenzyme:
The breaking of peptide bonds in the protein, generally by proteolytic enzymes, giving rise to fragments of the apoenzyme that may be partially or completely inactive.

(2) Reaction of inhibitor with substrate:
The binding or the subsequent transformations of the substrate are hindered.

(3) Reaction of inhibitor with coenzyme:
Generally the ability of the coenzyme to participate in the reaction is reduced, although the affinity for the apoenzyme may also be decreased.

(4) Reaction of inhibitor with activator.

(5) Reaction of inhibitor with enzyme complex:
Combination with the enzyme-substrate, enzyme-coenzyme, or enzyme-activator complex, although there is not necessarily a reaction with any of the individual components.

(6) Entry of inhibitor into the reaction sequence:
The inhibitor may be acted upon by the enzyme so that it undergoes reactions similar to the substrate, thus reducing the amount of substrate reacted or causing a subsequent block if a normal transfer reaction is slowed.

(7) Reaction of inhibitor with linking components in an enzyme aggregate:
Dissociation of enzyme units in a complex system by interaction with substances, perhaps nonprotein in nature, functioning structurally in the spatial orientation of these units.

Fig. 18 shows several ways in which inhibitors can interfere with an enzyme reaction. Molecules X and Y are either two substrates or a substrate and coenzyme. Z is a co-factor (not always necessary) such as a metal ion, functioning in the binding of a substrate on the enzyme surface. The inhibitor in every case is represented by the solidly shaded molecule. Case A illustrates the unhindered enzyme reaction. Example B shows an inhibitor with substrate-analogous structural and binding properties. In example E only a part of the substrate structure is occupied by the inhibitor. It is not necessary for the sites of the enzyme to be directly blocked. In example G, the inhibitor functions in the spatial vicinity of the enzyme and inhibits the reaction of enzyme with substrate by steric or electrostatic repulsion (the inhibitor breaks, for example, the hydrogen bonds needed for unhindered function). The reaction types H-L are examples of an interaction between inhibitor and enzyme-substrate complex.

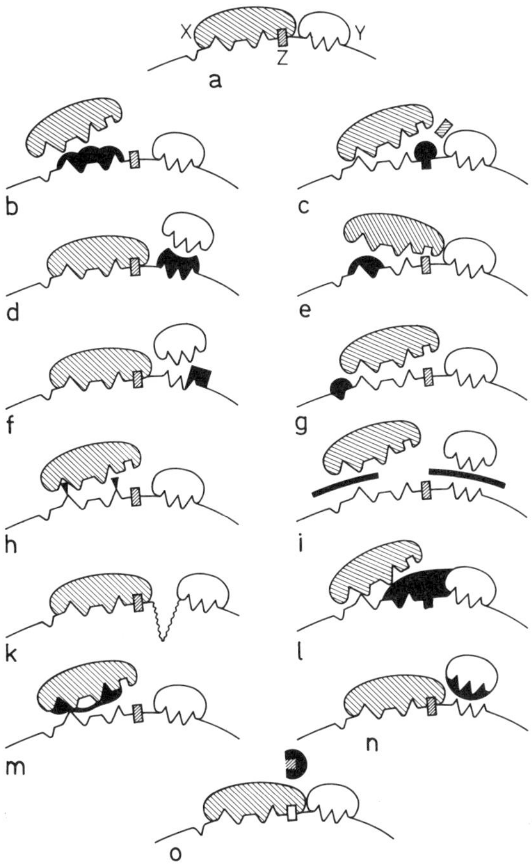

Fig. 18. (From J. LEYDEN WEBB: Enzyme and metabolic inhibitors, Vol. I, p. 53. New York-London: Academic Press 1963). Schematic possibilities of blocking enzyme function. X and Y signify two substrates or one substrate and a co-enzyme, Z a co-factor, e.g. a metal ion. The molecule inhibiting is shaded black

These possibilities may induce partial or complete inhibition. In a molecule, the inhibitor itself is capable of satisfying the structural conditions for substrate + coenzyme + cofactor. Inactivation may be reversible or irreversible. Therefore, these examples do not specify conditions of an "all-or-none" character, and even the enzyme surface has only a schematic character. Furthermore, it must be realized that to discuss complicated problems two-dimensionally is in itself problematic, since conformational changes in the apoenzyme may be a partial reaction of the enzyme-substrate reaction.

A detailed discussion of the different types of enzyme inhibitors and their kinetics will be found in the monograph by WEBB [547].

CLELAND [192, 193, 194] investigated the kinetics of enzyme-catalyzed reactions with two or more substrates or products. He proposed a simplified nomenclature

of possible mechanisms for enzyme-catalyzed reactions, especially for inhibition. Further, CLELAND suggests a shorthand formulation for such mechanisms. There is not, however, room in this book to discuss this matter in greater detail.
According to the older template hypothesis, it was assumed that the active site had a rigid structure to which the substrate had to be fitted. But according to KOSHLAND [507] it is now assumed that a dynamic interaction is involved in which the substrate itself induces a conformational change in the protein structure of the active site, leading to an appropriate alignment of catalytic groups ("induced fit"). This means that only a few amino acids of a sequence need be directly involved in the catalytic process ("contact" amino acids), and that a single amino acid, e.g. serine, will be sufficient as the "transfer" amino acid.
It also implies that although an enzyme preferably transforms its specific substrate, a partial action is maintained on substrates not differing too greatly sterically.
BELLAU and LAVOIE [82] came to similar conclusions in a thermodynamic study using AChE as a model receptor. They found that "ligand binding will occur only at the expense of an actual physical change in the molecular species concerned ('ligand-induced perturbation theory of drug action')".
KRUPKA, KRUPKA and LAIDLER [518, 519] were especially concerned with the structure of the active center and the kinetics of enzyme action and inhibition. Their results provide good evidence on the active site of acetylcholine esterase and its mode of action.
Enzyme E and substrate S form a complex ES which, in the case in question, splits off choline and changes to the acetylated enzyme EAc. In the final stage, hydrolysis takes place, with re-formation of enzyme and acetate, which, together with choline, return to the acetylcholine synthesis.

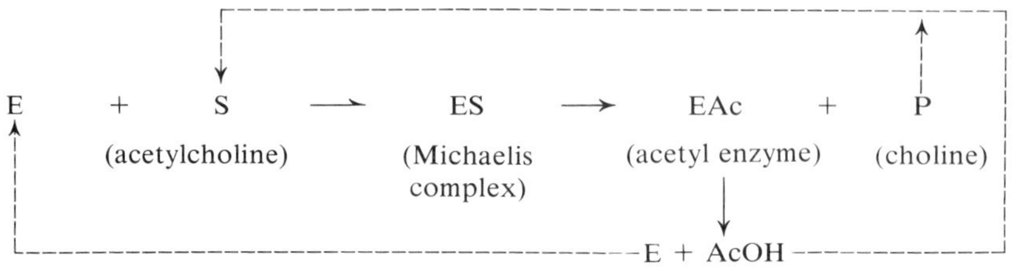

Fig. 19. Scheme of acetylcholinesterase function

The active center of the enzyme has two active sites. The "anionic site" binds the cationic part of the substrate by Coulomb forces; it is presumably the carboxyl group of an amino dicarboxylic acid such as glutamic acid. In other serine enzymes, aspartic acid was also found in the amino acid sequence of the active center.
The "esteratic site" contains the primary alcohol group of the transfer amino acid serine, together with activating acid and basic groups. The basic groups are very probably the imidazole rings of histidine molecules. By taking charge

of a proton, one imidazole ring activates serine alcohol to a form capable of being acetylated; after a conformational change in the active site, a second imidazole ring facilitates the analogous reaction with a water molecule. The resultant hydroxyl ion serves the purpose of hydrolyzing the acetylserine. The acid group in the esterase part has not yet been identified; its role is presumably to protonate the ester oxygen in acetylcholine. Fig. 20 illustrates the schematic structure of acetylcholine esterase. Fig. 21 is a schematic representation of the enzyme substrate complex ES, and Fig. 22 shows the hydrolysis of acetylated AChE.

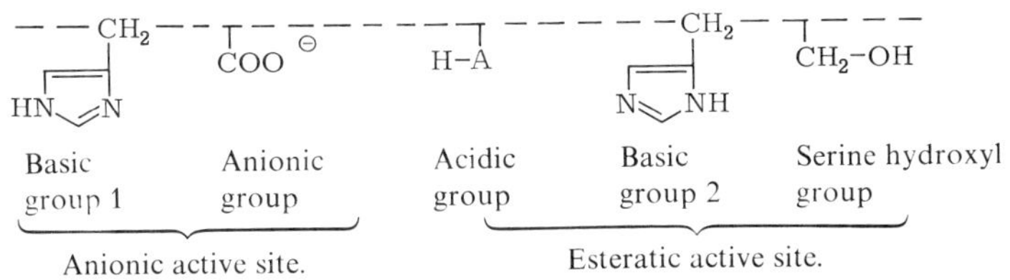

Fig. 20. Schematic construction of acetylcholinesterase

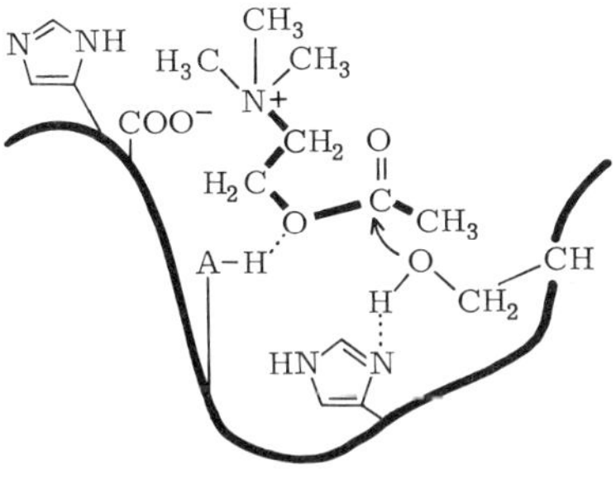

Fig. 21. Enzyme substrate complex (schematic) ("Michaelis complex")

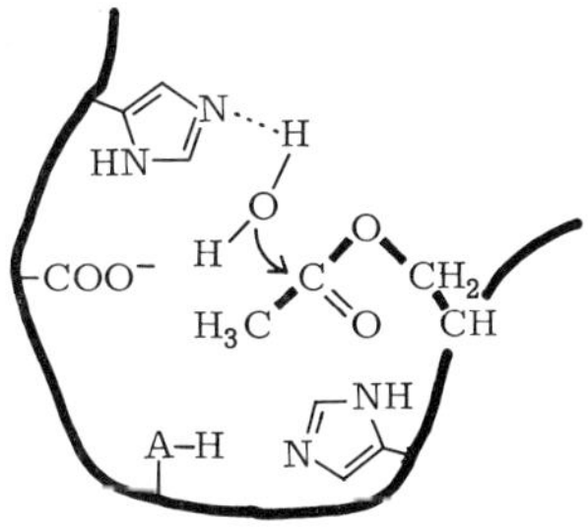

Fig. 22. Hydrolysis of acetylated AChE (schematic)

It is necessary to postulate this interaction between serine and histidine because the alcohol group of serine does not react with DFP alone.

The enzyme-substrate reaction represented by Fig. 21 may be described as the primary step of a trans-esterification of acetylcholine by the serine alcohol of AChE. An argument that may be put forward in support of this hypothesis is that AChE inhibitors of the *paraoxon* type can be transesterified, under mild conditions, with almost any alcohol (c.f. Wolfen method for synthesis of *demeton* on page 121). An older mechanism proposed by CUNNINGHAM [226] is based, on the other hand, on a trans-acylation as shown by Fig. 23 (see p. 173). Acetyl imidazole is primarily formed from acetylserine and is rapidly hydrolyzed. The hydrolysis of acetylserine can, however, be directly formulated on step (III), analogous to Fig. 22.

In this connection a proposal of PORTER, RYDON and SCHOFIELD [745, 785] is of particular interest. They postulate that the serine alcohol reacts with an adjacent aspartic or glutamic acid of the peptide to form a Δ^2-oxazoline (Eq. (1)). Oxazolines react directly with DFP, the rate of reaction increasing with their pKa value. This value should not be less than 4.

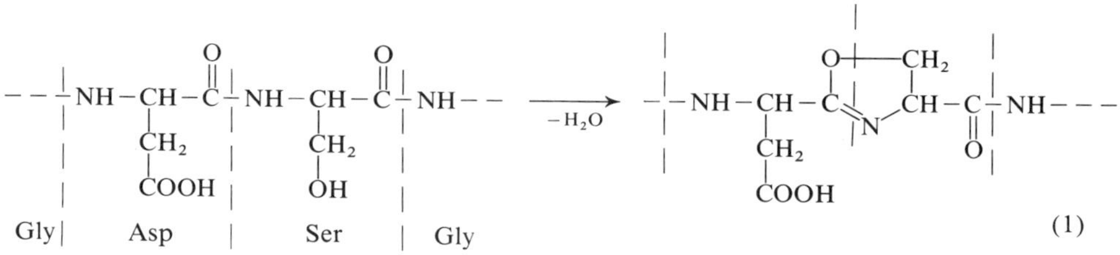

(1)

The primary step is a nucleophilic attack of the oxazoline nitrogen on the ester-carbonyl group, followed by a rearrangement with subsequent cleavage of the ester alcohol:

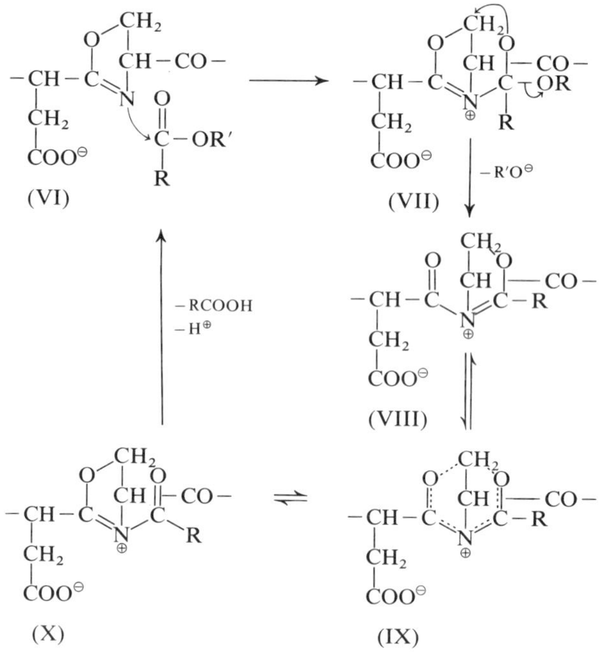

Scheme 12: Oxazoline hypothesis

Cation (VIII) is in tautomeric equilibrium with cation (X). In the case of carboxylic acid substrate, (X) further reacts to (VI), whereas in the case of the phosphoric acid substrate (inhibition), the form (VIII) is favoured in equilibrium. It is not rapidly hydrolyzed in accordance with Scheme 12 but reacts, preferably in accordance with Eq. (2), to form seryl phosphate:

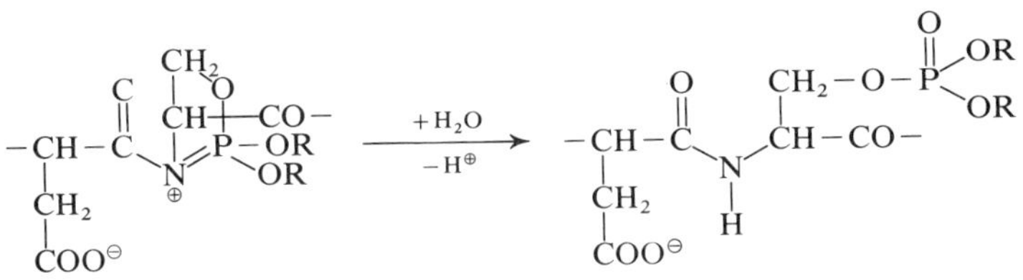

(2)

Some authors, for example BENDER [83] and HEILBRONN [394], disagreed with the validity of this mechanism mainly on the grounds that such a drastic change of the peptide chain under physiological conditions is unlikely and also that it is improbable that the carboxylation of aspartic acid, due to its slight nucleophilicity, functions in the nucleophilic hydrolysis of the acetylated enzyme. The established dependence of hydrolysis upon the pH value is a further opposing factor. The participation of histidine in the deacetylation is, therefore, assumed to be certain, although in the amino acid sequence no histidine has yet been found in spatial proximity to serine. On the other hand, the hypothesis of PORTER, RYDON and SCHOFIELD [745] has the advantage that the reactivity of DFP to serine in the form of oxazolines has been proved by model studies.

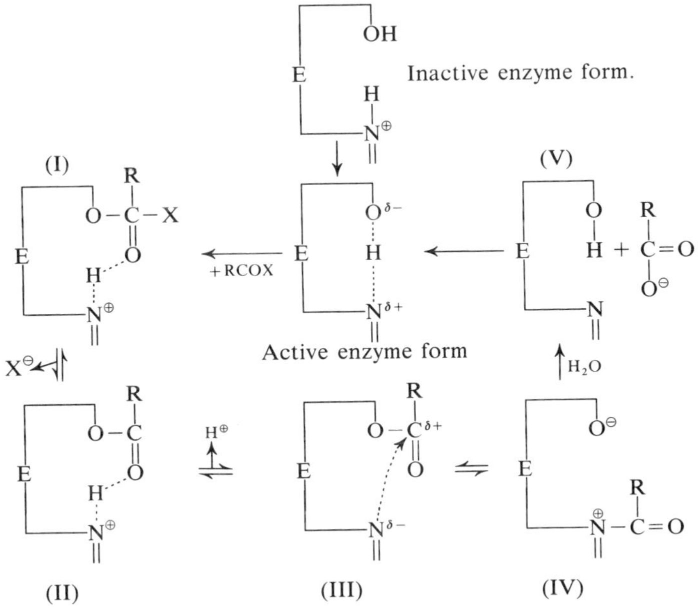

Fig. 23. Enzymatic hydrolysis (CUNNINGHAM [226])

SHALITIN and BERNHARD [900] studied the hydrolysis of O-acylserine derivatives on models of aspartylseryl peptides and assumed that non-polar conditions prevail at the active site, under which the carboxyl group of the aspartate might very well be present in a non-ionized form (XI). Hence, it is possible for imide intermediate (XII) to be formed, which, in an imidazole-catalyzed reaction, may undergo faster hydrolysis to (XIII) than other O-acyl serine compounds, since the action of β-imide as an electron acceptor raises the acidity of serine hydroxyl (Eq. (3)).

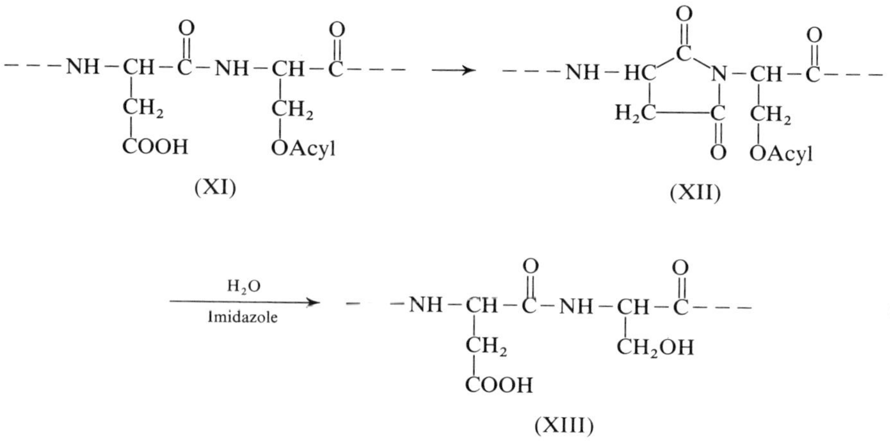

(3)

The misgivings of BENDER [83] and HEILBRONN [394] concerning the formation of oxazoline in the peptide chain under physiological conditions tend to lose weight when it is considered that O-acylserine derivatives may form oxazolines in alcohol and in the presence of weak bases such as potassium acetate, in other words, under relatively mild conditions [84, 819] (Eq. (4)):

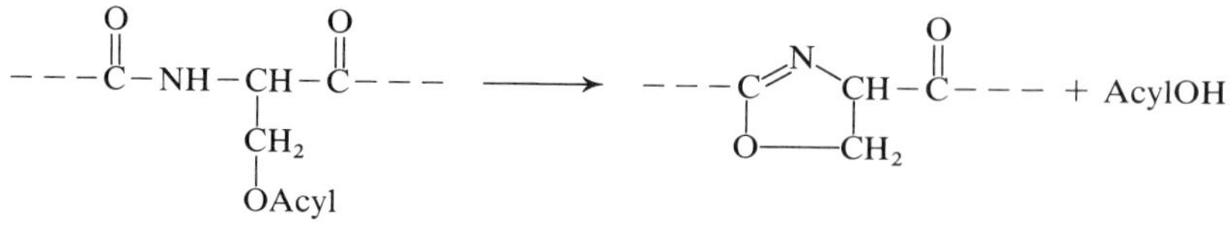

(4)

Accordingly, the mechanisms of inhibition and hydrolysis cannot be considered as having been completely explained for the serine enzymes. Further experiments in this direction, such as those carried out by CRAMER and MACKENSEN [221] using bis-imidazolyl cyclodextrines as chymotrypsin models, are most certainly desirable and, in fact, necessary.

Important contributions to the theory of AChE may also be expected from the Leuzinger four-chain model of AChE. If it could be confirmed that chain A containing the transfer-aminoacid serine must be combined with chain B in order to become activated, as was suggested by LEUZINGER *et al.* [544, 545, 546], an improved theory of AChE functioning should result. At present it is difficult to understand the different mechanisms of AChE inhibition by different inhibitors (e.g. interaction by Coulomb forces, by hydrophobic interactions etc.), if only one structural type of the active center in AChE, or changes in the structure of the peptide chain may be presumed.

The molecular dimensions of the active site of an esterase can be determined by indirect methods. O'BRIEN [688] elegantly used specific inhibitors of certain active sites of the enzyme. Whereas DFP, as a phosphoryl halide, is an almost selective inhibitor of the esteratic site in cholinesterases, other phosphoric esters (e.g. phosphoryl cholines of the Tammelin ester and the *amiton* type) are bound to the anionic site of the enzyme analogously to acetylcholine by Coulomb forces, van der Waals' forces or hydrogen bonds. Interference with binding can be effected with such compounds as $(C_2H_5)_4N^{\oplus}\,Br^{\ominus}$ (TEA) or $(C_3H_7)_4N^{\oplus}\,Br^{\ominus}$ (TPA). A measure of this effect is the decrease in the pI_{50} values under the influence of TEA or TPA (pI_{50} is the negative logarithm of I_{50}, i.e. the molar inhibitor concentration necessary for 50% enzyme inhibition).

The difference between the inhibiting effects of TEA and TPA must have a maximum if TEA covers only the anionic site and TPA, because of its greater molecular dimension, is capable also of overlapping the esteratic site. Thus, the distance between the anionic site and the esteratic site is limited to a range between 4.5 Å (radius of TEA) and 5.9 Å (radius of TPA). For fly head cholinesterase, such a maximal difference in the pI_{50} values is found. If, at the other extreme, there is no difference in effect between TEA and TPA and the overall reduction of the pI_{50} values is slight, the distance between the anionic site and the esteratic site would be greater than 5.9 Å. A third possible case arises if, on the one hand, the effect of a selective esterase inhibitor such as a phosphoryl choline is markedly reduced and, on the other hand, there is no difference in effect between TEA and TPA.

The distance between the anionic and esteratic sites would then be less than 4.5 Å (example: plasma and erythrocyte cholinesterases). As an important consequence for future organophosphate synthesis, O'BRIEN recommends that, apart from the phosphorylating effect of a molecule, emphasis should be given to that part of the structure of an inhibitor which corresponds to the anionic site of the enzyme.

The first simplified kinetic theory of the reaction between substrate and enzyme was suggested by MICHAELIS and MENTEN in 1913 [646]. In their theory, it is assumed that enzyme (E) and substrate (S), in a reversible reaction, form the non-specific adsorption complex ES ("Michaelis-Menten complex") which then decomposes to enzyme and the products:

$$[E] + [S] \underset{k_{-1}}{\overset{k_1}{\rightleftharpoons}} [ES] \xrightarrow{k_2} [E] + [P]\,. \tag{5}$$

A series of differential equations are derived therefrom, which cannot be solved in a closed form. Assuming that the initial concentration of the enzyme (to be more precise, the initial concentration of active centers) $(E)_0$ is very much smaller than the initial concentration of the substrate $(S)_0$ and that the steady-state approximation is $d(ES)/dt = 0$, one arrives from the equation $(E)_0 = (E) + (ES)$, at the expression

$$-d[S]/dt = d[P]/dt = k_2[E]_0[S]/(K_m + [S]) \tag{6}$$

where K_m is the so-called Michaelis-Menten constant

$$K_m \equiv (k_{-1} + k_2)/k_1 \tag{7}$$

If $[S] \ll K_m$, the reaction will be of the first order with respect to E and S, whereas if $[S] \gg K_m$ the reaction will be a zero order reaction with respect to S. If $[S] \gg K_m$ (i.e. all the enzyme is used up to form the complex ES: the enzyme is saturated with the substrate), the maximal specific activity of the enzyme is observed to be:

$$(1/[E]_0)(d[P]/dt)_{\max} = U \tag{8}$$

where U having the dimension $\sec^{-1}$ is the "turnover number". Generally, the turnover number is a function of many experimental parameters, including the identity of the enzyme itself. Turnover numbers normally range from 10 to 100 $\sec^{-1}$ although occasionally they have values of up to 10^6 $\sec^{-1}$ (Review see [931]).

Krupka and Laidler [519] described the competitive inhibition of acetylcholinesterase by substrate and reversible inhibitors, using Michaelis-Menten kinetics. Winteringham and Disney [1028] studied the inhibition of cholinesterases by carbamates using an integrated Michaelis-Menten equation (Eq. 9)), and assuming $[S] \ll K_m$ and $[S] \ll K_s$ (K_s is the dissociation constant of the reaction $ES + S \rightarrow ES_2$):

$$\ln([S]_0/[S]_t) = V \cdot t/K_m \tag{9}$$

(V is the maximal theoretical rate for substrate excess without inhibition). If $[S]$ is relatively large, the overall rate of product formation v is expressed by:

$$v = V/(1 + [S]/K_s) \tag{10}$$

Assuming the primary step in the reaction between esterases and P-ester inhibitors to be reversible. Main [587] developed equations for the bimolecular reaction rate constant which contain an affinity constant and a phosphorylation constant. His kinetic scheme is a follows:

$$E + I \underset{k_2}{\overset{k_1}{\rightleftharpoons}} EI_{rev} \xrightarrow{k_p} EI_{irrev} \tag{11}$$

The reversible step depends upon the affinity of the inhibitor for the active center and is governed by the affinity constant K_a:

$$K_a \equiv \frac{k_2}{k_1} \tag{12}$$

The irreversible inhibition, on the other hand, is determined by the phosphorylation constant k_p. The total inhibition potential depends, therefore, upon the affinity of the inhibitor for the active center and upon the phosphorylation potential of the inhibitor.

The inhibition potential is often expressed also as I_{50}, i.e. as the molar inhibitor concentration necessary for 50% enzyme inhibition, or as pI_{50}, i.e. as the negative logarithm of I_{50}. This quantity is obtained in *in vitro* experiments and is usually only a rough approximation to k_p.

Under conditions similar to those in the simple Michaelis-Menten scheme, it is possible to use a steady state approximation in MAIN's scheme, and hence:

$$\frac{d[\mathrm{EI}]_{\mathrm{irrev}}}{dt} = \frac{[\mathrm{I}]}{([\mathrm{I}] + K_a)} \cdot k_p [\mathrm{EI}]_{\mathrm{rev}} \tag{13}$$

If Eq. (13) is integrated between $[\mathrm{EI}]_{\mathrm{rev}1}$ and $[\mathrm{EI}]_{\mathrm{rev}2}$, t_1 and t_2, taking into account that interval Δt is:

$$\log \frac{([\mathrm{E}] - [\mathrm{EI}]_{\mathrm{irrev}\,1})}{([\mathrm{E}] - [\mathrm{EI}]_{\mathrm{irrev}\,2})} = \log\left(\frac{v_1}{v_2}\right) = (\Delta \log v)\,,$$

then

$$\frac{1}{[\mathrm{I}]} = \frac{\Delta t}{2.303\,\Delta \log v}\,\frac{k_p}{K_a} - \frac{1}{K_a} \tag{14}$$

k_p/K_a has the dimension of a bimolecular reaction constant [min^{-1} (mole litre)$^{-1}$]. Since $k_1 = k_p/K_a$, it follows that

$$\frac{1}{[\mathrm{I}]} = \frac{\Delta t}{2.303\,\Delta \log v} \cdot k_1 - \frac{1}{K_a} \tag{15}$$

[I] may be varied for different reactions without the constancy of [I] changing in each reaction, i.e. the inhibition reaction will be of the first order. Eq. (15) may then be utilized experimentally since the corresponding $\Delta t/2.303\,\Delta \log v$ values for different 1/[I] values can be determined by plotting $\log v$ versus t at constant [I].

According to Eq. (15), $1/[I]$ plotted against $\Delta t/2.303\,\Delta \log v$ gives a linear relationship, the gradient of the straight line is k_1, the intercept on the $1/[I]$ axis is $-1/K_a$, the intercept on the $\Delta t/\Delta \log v$ axis $1/k_p$. If such axial intercepts are obtained after extrapolation of the straight line, it is indicative of reversible intermediate complexes EI_{rev} (example: Malaoxon; see Table 9) and there is thus the possibility of estimating K_a and k_p. With direct irreversible phosphorylation, the straight line passes through the origin (Example: DFP; Table 9). If Eq. (15) is solved with respect to k_1, then k_1 can be determined from any point of the straight line according to:

$$k_1 = \frac{2.303\,\Delta \log v}{\Delta t}\left(\frac{1}{[I]} + \frac{1}{K_a}\right). \tag{16}$$

Inhibition of the human serum cholinesterase with Malaoxon (Conc. 10^{-3} mol.) and with DFP (conc. 10^{-6} mol.) at 37°C and pH 7.6 gives the following numerical examples:

Table 8. K_a, k_p, and k_1 of Malaoxon and DFP

	Malaoxon	DFP (line near the origin)
K_a	$7.7 \cdot 10^{-4}$ Mol	$1 \cdot 10^{-5}$ Mol
k_p	11 min^{-1}	30 min^{-1}
k_1	$1.42 \pm 0.11 \cdot 10^4$ Mol^{-1} min^{-1}	$3 - 4 \cdot 10^6$ Mol^{-1} min^{-1}

Braid and Nix [131] obtained corresponding values with *mevinphos*, Sumioxon, DDVP and *phosphamidon* for the inhibition of bovine erythrocyte cholinesterase. As with DFP the curve for DDVP runs near the origin so that no meaningful value of k_p nor, therefore, of K_a could be derived (Values in brackets in Table 9).

Table 9. Values of kinetic constants for inhibition of purified acetylcholinesterase

Inhibitor	$k_1 \pm$ s.e.* (M^{-1} min^{-1})	$k_2 \pm$ s.e. (min^{-1})	$K_a \pm$ s.e. (M)
Phosdrin	$1.36 \pm 0.05 \times 10^5$	9.24 ± 0.58	$6.90 \pm 0.82 \times 10^{-5}$
Sumioxon	$4.02 \pm 0.01 \times 10^4$	5.58 ± 0.35	$1.38 \pm 0.21 \times 10^{-4}$
DDVP	$1.56 \pm 0.01 \times 10^4$	(50) ÷	(3×10^{-3})**
*Phosphamidon***	$6.30 \pm 0.41 \times 10^2$	9.36 ± 1.98	$1.90 \pm 0.50 \times 10^{-2}$

* Standard error of the mean.
** Estimated values.

Parameters k_p and K_a might be expected to yield improved information on structure-activity relationship in comparison to the simple bimolecular rate constant for the following reasons:

1) K_a represents the mass law proportion of inhibitor reacting with the enzyme.
2) K_a comprises the steric relationship between inhibitor and enzyme and, furthermore, the pH-dependent ionic state of the active site.
3) The rate term k_p includes both the bond energy of the leaving group in the inhibitor and environmental factors.

The application of MAIN's analysis (see page 176) to the inhibitory effects of *parathion-methyl*, *fenitrothion* and its higher *m*-alkyl substituted homologues against fly head and bovine erythrocyte cholinesterases, produced several points of interest concerning the mechanism of the selective insecticidal activity of organophosphates of the *parathion* type having alkyl groups of increasing size at the *m*-position in the benzene ring. HOLLINGWORTH, FUKUTO and METCALF [414] found, as was also to be expected chemically, that mesomerism of the nitro-group with the benzene ring is increasingly disturbed, i.e. MAIN's k_p values, $\Sigma\sigma$ of the ring substituents, and the corresponding hydrolysis rate constants change in the same way within a series of *m*-alkylated phenolphosphates. Differences in toxicity must, therefore, be due to differences in the affinity of the inhibitor for the various enzymes studied, i.e. K_a (p. 177, Eq. (12)) must change from ester to ester. The lower the K_a the stronger will be the binding of the ester to the enzyme. On account of the distance between the electron donor in the anionic center and the esteratic site (4.3 to 4.7 Å for bovine AChE, 5.2 to 5.5 Å for fly AChE), the *m*-alkyl group at the benzene ring (distance between P and C of 5.2 to 6.5 Å in the meta-position) fits poorly on the anionic site of bovine AChE, whereas the distances in *fenitrothion* and in fly AChE are in good agreement, i.e. *fenitrothion* has a very low order of toxicity to cattle (and other mammals) and high toxicity to flies (and other insects). This concept related to dimethyl esters is, however, of limited validity; the corresponding diethyl phosphates, and phosphonates are rather toxic to mammals.
Another example of hydrophobic interaction with the active site of AChE is the fact that alkyl groups on the phosphorus lead to high toxicity if they are isosteric to the choline group of *amiton* [124]. It has so far been argued that the toxicity of *amiton*, a phosphoryl analogue of acetylcholine, depends upon its excellent affinity for the enzyme surface, which again is a result of Coulomb forces between an anionic site in the enzyme and the quarternary nitrogen atom of *amiton*. BRACHA and O'BRIEN [125] calculated the effective part of the charge on the nitrogen atom for binding to AChE to be only about 18%. The affinity is also due to hydrophobic interaction, as in the series of O,O-diethyl S-alkyl phosphorothiolates (XIV). With a six-carbon chain

$$(C_2H_5O)_2\overset{\displaystyle O}{\overset{\|}{P}}-SCH_2CH_2-C(CH_3)_3 \qquad \text{(XIV)}$$

it reaches a maximum which is about the same for both classes of compounds.
The problem of hydrophobic interaction was very recently discussed by a group of Russian authors. KABACHNIK *et al.* [463] investigated the correlation between

inhibition of the enzymes AChE [3.1.1.7] (bovine erythrocyte ChE) and BuChE [3.1.1.8] (horse serum ChE), and the structure of compounds obeying the general formula

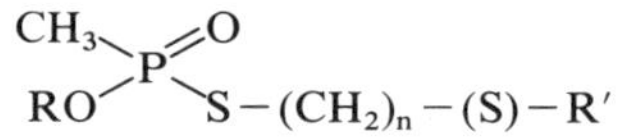

By a systematic variation of R, R′ and *n* with respect to chain length and branching of the alkyl groups, the authors tried to evaluate the existence and steric location of lipophilic surfaces at the active site of the enzymes used. They concluded from the rate of hydrolysis and from the ChE inhibition rate of their compound series that there exist in the vicinity of the anionic centers two hydrophobic regions, one directly surrounding the anionic center and of greater importance with BuChE, the other hydrophobic region in both enzymes being situated somewhat distant from the anionic centers. The best fitting complements are compounds containing a tert. butyl group. The size of this second lipophilic region corresponds in BuChE to a six-carbon chain, in AChE to an eight-carbon chain.

Near the esteratic site of BuChE, separated by a hydrophilic group there are two hydrophobic regions, the extent of which corresponds to a seven-carbon chain. In AChE only one hydrophobic group exists near to the esteratic site and is rigorously correlated to the isohexyl moiety. The hydrophobic regions of the esteratic site in both the enzymes are completely unable to accept compounds containing tert. butyl groups.

The different properties of AChE and BuChE are, to a large extent, due to these differences concerning the hydrophobic regions. For their biological interpretation, the Russian authors put forward a very interesting hypothesis. Because the interaction between substrates and anionic centers in AChE is governed by Coulomb forces, and in BuChE additionally by van der Waals forces, the hydrophobic regions prevent a "parasitic" sorption of acetyl choline by other active sites in the peptide chain. Acetyl choline contains a highly hydrophilic group which by sorption to other sites would retard the enzymatic hydrolysis of acetyl choline. The hydrophobic regions push this substrate to the active center of AChE, thus ensuring maximum turnover. As was suggested by Kabachnik *et al.*, such lipophilic regions play the role of an "energy hill" from which the substrate "rolls down" to the active center of AChE.

Similar conditions may exist also at the choline receptor site, i.e. sorption and desorption rates of the substrate are influenced by the size of the alkyl chains; hence the cholinomimetic effect of acetyl choline changes to a blocking effect.

It might be of special importance if the hypothesis of hydrophobic regions in AChE could be introduced into the Leuzinger model of a four-chain AChE.

A further interesting relationship between structure and activity was found with *amiton*: phosphorylation of AChE, i.e. an electrophilic attack on the enzyme by phosphorus, is favoured by electrophilic substituents attached to the phosphorus. This is also indicated by the Schrader rule and the P-XYZ scheme. For the same reason, the phosphorylating and inhibiting action is greatly diminished when a tri-ester is converted hydrolytically into the di-ester, for under physiological conditions the resulting hydroxy group on the phosphorus is present as a $^{\ominus}|\overline{O}$-P group, i.e., as a donor. As AHARONI and O'BRIEN [8] found, the anticholinesterase activity of various esters is reduced by O-dealkylation by a factor of between 6,000 and 665,000 with one exception: when there is possibility of internal salt formation between the $^{\ominus}|\overline{O}$-P group and one of the remaining substituents on the phosphorus. An example is *amiton*, the activity of which, after dealkylation to (XV) is reduced only 179-fold (AChE fly head) or 292-fold (red cell AChE). With *amiton* in the ammonium form (XVI) an internal salt is no longer possible, so that the activity is once more reduced by a factor of 80,000. The permissible structure variance is narrow, so that the P-N distance cannot be varied. Little is known about the mechanism of action of such inner salts, or whether this effect is to be observed in the *demeton* series.

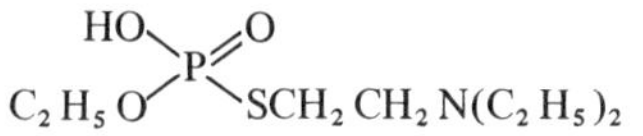

(XV)

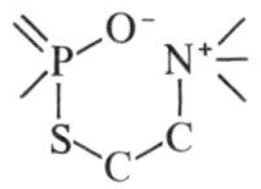

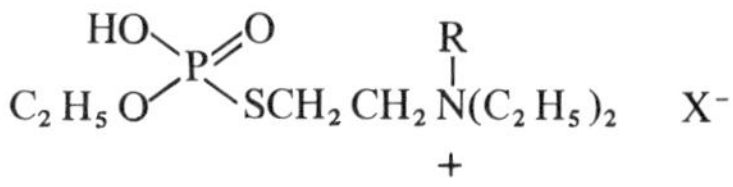

(XVI)

b) Mechanism of Action in Arthropods

Knowledge of the nervous system of insects, of its chemical function and of the biochemical injuries caused by insecticides, is not as soundly based as it ought to be, considering the economic importance of noxious insects; this deficiency is due not least to the very great experimental difficulties involved. Cockroaches and locusts have been relatively well studied.

The nervous system of cockroaches consists of a series of ganglia which are comparable, not with the peripheral ganglia of a mammal, but rather with the brain and spinal cord, in other words, with the central system. The insect ganglia are, for example, the seat of coordination. The blood-brain barrier, which protects the mammal against exogenous substances, corresponds to an unbroken protective membrane in insect ganglia. In contrast to mammals, the myelin sheath is absent in all insect nerves. In mammals, the muscles have one neuromuscular junction to an axon on every fibre (end plate); in insects, on the other hand, a muscle is innervated by only a few axons, every muscle fibre having many end plates slightly separate from each other.

The autonomic system controlling heart, intestine, spiracles, etc., in insects does not consist of a cybernetic system formed by antagonistically acting factors (sympathetic-parasympathetic) as in mammals, but only of a sympathetic (visceral) system.

The visceral system in insects has no peripheral synapses and consequently no peripheral ganglia which, in mammals, are especially sensitive to organophosphates [820].

The transmission of impulses in the axon of the insect is comparable to that of vertebrates in as much as it is also dependent upon the ratio of Na^+ to

K^+ ions. Herbivorous insects with high K^+ and low Na^+ concentrations in the haemolymph which might impair nervous transmission, differ from carnivorous insects in the anatomical structure of the nerve endings where a high Na^+/K^+ ratio is encountered [699, 700, 967].

Chemically, the neuromuscular mechanism of impulse transmission in insects is relatively unknown; the cholinergic system of neuromuscular function in vertebrates is completely lacking, for example, in cockroaches. The fact that there is no cholinergic system would explain the ineffectiveness or poor blocking effect of acetylcholine injected into insects better than the older hypothesis that acetylcholine in quaternary form is unable to penetrate into the neuromuscular junction. The fact that atropine is ineffective in locusts is difficult to interpret if impulse transmission is attributed to a cholinergic system. The neuromuscular effect of eserine is perhaps due to the attack of eserine on other receptor sites at the non-cholinergic end plates. The fact that some organophosphate insecticides exert an influence on the central nervous functions, but not upon the muscular activity of cockroaches may likewise be accounted for by the absence of cholinergically controlled end plates.

It is generally accepted today that chemical transmission at the insect neuromuscular junction is non-cholinergic, USHERWOOD and MACHILI [989] suggested L-glutamic acid as the chemical transmitter at the excitatory synapses in the locust, grasshopper and cockroach, because L-glutamate is the most active excitatory substance among numerous amino acids, active even at concentrations as low as 10^{-6} w/v.

USHERWOOD and MACHILI [989] review the very extensive literature on the pharmacological problem of nerve conduction in insects.

A review of the properties of the nerve axon (without, however, giving consideration to organophosphates) has been published by NARAHASHI [675]. DAVEY [228] describes the autonomous system of insects. Functional aspects of the organization of the nervous system of insects are discussed by SMITH and TREHERNE [909] and the comparative anatomy of the nervous system of insects by SCHMITT [820]. OSBORNE [699, 700] investigated the fine structure of certain neuromuscular junctions of blowfly larvae.

BURT [148] studied the biophysical aspects of nervous activity in relation to the mode of action of insecticides. Principally, there exist two different mechanisms of interaction with the nervous system. Insecticides of the DDT group affect the transmission of nerve impulses within the nerve cells (intracellular nerve poisons), organophosphates affect the transmission of nerve impulses from one cell to another (inter-cellular nerve poisons). As information about the organization of the central nervous system of insects accumulates, additional effects will probably be identified.

A problem which has received little attention is the active transport of insecticides to the nervous system of insects. At least three steps must be considered in the process of penetration to the active site: penetration through the cuticle; spread through the tissues to the nervous system, and penetration into the nervous system. GEROLT [330] suggested that insecticides like *dieldrin*, after application to the cuticle, penetrate via the respiratory system (spiracles, tracheae, and tracheoles) into the nervous system (cf. [728]). The main carrier, however, in the transport of insecticides from the innermost layers of the cuticle to the active site seems to be the insect haemolymph. The insecticide-carrying potency of the haemolymph depends, next to its circulation rate round the body, on its sorption properties.

Acetylcholine, acetylcholinesterase and choline acetyltransferase have, however, been found in many insect species and, by analogy, mechanisms similar to those in vertebrates are held responsible for the insecticidal effect of organophosphates [202]. AChE has frequently been demonstrated in numerous organs of insects. It is not possible here to review all the papers on this subject, but the reader is referred to a summary by COLHOUN [202] and a paper by EDWARDS and GOMEZ [258] containing an extensive bibliography. The bound acetyl cholinesterase of the central nervous system of *Acheta domesticus* was analyzed in an zymogram. For this purpose, EDWARDS and GOMEZ chose an acrylamide gel electrophoresis technique and showed that in three of the eight cases of demonstrable esterases, true acetyl cholinesterase was involved.

There is, however, no doubt that in future far greater attention must be devoted to this problem in experimental work, especially as there are increasing indications that, even in vertebrates, the inhibition of acetylcholinesterase cannot be the sole mechanism of action.

BRADY and STERNBURG [128] evaluated the amount of *in vivo* thoracic ChE inhibited in house flies at the time of knock-down after the application of six different organophosphate anticholinesterases. For each compound, a specific level of inhibition was necessary for knock-down (see Fig. 24).

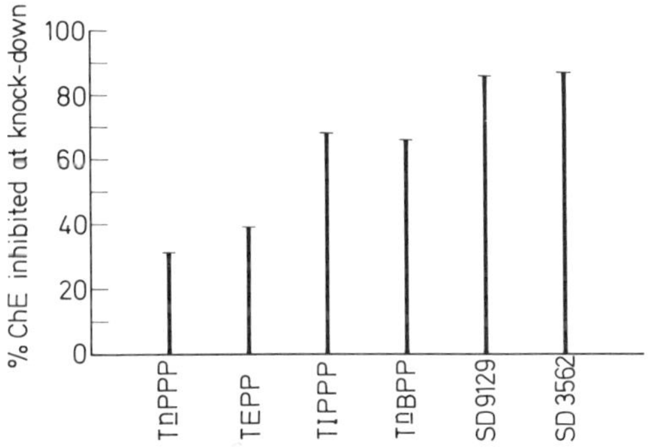

Fig. 24. Percentage of housefly thoracic ChE inhibited at the time of knock-down after treatment with several different organophosphates [128]

TEPP = Tetraethyl pyrophosphate
TiPPP = Tetraisopropyl pyrophosphate
SD 3562 = α-Isomer of *dicrotophos*
SD 9129 = N-monomethyl analogues, *monocrotophos*
TnPPP = Tetra *n*-propyl pyrophosphate
TnBPP = Tetra-*n*-butyl pyrophosphate

As Fig. 24 shows, the cholinesterase inhibition levels vary between 31.1% inhibition with TnPPP and 86.7% with SD 3562. This means that there is no one level of total cholinesterase inhibition which results in poisoning symptoms. These data must be regarded as evidence against the exclusive significance of cholinesterase inhibition in organophosphate poisoning. As early as 1960 VAN ASPEREN had found only 27% cholinesterase inhibition in house flies with DDVP.

It is interesting that BRADY and STERNBURG could establish a correlation between cholinesterase inhibition levels at knock-down and reactivity of organophosphates as measured by stability to hydrolysis in water (see Fig. 25), and bimolecular rate constants for the reaction with ChE (see Fig. 26).

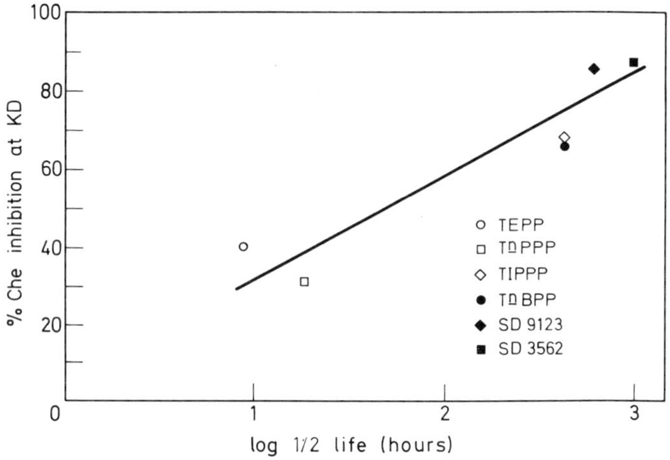

Fig. 25. Relations of percentage ChE inhibited at knock-down (KD) to the stability of each of several organophosphates [128]

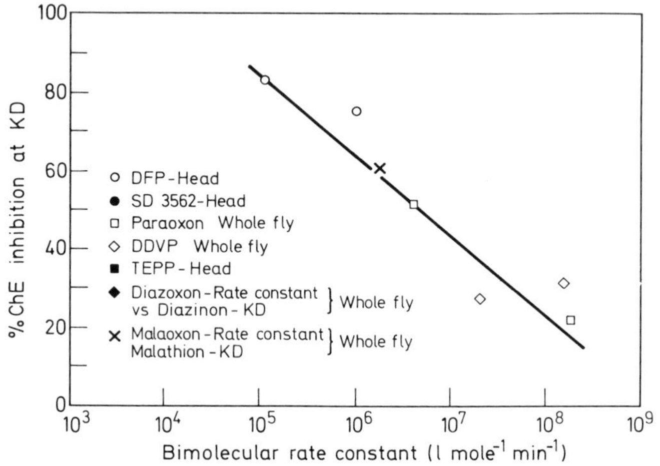

Fig. 26. Relation of percentage ChE inhibited at knock-down to the bimolecular rate constant of each of several organophosphates ([128] there also data sources)

An important conclusion to be drawn from these data is that the inhibition level at knock-down appears to be inversely related to the reactivity of the organophosphate [128]. In an attempt to clarify this effect, BRADY and STERNBURG cite numerous authors. The most important hypotheses are the following:

1) The more reactive inhibitors effect a rapid accumulation of acetylcholine in the synapses. This high acetylcholine concentration, on the one hand, protects cholinesterase against further inhibition, and on the other, increases the nervous activity and knock-down.
2) Organophosphates react not only with the enzyme acetylcholinesterase but also with the receptor site, whereby cholinesterase inhibition levels at knock-down can be influenced. Reaction at the receptor would antagonize the effect of ChE inhibition, i.e., the accumulation of ACh.
3) Organophosphates evoke the liberation of endogenous substances which stimulate electrical activity and which effect an excess of acetylcholine and subsequently knock-down.

A relationship between ChE inhibition levels and penetration, such as lipid or water solubility, is not evident. In this connection there is agreement also with the work of HANSCH and DEUTSCH [368].

VOSS [992] found cholinesterases of the insect ChE type in spider mites *(Tetranychus urticae)*. LEE and HODSEN [541] established their presence in parasitic nematodes *(Haemonchus contortus)*.

The *Haemonchus* AChE differs, however, from the acetylcholinesterase of vertebrates because the reaction of *Haemonchus* ChE with ®Haloxon (see page 105) is irreversible, whereas the inhibition by ®Haloxon in sheep (host animal) is reversible. Aliphatic esters are not hydrolyzed by homogenates of *H. contortus*, which suggests that no carboxyl esterase [3.1.1.1] (formerly "ali-esterase") is present. A more recent study conducted by KNOWLES and CASIDA [502] deals with the inhibition of cholinesterase in *Ascaris lumbricoides* by organophosphates. The ascarid acetylcholinesterase degrades acetyl 2-methyl choline faster than acetylcholine itself. In this respect, it resembles the bee brain cholinesterase. Using DFP as the selective cholinesterase inhibitor (it blocks only the esteratic site), no cholinesterase can be found in the homogenates and carboxyl esterases are also absent. The mechanism of the anthelmintic effect is the inhibition of the acetylcholine esterase, which, however, proceeds at a much lower rate than in mammals and insects.

The presence of acetylcholinesterases in helminths, mites and insects has thus been proved, but there is no positive evidence of their function in nervous transmission. Atropine and 2-PAM, the principal antagonists of organophosphates in mammals, are only slightly or not at all effective in insects. In the development of useful working hypotheses for the synthesis of new insecticides, we have, with our existing knowledge and concepts, temporarily arrived at a barrier which can only be surmounted by greatly increased experimentation in all disciplines concerned with crop protection.

The extent to which other biochemical processes in insects are correlated with the insecticidal activity of phosphoric esters still remains to be clarified. There are, however, some indications that the insect metabolism is greatly affected by organophosphate insecticides. For example, the concentration of the acid-soluble fraction of hydrolyzable phosphates in the haemolymph of lepidopterous larvae was greatly raised following application of *paraoxon*. At the same time, the activity of the alkaline phosphatase [3.1.3.1.] dropped at the stage of total paralysis to approximately 50% of the normal enzyme activity [439]. A decrease in the activity of the acid and alkaline phosphatases in the haemolymph following application of DDVP was reflected in an increase in the corresponding enzyme activities in the intestine [440].

Of the haemolymph transaminases, the activity of alanine-ketoacid-amino-transferase [2.6.1.12] was greatly reduced by DDVP up to the time of total paralysis, whereas D-aspartate-aminotransferase [2.6.1.10] remained largely uninfluenced. In parallel experiments it was found, that following electrophoretic analysis of the free amino acids of the haemolymph, there was a continuous increase in the concentration of L-(+)α-alanine, whereas the concentration of α-ketoglutaric acid decreased [441]. The consequence of *paraoxon* intoxication is also of interest: Methionine is the only amino acid that is no longer detectable in the total hydrolyzate of the haemolymph of lepidopterous larvae. This offers a new field for study, particularly as methionine represents the most important component of methylation in metabolism and is closely linked with Vitamin B_{12} through d,1-homocysteine. It has not yet been established to what extent these results of JARCZYK may be directly correlated with the lethal effect of insecticidal organophosphates. It is to be expected that, in view of the development of special inhibitors, greater experimental study will be devoted to the C_1 metabolism in insects.

While in the tests so far discussed a particular inhibitor was taken as starting point, and the enzymes concerned were sought, in experiments such as those of WATANABE, KOBARA [1001] and MATSUMURA and SAKAI [606] a rather different approach is used: given enzyme combinations are examined for their special activities.

Using an agar gel electrosphoresis technique, MATSUMURA and SAKAI obtained a zymogram from homogenates of the American cockroach *(Periplaneta americana)*. At least 14 bands were found which hydrolyzed α-naphthyl acetate. Every band was isolated and tested against other esters, insecticidal organophosphates and carbamates for their hydrolyzing properties. It was found that for the hydrolysis of each of the ester types investigated, a special enzyme combination is responsible. The cockroach enzymes belong to two groups:

1) to the A esterases, these are aryl esterases that hydrolyse *paraoxon* but are not thereby inhibited,
2) to the B esterases, these are aliesterases that are inhibited by *paraoxon*. C esterases, which neither degrade organophosphates nor are inhibited by such, were not, however, isolated.

For example, *parathion* and *diazinon* were hydrolyzed by the enzymes 3 (A), 7 (A), 11a (A) and 12 (A), *malathion* at the CH_3O group by the enzyme 3 (A), 11a and 12, at the carbethoxy group by enzyme 8 (B) and 11 (B), DDVP by enzyme 7 (A) and 12 (A), while DFP is specifically degraded by enzyme 12 (A). Enzyme 2 belongs to the cholinesterases.

It would be important and desirable to obtain zymograms from as many insect species as possible, to know the susceptible and resistant strains, and also to know the hydrolyzing properties for many organophosphates. Such *in vitro* tests are very suitable for the analysis of action *in vivo* and conversely would permit conclusions for the synthesis of new products.

4.2. Structure and Activity

Some of the common factors relating structure and activity of substituents at the phosphorus atom were dealt with from a qualitative aspect in section 2.1.b "Bond properties" (page 22). The phosphorylating potential of esters with a complex substituent was derived according to the way they could be fitted into the P-XYZ-scheme (see page 41). Repeated reference was made to Schrader's rule (see page 40), linking structure and biological activity of the organophosphates. Finally, numerous substantial arguments were put forward showing that the biological activity of organophosphates is to be considered as an inhibition of cholinester-splitting and other serine enzymes which, chemically, undergo phosphorylation of the serine-alcohol group at the esteratic site.

For the synthesis of new insecticides, it would be extremely interesting if the physical and physicochemical properties of a substance could be correlated to biological activity. Such easily determined properties are, for example, the *p*Ka values of unphosphorylated molecules, the hydrolysis rate of an ester in certain pH ranges, the solubility properties (e.g. partition coefficients in oil-water systems) etc.

The dependence of the inhibitory effect both upon the affinity of the inhibitor for the active center of the enzyme as well as upon the phosphorylation constants means that, due to the complex nature of the inhibition potential, it is only in very favourable cases that there is a direct correlation between inhibition and phosphorylating action. Perhaps the most consequential experiments in this direction were those carried out very early by ALDRIDGE and DAVISON [13], by FUKUTO and METCALF [311] primarily in the series of phenyl dialkyl phosphates. FUKUTO [307] determined the bimolecular rate constants of the inhibition of erythrocyte acetylcholinesterase by different *paraoxon* analogues, and correlated the values found to the rates of hydrolysis in water. In the series of O,O-diethyl O-(substituted phenyl) phosphates, FUKUTO and METCALF [311] demonstrated that there are correlations between the inhibition of fly brain acetylcholinesterase by an ester and the influence of the substituent on the phenyl ring on the lability of the P-O-phenyl bond as measured by Hammett's σ-constants, shifts in the P-O stretching frequencies and hydrolysis rates. If, for example, the logarithm of I_{50} (molar concentration at which 50% inhibition of the fly brain acetylcholine esterase is caused) is plotted against the σ-values of the substituent, a relationship is obtained that shows an astonishingly linear tendency, considering the biological-statistical complexity of I_{50}. The same holds for the relationship between log I_{50} (molar, fly brain acetylcholinesterase) and the P-O-C_{arom} stretching frequencies. It has been shown by ALDRIDGE and DAVISON [13] that a linear relationship is also obtained when if the log I_{50} is plotted against the log of the hydrolysis

constants for a series of compounds. On the other hand, we would draw attention to the fact that reactivity and ChE-inhibition level at knock-down, i.e. *in vivo*, may correlate negatively as was found by BRADY and STERNBURG [128] (see page 184f.).
MURDOCK and HOPKINS [c.f. 13, 670] also investigated a series of O,O-dialkyl S-aryl phosphorothiolates for their anti-cholinesterase action from the point of view of their possible role as environmental degradation products or impurities in formulations of insecticides. They also investigated the hydrolytic properties of these compounds in relation to their structure. Their toxicity against *Musca domestica* was used as the only measure of insecticidal activity. In compounds of the formula:

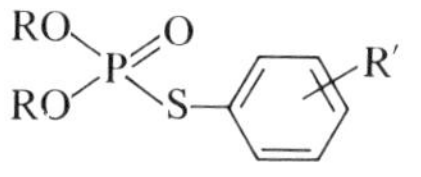

R = CH_3-, C_2H_5-, i$-C_3H_7-$

R′ = H, 4–Cl-, 2–Cl-, 2,4,5,$-Cl_3$-,
4–Br-, 4–F-, 4–CH_3-, 4–NO_2-

the same correlations were found as mentioned above [311]. The thiolphosphates are active AChE inhibitors and relatively labile to alkaline hydrolysis; the parallel is, however, not always very close. There is a correlation between house-fly toxicity and ChE inhibition (*in vitro*), i.e. the mechanism of action corresponds to that of the insecticidal phenol esters. The toxicities to *M. domestica* and the I_{50} values (fly head ChE inhibition) for the most active and the weakest compound do, however, lie closer than for the phenol esters. As regards their actual insecticidal potency, i.e. the spectrum of activity against arthropods, experience has shown that the phosphoric acid esters of the thiophenols are inferior to the analogous phenol esters. In numerous cases the direction of their activity is shifted towards herbicidal and fungicidal action. This effect is probably no longer attributable to the phosphorylating action of the thiolesters but rests rather on the hydrolytically released thiophenol.
In later studies, FUKUTO and METCALF [312, 316] investigated the relationship between the structure and activity of O-alkyl O-*p*-nitrophenyl alkanephosphonates (I) and O-*p*-nitrophenyl dialkanephosphinates (II), from similar aspects. They compared the pseudomonomolecular rate constant of alkaline hydrolysis K_h

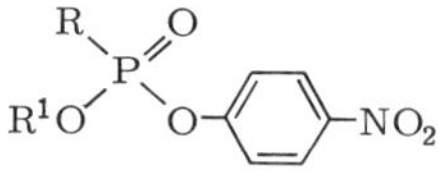

(I)

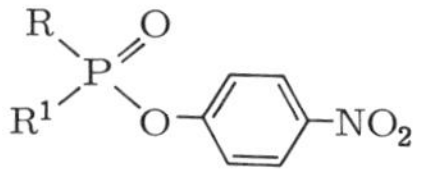

(II)

with the bimolecular rate constant of enzyme inhibition K_e. Whereas ALDRIDGE and DAVISON [13] obtained a straight line for log I_{50} plotted against log K_h, FUKUTO and METCALF [312, 316] did not find a linear trend when log K_e was plotted against log K_h of phosphonic esters with different alkyl groups R. Here, several points arise that are of importance for the synthesis of phosphonic esters, since the phosphonyl radical itself is varied and not the phenyl substituent. Branching on C_1 or C_2 in the alkyl group R reduces the rate of inhibition, whereas branching on C_3 and C_4 (e.g. R = isopentyl, isohexyl) raises the rate of inhibition. If, on the other hand, log K_e is plotted against the log of toxicity (*LD*$_{50}$ *Musca domestica*) [316], such a wide scattering of points results that no definite trend is perceptible.

With phosphonic esters of the *parathion* series there is a similar correlation between toxicity and the branching of the alkyl group on the phosphorus (increasing steric hindrance, decreasing affinity to) the active center, decreasing toxicity) [309]. On the other hand, no correlation of this kind could be established, in flies, for nitro-substituted diethyl naphthyl phosphates. The reason for this is probably that the mechanisms of detoxication are different from those of the corresponding phenol derivatives [313].

A point is thus reached at which the steric factors operating at the site of action begin to play a part. They can no longer be derived from the hydrolytic properties of a compound. Similar results have also been obtained by Russian authors [589]. Experiments to discover a correlation between K_e, Taft's σ^*-values and toxicity to *Musca domestica* in the series of amido-esters of 2,4,5-trichlorophenol (III) [317] likewise proved unsuccessful. Log K_e plotted against σ^* gave a straight line but log K_e plotted against the log of dose *(LD*$_{50}$ *M. domestica)* showed no uniform trend. The correlation for the

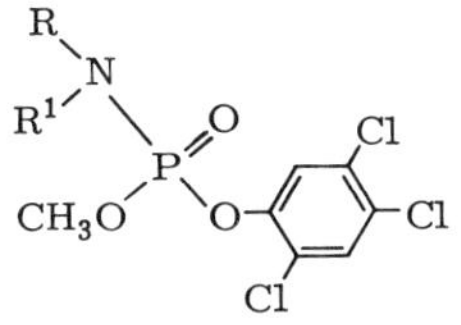

(III)

N-monoisopropyl and the N-mono-tert. butyl derivative of (III) was especially poor. Here, the toxicity is considerably higher than the K_e values suggest. Different mechanisms of activation in the organism were postulated as an explanation.

In this connection, it is not without interest to note that the toxicity of phenols to plants and bacteria, as well as their uncoupling effects on oxidative phosphorylation, has been examined in relation to Hammett's σ values of the substituents on the phenol (see [302]; further references are listed there). HANSCH, FUJITA *et al.* used the σ values with a substituent constant π, which is a free-energy parameter evaluating the lipophilic or hydrophobic character of a substituent on the phenol. The decisive factor here is the dissociation of phenol at physiological pH, i.e. the ratio of phenolate to phenol varies with the substituent at the ring. The following Eq. (17) holds for many cases and also for highly specific enzymatic reactions:

$$\frac{1}{C} = a \cdot \pi + \rho \cdot \sigma + c \tag{1}$$

In this equation, C is the molar concentration of a compound at which a 50% effect occurs (e.g. the LD_{50}, the ED_{50} the isonarcotic concentration etc.), a and c stand for constants. π is defined as $\log P_x - \log P_H$, where P_x and P_H are the partition coefficients determined in a 1-octanol-water system of the substituted and unsubstituted phenol, respectively.
HANSCH and DEUTSCH [368] continued their investigation into the structure-activity relationship of cholinesterase inhibitors. Using the substituent constants already mentioned and a regression analysis, they attempted to deduce the influence of substituents on cholinesterase inhibition from three factors:

1) electronic
2) steric and
3) hydrophobic factors

which are formulated as Hammett's σ constants, Taft's steric constant E_s and Hansch's constant π. Using the preliminary investigations of FUKUTO and METCALF [311] they examined the correlation between the I_{50} and the three substituent constants for numerous carbamates and phosphates of the structures (IV)–(VII):

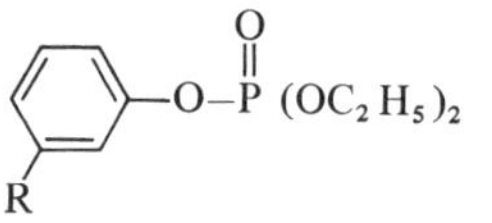

(IV)

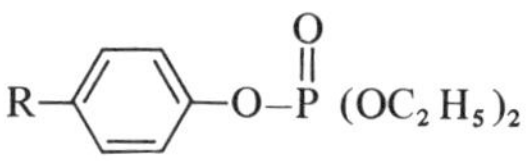

(V)

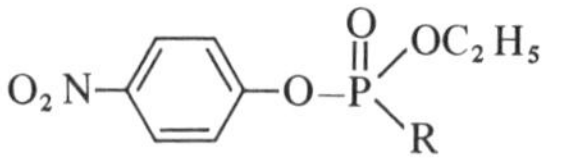

(VI)

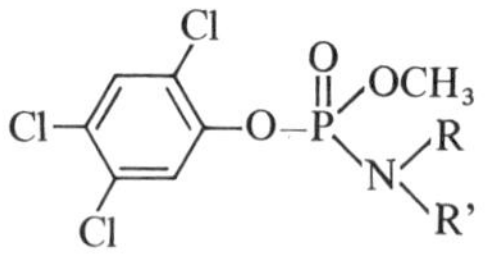

(VII)

HANSCH and DEUTSCH found that the substituent constants for these four types did not significantly influence activity. In the series of *m*-substituted phenyl phosphates this correlation accounts for only 7% of the variance in the data, in *p*-substituted compounds (V) about 50%. In some cases hydrophobic bonding does not play a significant role in the action of the compounds (IV) and (V), nor in the case of the phosphonic acid esters (VI). Conditions are more favourable for the carbamates. BRADY and STERNBURG [128] came to similar conclusions.
BRADY and ARTHUR [126] proved decreasing toxicity in flies with increasing stability to hydrolysis in a series of *dimethoate* analogues, since by increasing the size of the alkyl groups at both phosphorus and nitrogen, the rate of hydrolysis and the toxicity are reduced. Biological Hammett series have also been set up for the activity of substituted G-penicillins against *Staphylococcus aureus*, sub-

stituted benzoic acids against *Aedes aegypti*, and chloramphenicol derivatives against *Escherichia coli*, *Staphylococcus aureus* and *S. haemolyticus* (see [369]).

The attempts described to establish a correlation between the physiological and the physicochemical properties of a compound have contributed enormously to an understanding of the biological activity of organophosphates. Both advantage and disadvantage of such concepts are an over-simplification, as is indicated by the comparison between K_h and the more complex quantity K_e (see p. 188). A further disadvantage is that, in such experiments, the possibility of a chemical change in a molecule between its transport form and active form on the way to the site of action (lethal synthesis) is not taken into consideration, i.e. the construct is not relevant. The conditions are still more difficult to survey when one leaves the series of the phosphorylated phenols for the aliphatic, fused or heterocyclic hydroxy compounds. Furthermore, experience has shown that changes in the spectrum of insecticidal activity are to be expected, in which case the evaluation of cholinesterase inhibition *in vitro* does not improve the relevance of the construct.

All these parameters, like the chemical changes of a compound on its way to the site of action, enter into the overall effect. Here, one is still dependent upon empirical methods, e.g. the species-specific and strain-specific responses of test organisms to the series of insecticides must be determined in relation to the systematic variation of all substituents and structural variables. This procedure offers, firstly, the greatest opportunity of finding new classes of compounds and secondly, the best possibility – in view of our limited knowledge of the biochemistry of insects – of chemically approaching optimal activity. Articles in periodicals such as the Journal of Economic Entomology and the monograph of SCHRADER [863] demonstrate this very clearly.

A very instructive example is given by a comparison of some esters derived from 2,4-dichloro- and 2,4,5-trichlorophenol:

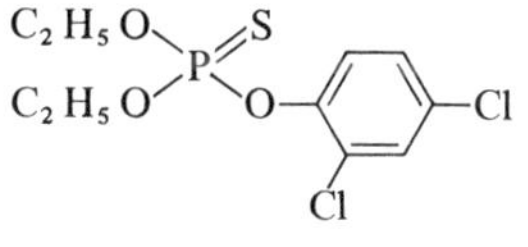

nematicide
dichlorfenthion

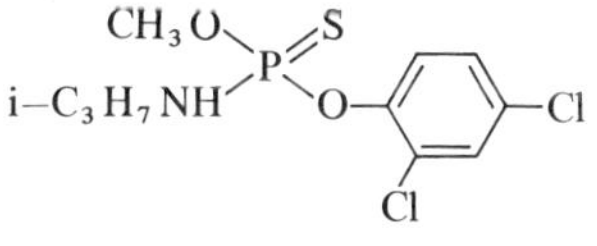

herbicide
DMPA

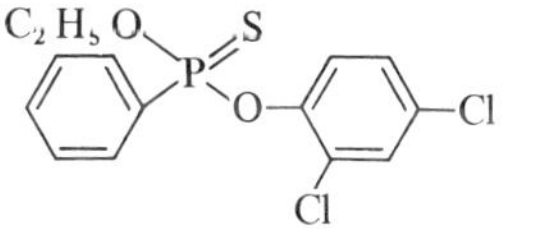

acaricide
S-Seven

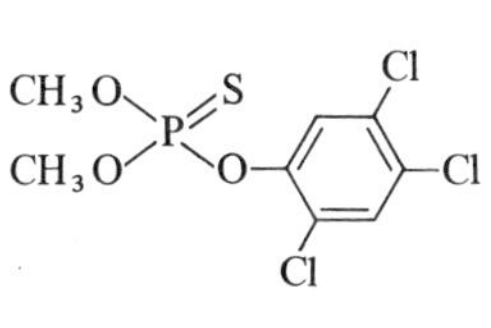

hygiene, vet. med.
fenchlorphos

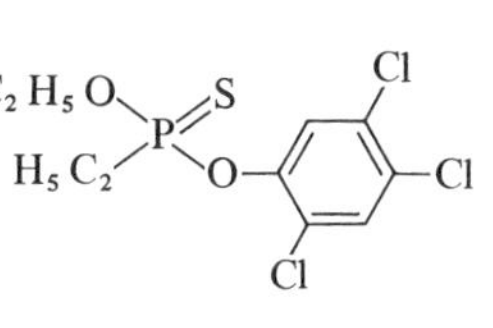

soil insecticide
trichloronate

Very closely related compounds exhibit quite different types of activity. It would be extremely difficult to forecast the practical application of these compounds from their activity, for example, against *Musca domestica.* A further argument emerges from this comparison: the opinion is frequently expressed that the synthesis of many chemically closely related compounds is unnecessary, makes the market difficult to survey and complicates their application. If the aim is to produce pesticides with specific action, then there is no choice but to possess a diverse armament to combat pests. If chemical crop protection were pursued with a few products produced in large quantities, considerable gaps would exist in the protection program.
A comparison between the products ®Potasan and *coumaphos* further substantiates our claim. The introduction of a single chlorine atom increases the activity against mosquito larvae by the factor of 10,000, although neither the chemical nor the physical properties differ by similarly high factors. The biocidal effect caused by introduction of one chlorine atom was not predictable and could only by established empirically.

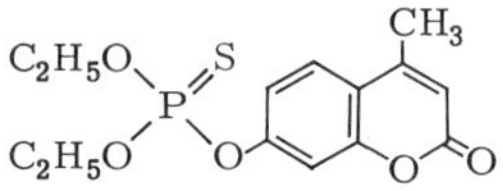

Potasan

LC_{100} (*A. aegypti*, larvae) 100 ppm

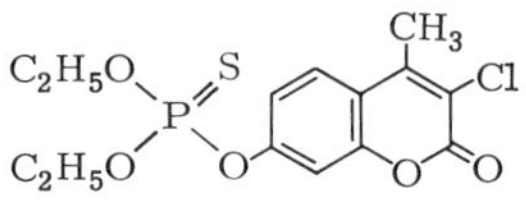

coumaphos, Asuntol (Muscatox)

LC_{100} (*A. aegypti*, larvae) 0.01 ppm

The favourable properties of *malathion* have stimulated intensive investigation of this group.
In 1966 DAUTERMANN and MAIN, for instance, investigated the toxicity, cholinesterase inhibition, carboxyl esterase inhibition, and hydrolysis of *malathion* analogues by carboxyl esterase, in relation to the alkyl group in the carbalkoxy moiety [227]. HASSAN and DAUTERMANN [379] showed that *d-malathion* is more toxic to mice and house flies; *d-malathion* is superior to *l-malathion*, also as inhibitor of cholinesterase and liver carboxylesterase. The α-carbethoxy group of *malathion* is said to be preferentially hydrolized in the liver. CHIU, HASSAN, GUTHRIE and DAUTERMANN [180] investigated structural analogues of *malathion* (Fig. 27).
Their investigations indicate that, for the inhibition of cholinesterase and carboxylesterase, the distance between the α-carbethoxy group and the phosphorus atom plays an important role. The α-glutarate analogue possesses the highest toxicity to mice and the β-glutarate analogue the lowest. In the thiono-series *malathion* itself is the most active against house flies. In the oxon series, malonate malaoxon and α-glutarate malaoxon are somewhat more active.
The differing toxic properties of cis- and trans-*mevinphos* (VIII, IX) and of cis- and trans-®Bomyl (X, XI) are also explained in terms of differing fit of the molecule to the anionic and esteratic site of cholinesterase [658]. It is known that cis-*mevinphos* is about 50 times more toxic to flies and about 20 times more toxic

to the mouse than trans-*mevinphos*. The cis-compound inhibits AChE somewhat more effectively and, in the mouse liver, is more slowly degraded than trans-*mevinphos*.

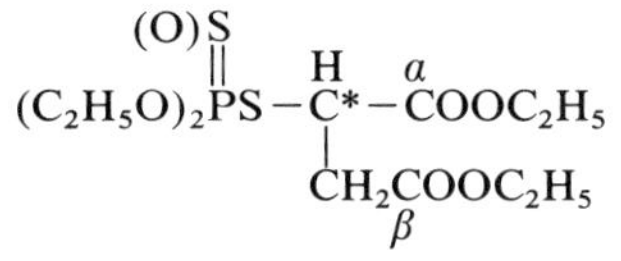

Succinate malathion (malaoxon)

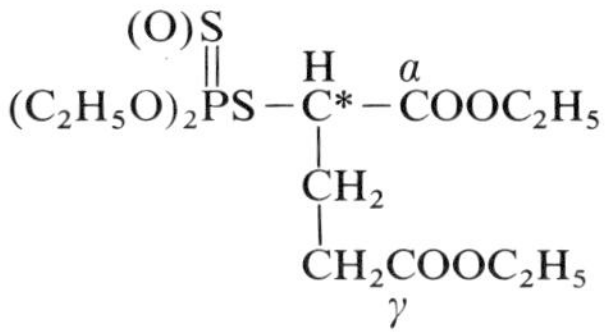

α-Glutarate malathion (malaoxon)

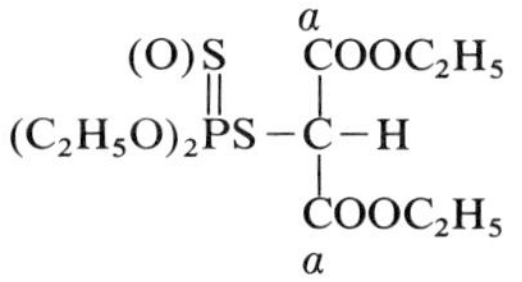

Malonate malathion (malaoxon)

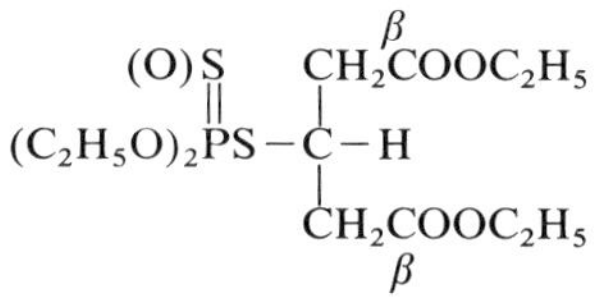

β-Glutarate malathion (malaoxon)

* Asymmetrical carbon.

Fig. 27. Structure of branched-chain analogues of diethyl malathion and malaoxon [180]

The distance of 4.5 – 5.9 Å between anionic and esteratic site of insect ChE [688] is related to the $P-C_{carbonyl}$ distance for

cis-mevinphos of 4.3 – 5.2 Å (good fit),
trans-mevinphos of 2.2 – 4.4 Å (poor fit),
cis Bomyl of 4.4 – 5.2 Å (good fit),
trans-Bomyl of 4.4 – 5.2 Å (good fit).

In fact the toxicities of cis- and trans-Bomyl are comparable; both isomers are more toxic for flies than might be expected from a comparison with cis-*mevinphos*, as was also found by NEWALLIS *et al.* [677].

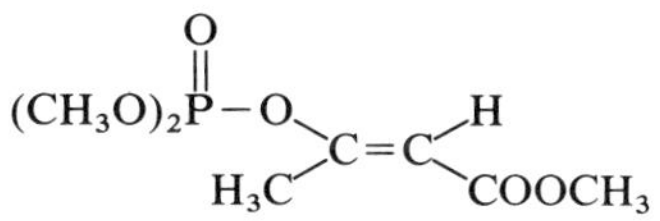

cis-mevinphos (VIII)

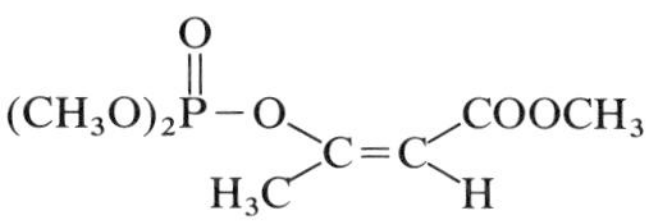

trans-mevinphos (IX)

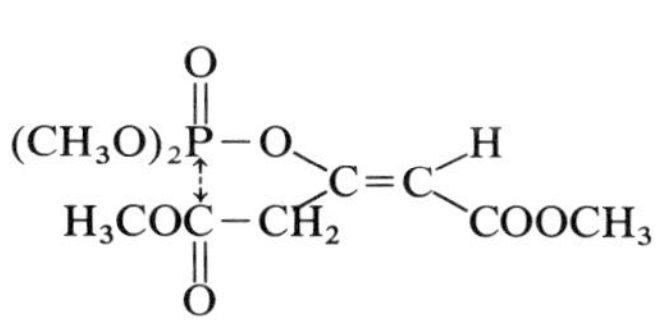

cis-Bomyl (X)

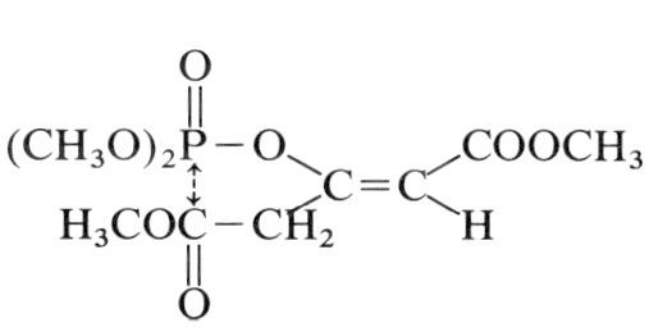

trans-Bomyl (XI)

In the following tables, structural variations of *parathion* are compared with respect to their activity against aphids *(Doralis fabae)* at threshold concentrations, and the acute (single dosage) oral LD_{50} values for rats expressed in mg/kg [866, 869]. Similar comparisons, but of systemic activity have been published, for example, by Menn and Szabo [626], Coe *et al.* [199], Flynn and Eden [292], and Metcalf, Reynolds, Fukuto and Collins [635] for variants of 2,4,5-trichlorophenol esters, by Metcalf and Fukuto [630] for Phosphono-demeton compounds, by Schrader [866] and Francis and Barnes [297] for methylthiophenols and by O'Brien and Hilton [689] for the *amiton* series. Bracha and O'Brien investigated carbon isoesters of *amiton* [124, 125] (see p. 179).

Table 10. Biological properties of structural variations of *parathion-methyl* and *parathion* [866, 869]

Compound no.	Name	Structural formula	Toxicity: LD_{50} rats oral mg/kg	*Doralis fabae* % conc.	% mortality
1	*parathion-methyl*	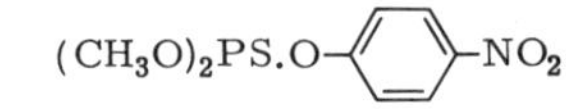	14	0.0008	100
2		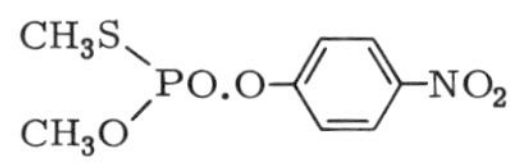	50	0.0008	30
3	*paraoxon-methyl*	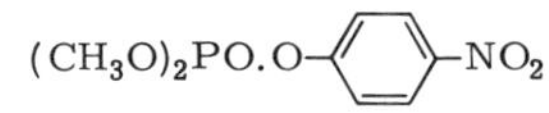	2.5	0.005	100
4		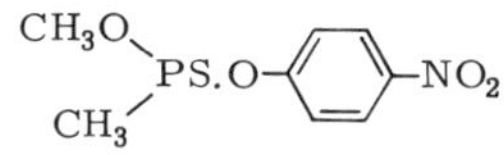	1	0.001	100
5		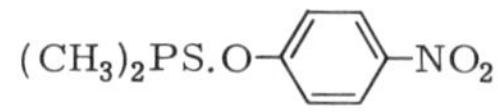	100	0.1	98
6		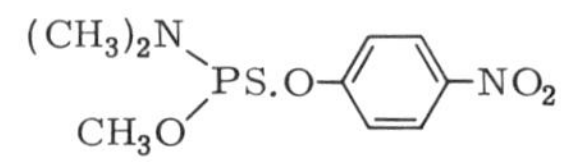	250	0.01	90
7		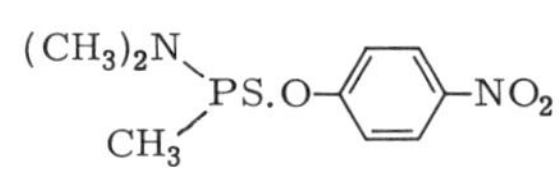	500	0.1	0
8			1000	0.1	80
9			6.8	0.00016	20
10		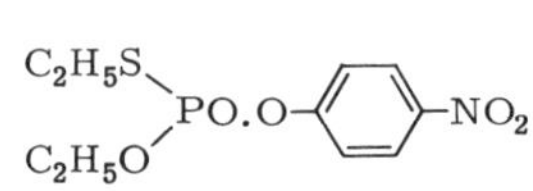	50	0.01	100
11		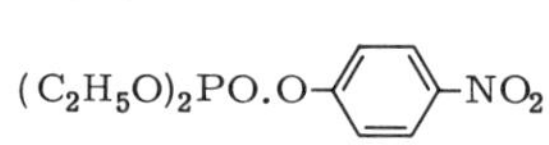	2.5	0.00016	10

Table 10. (continued)

Compound no.	Name	Structural formula	Toxicity: LD_{50} rats oral mg/kg	*Doralis fabae* % conc.	% mortality
12		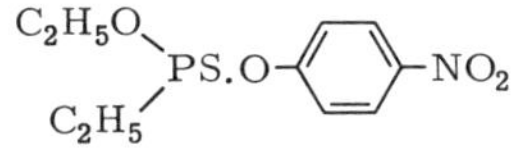	2.5	0.001	100
13		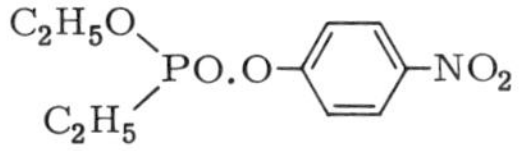		0.001	100
14		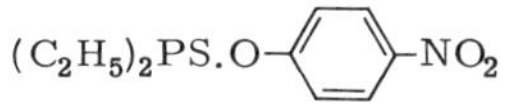	5	0.01	100
15		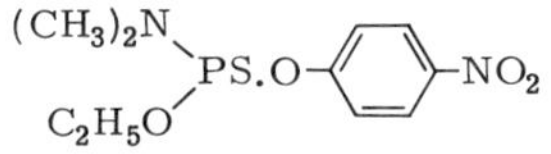	50	0.001	90
16		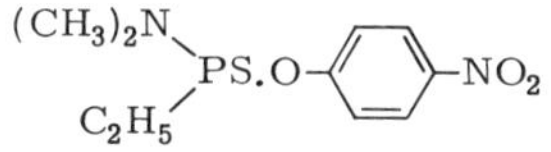	250	0.01 0.1	0 100
17		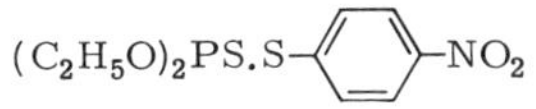	10	0.1	90
18		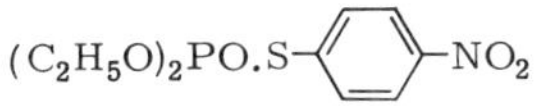	2.5	0.1	100
19		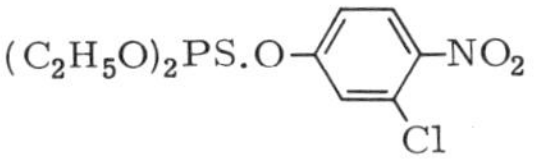	50	0.001	100
20		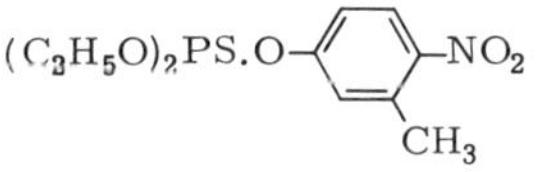	10	0.02	100
21	®Chlorthion	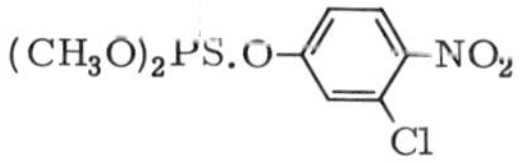	625	0.001	100
22	*fenitrothion* ®Folithion ®Sumithion	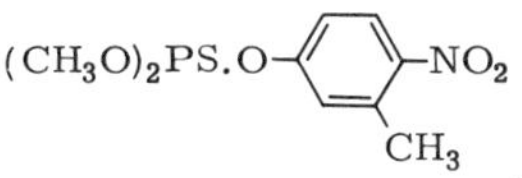	250	0.004	100
23	®Fluorthion	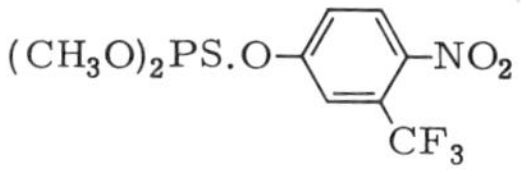	250	0.02	100

An analysis of the results given in Tables 10 and 11 shows that mammalian toxicity and activity against aphids do not run parallel (this also holds for other insect and mite species). Conversion of P=S to P=O results in increased toxicity (Table 10: 1 to 3, 9 to 11). Phosphonates are usually more toxic than phosphates (Table 10: 4 to 1, 12 to 9), except amides (Table 10: 7 to 4, 16 to 13). Ester amides are usually less toxic and also less biologically active than the esters themselves

Table 11. Biological properties of structural variations of methyl isopropyl parathion [866, 869]

Compound no.	Name	Structural formula	Toxicity: LD_{50} rats oral mg/kg	*Doralis fabae* % conc.	% mortality
1		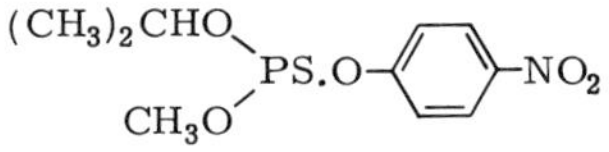	5	0.00016	100
2		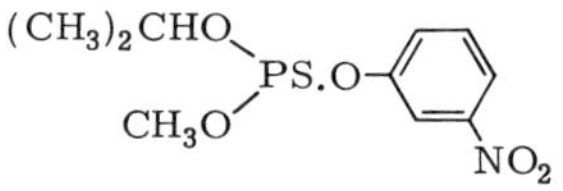	5	0.001	90
3		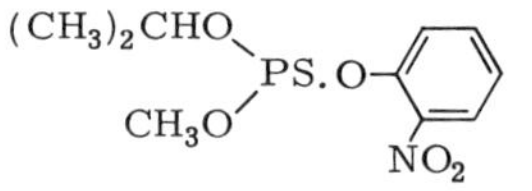	37.5	0.001	40
(4	*parathion-methyl)*	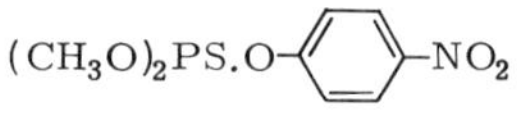	14	0.0008	100
5		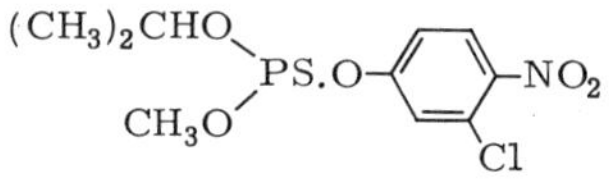	50	0.004	100
(6	®Chlorthion)	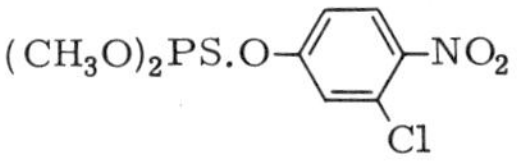	625	0.001	100
7		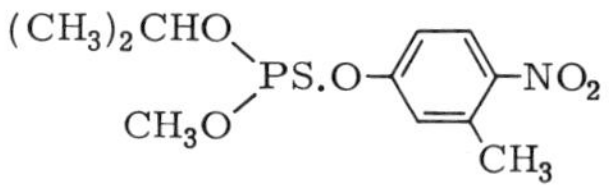	25	0.0008	40
(8	®Folithion)	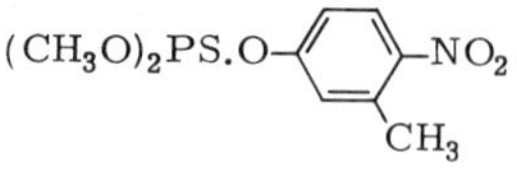	250	0.004	100

(Table 10: 6 to 3, 15 to 9). Thiol esters are less toxic and less active than the corresponding thiono-compounds (Table 10: 2 to 1, 10 to 9). Methyl esters are less toxic (Table 10: 1 to 9, 21 to 19, 22 to 20) and sometimes less active (biting insects) and sometimes more active (sucking insects) than the ethyl esters. This very largely depends upon the insect species.

An example is the data of KERR [478] on the dependence of the action against the Southern chinch bug (*Blissus insularis* Barber) upon the structure of various organophosphates. Here it applies that the dimethyl esters are inactive and that insecticidal activity increases by way of the diethyl esters to the diisopropyl esters. Furthermore, with certain exceptions the following holds true:

a) Aliphatic esters such as *malathion*, *demeton*, *phosphamidon* and *dimethoate* are inactive,
b) Aryl esters, like *parathion*, *dichlorfenthion* or *fensulfothion* possess very good activity,
c) Esters of heterocyclic compounds such as ®Dursban, *thionazin* or *diazinon* occupy an intermediate position.

Phosphinates are less toxic and usually less active as insecticides than phosphates and phosphonates (Table 10: 5 to 1, 14 to 12). Substituents at the *m*-position of the phenyl ring have a strong detoxifying action (Table 10: 19 and 20 to 9, 21, 22, 23 to 1; Table 11: 5 and 7 to 1), but the biological activity need not necessarily decrease. Branching in an alkoxy group on the phosphorus often increases toxicity, with activity remaining constant (Table 11: 4 to 1, 6 to 5, 8 to 7). In very many cases, systematic variation enables a phosphoric ester to be matched to special practical requirements, it being possible to reduce toxicity levels or to alter spectra of activity. The tables show that this approach has led to some of the most important commercial products.

On the other hand, this empirical method and also SCHRADER's rule (see page 40) or the P-XYZ scheme (see page 41) may result in too much importance being attributed to phosphorus as the central atom. With a given 'acyl' or X-Y-Z, the preparative variation of the other two groups can no longer decisively change the action of the complete molecules, although it does permit a certain "tuning". The real is question not, for example in the case of *azinphos*, why the ethyl ester and not the methyl ester is active against resistant spider mites, while the reverse is true for *omethoate*. The question to be asked is: To what extent do the expected metabolites, perhaps benzazimide or thioglycolic acid, differ in their biochemical action within the spider mite. A more promising model for special problems such as acaricidal action, resistance, insect or type-specific control agents could be made available by providing an insect or mite-specific inhibitor with the phosphoryl group as "vehicle". In this way insecticidal compounds would be obtained which reveal a "bi-toxic" action, since primary degradation of the ester in the organism produces a physiologically active substance which can evoke further specific inhibition reactions. At the present time, however, this model which is orientated to the biochemistry of insects has the decided disadvantage that a special toxicology of insects in contrast to mammals virtually does not exist. For the more important commercial products from the series of insecticidal organophosphates the threshold concentrations of activity are known for numerous species of insects and mites, although we are not able to refer to them in any great detail in this work. Basically, however, it is difficult to establish the details of the correlation between spectrum of activity and structure of an insecticidally active compound, since this largely involves the know-how of the crop protection industry. On the other hand, numerous papers on structure and activity are published; they are restricted to the biological parameters: toxicity (mouse or rat, usually oral), AChE inhibition *in vitro* and, at the most toxicity for flies (usually to *Musca domestica* by topical application). Since it is useless to discuss a potential insecticide unless its spectrum of activity is known, investigations into structure and activity should begin by testing insecticidal activity against typical representatives of various classes and orders of arthropods. The action of a compound on *Musca domestica* reveals little about its insecticidal potential. In particular, it gives no information on type-specific activity, e.g., against spider mites, ticks or aphids.

To summarize, it should be emphasized that, for the synthesis of insecticidal organophosphates, the phosphorylating potential and the required detailed structure of the molecule are only two parameters, knowledge of which does not suffice for a prediction of activity. Other factors, that are hard to estimate, are involved in the complex term "activity", e.g. stability to climatic conditions, favourable formulating properties, resorption, penetration, partition, solubility, stability under hydrolytic, oxidative and reductive conditions etc.

a) Structure and Systemic Activity

Alongside a certain molecular size, polar centers appear prerequisite for systemic activity. It is known from organic chemistry that monoalkylamides of carboxylic acids (I) possess very high dielectric constants. This principle is realized in *dimethoate* types (II):

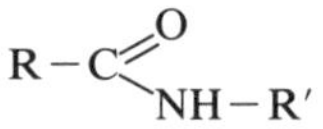

(I)

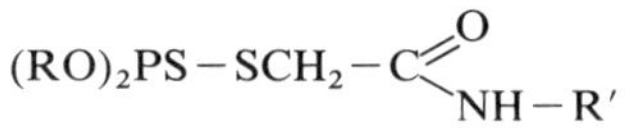

dimethoate types (II)

and is further enhanced in *omethoate* (III), which in addition contains the polar P=O group:

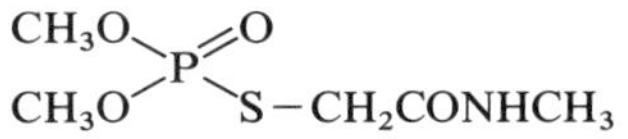

omethoate (III)

In many cases this is also applicable to the amido-group attached to phosphorus (IV–VI):

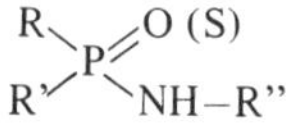

(IV)

e. g.

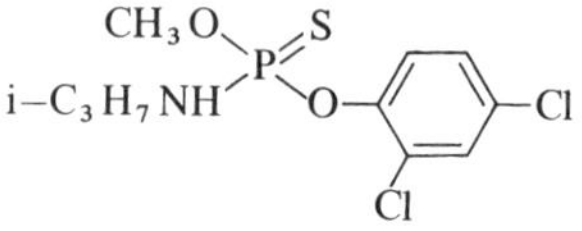

®Zytron (V)
DMPA

and

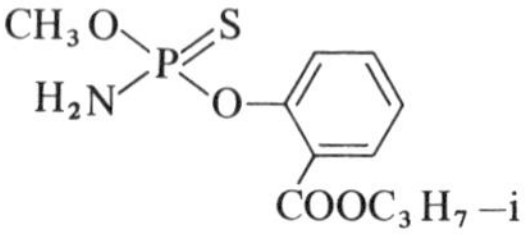

®Optunal (VI)

Here, however, there is an additional dependence upon R and R′, and systemic activity is not generally to be expected. A second type, which has a parallel in organic chemistry, is *demeton* (VII) and its analogues which in the lethal synthesis

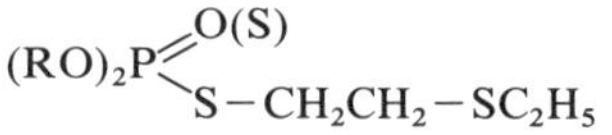

(VII)

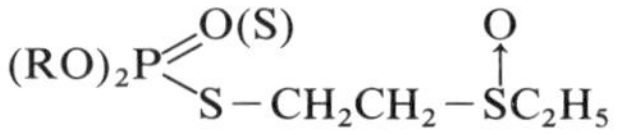

(VIII)

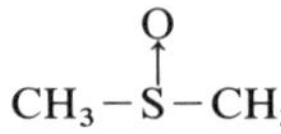

(IX)

are first converted to the sulfoxide (VIII). In the side chain these correspond to the dimethylsulfoxide (IX), whose penetrating and solvation properties are well known.
As was mentioned by UNTERSTENHÖFER [987], thiol compounds of the *demeton* series penetrate somewhat faster than the corresponding dithio-compounds, a feature established in other series of organophosphates, too.
O'BRIEN [686] suggested that systemically acting substances should have an 'intermediate liposolubility', which can be confirmed experimentally. However, it is very difficult to place this idea on a theoretical basis. In fact, the absorption and penetrating properties of a compound do not correlate in all cases with the partition coefficient in a two-phase system. One of the reasons may be that partition is an overall effect comprising the component solubilities of several groups in a molecule. It would be necessary to ascertain the absolute dimension of these component solubilities, e.g. the partial polarities within a molecule, because by suitable orientation to the surface or interfaces between two different phases, they could confer on a molecule an active penetration process. Without doubt, much effort has to be put into the search for a reliable basis for the experimental treatment of the problem of structure and systemic activity.

b) Structure and Activity of Fungicides

Because the Japanese authorities prohibited the use of mercurial pesticides in rice cultivation, active organophosphates were developed specifically against *Piricularia oryzae* (rice blast, Brusone disease). The relationship between structure and activity has therefore been relatively well investigated for these compounds. The most important of them are derived from thiophenol, *p*-chlorothiophenol or from benzyl mercaptan. For this type SCHRADER and SCHEINPFLUG proposed the following formula (I):

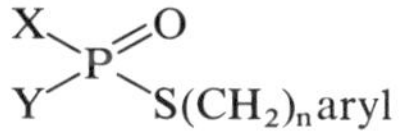

(I)

X: Lower alkoxy groups
Y: Lipophilic groups, in some cases X
n: 0, 1.

Possibly such compounds do not obey SCHRADER'S rule directly. It is notable that with few exceptions the compounds are oxon esters, which perhaps gives some indication of the type of mechanisms involved. Firstly, P=O compounds are strong phosphorylating agents which might attack certain *Piricularia* enzymes. Secondly, in the oxon form of an ester cleavage of the mercaptan or thiophenol groups is also favoured, which may perhaps disturb enzymatic processes in the fungus.
One can certainly accept that the mechanism of activity in the fungus is different from that of the insect. For example, the fungus is unable to activate the thiono-group to the oxon form by oxidation, for thiono-compounds are as a rule inactive against fungi. To effect penetration into the fungus, however, the polarity

of the P=O group must be balanced by a larger, lipophilic group, e.g. by a second thiophenol group, by cyclohexyl, butyl or phenyl radicals:

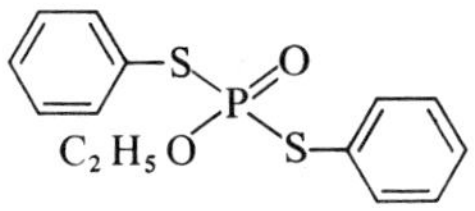

®Hinosan

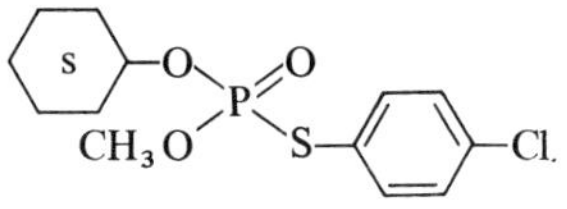

®Cerezin

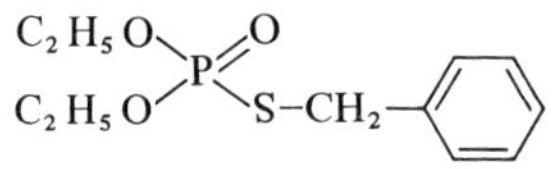

®Kitazin

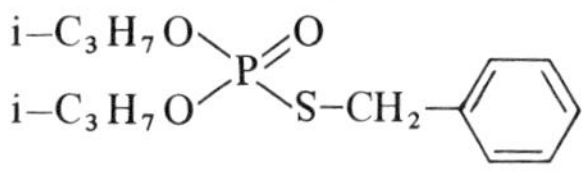

®Kitazin-P

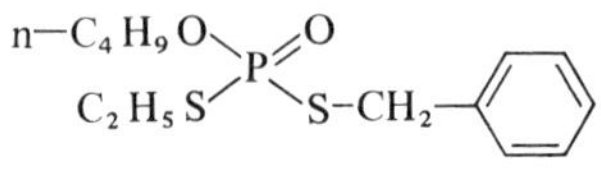

®Conen

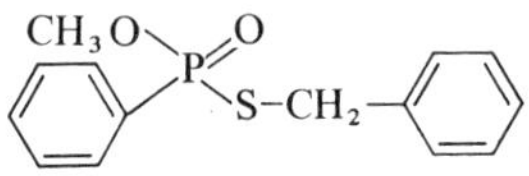

®Inejin F 254

Formula (I) above also covers compounds such as Kitazin or Kitazin P, in which X is identical with Y, the only prerequisite being that both groups together possess the required lipophilicity.
Other organophosphate fungicides belong to the group of NH-acidic compounds, for example Wepsyn *(triamiphos)* or Dowco 199. Here, too it can be assumed that the phosphoryl group merely serves as a vehicle to direct the active groups towards the correct metabolic pathway. Examples include the phthalimide, imidazole derivatives or, in the case of herbicidal organophosphates, also *s*-triazine, whose herbicidal activity has long been known.

4.3. Mixtures of Active Substances

(Synergism, Antagonism)

In practice, it often happens that several active substances are applied as a mixture. An intentional combination of sprays is frequent for technical reasons when different pests are to be controlled at the same time. Unintentionally, however, a mixture may result as a consequence of large-scale synthesis which often yields several isomers or by-products. An important example is the *demeton* group. Here, in addition to the relatively non-toxic *demeton-O methyl* the more toxic thiol isomer *demeton-S methyl* is produced. Other examples are the cis-trans

isomers of the enol phosphates such as *mevinphos*: the mixture resulting from large-scale manufacture contains up to 60% of the cis-compound. These by-products may appreciably alter the spectrum of toxicity of a substance.
CASIDA [165] described similar toxicity-potentiating effects for *dimethoate*. From a practical point of view, SYNNATSCHKE [938] described a transesterification of *dimethoate* in a formulation containing methyl cellosolve. The inital LD_{50} of the *dimethoate* formulation (150–250 mg/kg oral for the rat) may fall to 30–40 mg/kg after storage for 7 months and under tropical conditions to 15 mg/kg after 9 months. The question thus arises: what is the relationship of the toxicity of the mixture to that of the individual components? It is possible by a suitable combination of several components from the same or different classes of substance to gain insight into the mechanism of the biochemical action of a single component. Another problem of practical importance is whether an insecticide, against which resistance has developed, can be reactivated by mixing with suitable compounds which themselves do not necessarily possess insecticidal activity. In the case of *malathion* a considerable number of reports have been devoted to this problem (see [147]). The relationship between dose and insect mortality is generally assessed statistically. Toxicological problems in crop protection are mostly approached by the method of probit analysis, which was suggested by BLISS [102] as early as 1934, and by WADLEY in 1945 [994] and 1949 [995]. Meanwhile, computer programs are available for probit analysis, enabling a rapid statistical evaluation of the numerous experimental results and a direct answer to comparisons between activity and structure (e.g. FINK and HUND [281] and FINK, HUND and MEYSING [282]). As it is not possible to go into any great detail here, a reference to the comprehensive special literature of biological statistics must suffice ([283] and appendix). An important contribution to the evaluation of bioassays and field tests in crop protection was published by UNTERSTENHÖFER in 1963 [985].
Using probit analysis, the %-mortality is expressed in probits (PROBability unITs) [102] and a linear relationship (a straight line regression) is obtained instead of the sigmoid mortality curve. A prerequisite is usually the conversion of the dosage data (abscissa) into logarithmic form. FINNEY [283] formulates this linear relationship as

$$y = a + b x^{\alpha}$$

where y = the probit mortality,
x = the dose and α the variance index,
b = slope of the regression line.

Substances possessing similar mechanisms of action usually provide parallel regression lines. These substances can replace one another in a mixture, i.e. the toxicity of a mixture can be predicted if the ratio of the concentration of components having similar action is known, as BLISS [103] was able to show in a statistical treatment of mixtures of active substances. If a mixture of two substances affects different biochemical systems, then their regression curves vary in slope and form asymptotes to a hyperbolic dose-response curve. The same effect may also occur with similarly acting substances if, for example, they require a different time to reach the site of action. BLISS formulated four different types of action for mixtures of active substances:

1) Similar action, when the components act independently but similarly.
2) Independent action, when the components are both different and independent in action.
3) Synergistic action, when the toxicity of the mixture is greater than that of the sum of the individual components.
4) Antagonistic action, when a substance A (B) reduces the activity of substance B (A) in a mixture.

Considerable effort would be involved if such problems were subjected to an exact statistical treatment, which in most cases is not called for. SUN and JOHNSON [934] therefore suggested a simplified method of evaluating the insecticidal activity of mixtures against defined species. A difficulty here is that the mortalities of the components in a mixture can not simply be added to a total mortality, for the linear relationship between mortality and dosage does not follow an arithmetic scale but is probit-log dependent. SUN and JOHNSON overcame this difficulty by referring to the 50% mortality ($y = 5$) and the associated concentration. For this purpose, three dose-mortality curves were determined on log-probit paper for insecticides A and B and the mixture AB and from the individual LC_{50} values the toxicity indices (TI) were determined using A or B as standard. (Within a given test series the ratio of A and B in the mixture must remain constant).
The actual TI value of a mixture AB with compound A as standard is given by

$$TI_{\mathrm{AB}} = \frac{LC_{50} \text{ of A}}{LC_{50} \text{ of AB}} \cdot 100 \,. \tag{1}$$

The theoretical TI value of a mixture AB is equal to the sum of the toxicity indices which can be calculated from the percentage fractions A and B and their respective toxicity indices:

$$TI^{+}_{\mathrm{AB}} = TI \text{ of A} \cdot \%\text{A in AB} + TI \text{ of B} \cdot \%\text{B in AB} \,. \tag{2}$$

From the actual and theoretical toxicity of the mixture AB, the total toxicity can be calculated by the following equation:

$$CTC = \frac{TI_{\mathrm{AB}}}{TI^{+}_{\mathrm{AB}}} \cdot 100 \,. \tag{3}$$

CTC is the "*co*toxicity *c*oefficient" of the mixture. If the CTC is 100, the mixture probably exhibits a similar action. If the mixture AB shows a coefficient significantly higher than 100, this demonstrates synergistic action. Independent action is

characterized by coefficients with values less than 100, in which case the total toxicity should exceed that of the individual components.
Neglecting the acute toxicity of a synergist (or antagonist), then Eq. (3) simplifies to:

$$CTC = \frac{TI \text{ of A (in the mixture)}}{TI^{+} \text{ of A (alone)}} \cdot 100 = \frac{LC_{50} \text{ of A} \cdot 100}{LC_{50} \text{ of A (in the mixture)}} \tag{4}$$

For example, if *parathion-methyl* is tested alone and in a mixture with 1% *sesamex* against *Musca domestica*, then a *CTC* value of 37 is obtained (LC_{50} of *parathion-methyl* 0.0055%, LC_{50} of the mixture 0.015%, LC_{50} of *sesamex* at 1% can be neglected). This indicates an independent action, i.e. no synergistic effect. In an analogous manner GERSDORFF and MITLIN determined the *CTC* value of a *parathion parathion-methyl* mixture for house flies and found a value of 95, i.e. a similar action [332] which might, in fact, be expected . However, tests *in vivo* with *parathion*, *malathion* and mixtures of both showed different effects with rat esterases: in subacute doses, the substances act antagonistically, whereas at least a similar action can be demonstrated at acute doses [898].
In a later paper, SUN and JOHNSON [933] simplified the cotoxicity coefficient as the LC_{50} of a substance only with reference to the LC_{50} of this substance in a mixture. In this case *CTC* values significantly higher than 1 point to synergism. Values less than 1 indicate antagonism. The authors investigated a series of vinyl phosphates and vinyl phosphonates, each possessing one amido-group; alongside *amiton* and *schradan*, they studied dithioacid esters such as *azinphos* and *malathion*, monothio-esters, like ®Chlorthion, *parathion*, *parathion-methyl* and the oxygen analogue paraoxon-methyl and, finally, the monothiophosphonate EPN. In each case a mixture with 1% of the pyrethrum synergist *sesamex* were tested. It was clearly shown that *sesamex* acts synergistically with phosphates containing amino- or amido-groups. For susceptible house flies, *CTC* values of the order of 40 or more are found and, in the case of resistant strains, of up to 30. The toxicity of the thiono-compounds may be noticeably diminished by *sesamex* (*CTC* values down to 0.3), i.e. antagonistic action. Similar behaviour is found with aphids and spider mites. SUN and JOHNSON [933] explain the synergistic or antagonistic action of *sesamex* in mixtures with organophosphates in terms of an inhibition of the biochemical oxidation reactions which constitute the activity of thionophosphates. The fact that the inhibition of oxidizing systems by *sesamex* potentiates the action of esters containing amido-groups indicates that different mechanisms are involved.
Sesamex is also effective as a synergist with other oxidizable thionophosphates, as was reviewed by SUN, JOHNSON and WARD [935].
Another synergist is 2-(diethylamino)-ethyl 2,2-diphenyl valerate hydrochloride (SKF 525-A), which has a similar mechanism of action to *sesamex* [687]. In a mixture with organophosphates, SKF 525-A acts both synergistically and, depending upon the compound used, also antagonistically in *Musca domestica*. With many substances there is no notable effect. The structure-activity relationships are, therefore, difficult to define [294].

Triolo and Coon [969] found an antagonistic action between chlorinated hydrocarbon insecticides such as *aldrin* and organophosphates such as *parathion, paraoxon*, EPN, *azinphos-methyl*, TEPP or DFP. In mice, administration of 1 mg/kg *aldrin* markedly reduces the action of oral doses of the above-mentioned esters (not, however, of *schradan*). The protection lasts for several days. If the phosphoric acid ester is administered first, followed by *aldrin*, a synergistic effect is obtained. The protective action may be explained in part by the fact that *aldrin* activates liver A-esterases and plasma B-esterase (detoxication processes by enzyme induction) and, furthermore, that *aldrin* decreases the inhibition of brain ChE; *paraoxon* does not, however, decrease inhibition of plasma ChE.

Numerous combinations of insecticides have also been investigated by Keplinger and Deichmann [477]. They usually found normal additive effects (similar action), but found a marked antagonism (or protection) in the rat for combinations of *aldrin* with *diazinon, malathion, dichlorfenthion, dioxathion, carbophenothion* and a potentiation of activity in mice for combinations of ®Aramite with *dioxathion, diazinon* or *parathion.*

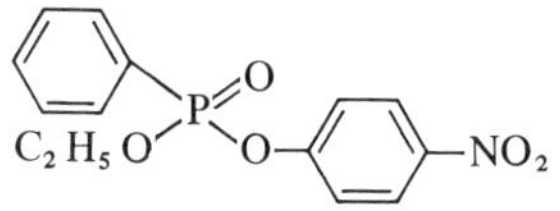

"EPN-oxide"

Synergistic effects are also known with mixtures in the phosphoric acid ester series itself [212]. For example, EPN-oxide inhibits the malathionases, i.e. the metabolism of *malathion* in the liver. The result is a synergistic action with EPN-O/ *malathion* mixtures. *Paraoxon*, on the other hand, inhibits the carboxyl esterases of the liver of humans and rats so strongly that its potentiating action on *malathion* can be neglected [588]. For a series of phenyl phosphates, Casida drew attention to a positive correlation between the synergistic action of *malathion* (toxicity to mice) and ability to inhibit carboxyl esterases [160]. Plapp and Eddy [741] therefore attempted to overcome *malathion* resistance in house flies and mosquitoes *(Culex tarsalis)* by using known carboxyl-esterase inhibitors. With triphenyl phosphate, S,S,S-tributyl phosphorotrithioate and -thioite, the resistance of house flies is reduced from 100-fold to about 5-fold and in favourable cases from 300-fold to about 5-fold [743]. In the case of susceptible house flies, a factor of only 2 is found. With mosquitoes it is possible with the known synergists to eliminate a 100-fold resistance.

Similar figures were also found for *Musca domestica* and *Chrysomya putoria* [81]. The ratio of synergist to insectide was, however, unfavourable; it was 5 : 1.

The most likely explanation of this effect may be that the synergists prevent the hydrolysis of the carbalkoxy group. In further experiments, Plapp *et al.* [737] found that synergists of the same type when used with other phosphoric acid esters, such as *parathion*, *fenthion*, *coumaphos*, *trichlorfon*, ®Ruelene or *ronnel* against *malathion*- and *parathion*-resistant strains of *Musca domestica* do not exhibit anything like the same synergistic potency as with *malathion*. Only when tri-ethyl, propyl or butyl trithiophosphites were used in a ratio of *parathion* to synergist of the order of 1 : 10 did a certain synergistic effect become apparent [742].

While it is possible by applying these results in reverse to obtain selective inhibitors of aliphatic esterases indirectly by way of their synergistic action against *malathion*, CASIDA *et al.* [162] provided another interesting approach. They investigated 112 phosphoric acid esters *in vitro* for their inhibitory action on mouse plasma esterases, which hydrolyze *malathion* and propionyl choline. Amongst the most active of the synergists they found were the trialkyl trithiophosphites and phosphates, diphenyl phosphates, neurotoxic compounds such as tri-*o*-cresly phosphate, and certain cyclic saligenin phosphates. The following generalization would, therefore, appear permissible [162]: a synergistic effect is to be expected if one compound interferes with the metabolism or detoxification reactions of another substance. There is, therefore, no necessity to postulate a direct correlation between synergistic action and neurotoxic effects.

More recent papers reviewing the problem of mixtures of active substances have been published by HEWLETT [405] and WILKINSON [1020].

After a critical examination of the literature, one is forced to conclude that the scientific interest of combinations of organophosphates with synergists is considerably greater than their practical importance. Furthermore, the increase in insecticidal activity is often accompanied by an increase in toxicity for mammals, as is shown from many examples of *malathion* synergists, so that one undesirable effect would be replaced by another.

4.4. Resistance

More and more frequently it happens that, in the case of certain pests, increasing quantities of insecticide must be applied in order to achieve the same effect, until either an economic limit or a toxicological threshold is reached. In extreme cases the pest becomes completely insensitive. The pest responds to the pressure of insecticides by the development of resistance, i.e. the ability to withstand doses of toxic substances which would be lethal to the majority of individuals in a normal population of the same species [1017].

Early examples are the resistance of San José scale to lime sulfur in 1913/14 and of California red scale *(Aonidiella aurantii)* to hydrogen cyanide in 1916 [982].

This resistance is the result of a selection which develops more rapidly the higher the rate of reproduction and the more rapid the succession of generations, in other words, the higher the biotic potential of a pest population. Another factor which is decisive for selection is the impracticability of 100% destruction of the pests. The survivors are therefore able to initiate a new population. Continued selection by means of an insecticide leads to resistant strains. The causes of resistance may be of a varied nature, for example, genetically induced changes in species

specific behaviour or of the morphological or physiological nature of an insect species.
The causes of resistance of insects to pesticides can be illustrated as follows:

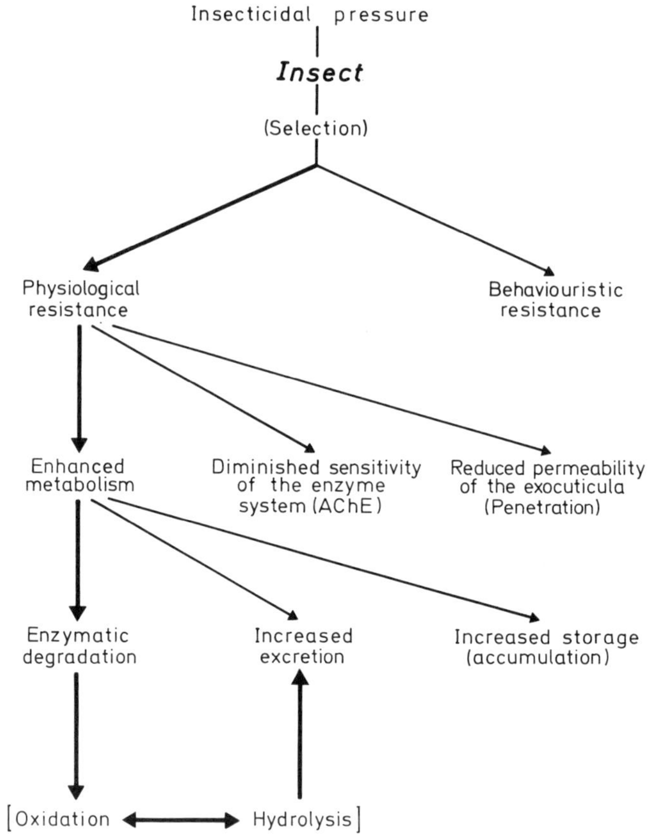

Scheme 13. Mechanisms of insect resistance

In the following paragraphs it will be seen that genetically controlled physiological resistance occurs along the paths indicated by the heavy arrows. It is almost certain that reduced permeability of the cuticula or an accumulation of the active substance in organs with markedly reduced metabolism have, so far, never been the main cause of resistance to insecticidal organophosphates.
SAWICKI [796] confirmed that the penetration-delaying factor alone confers little resistance in *diazinon*-selected strains of *Musca domestica*. When it is combined with the desethylating factor, however, resistance to many organophosphates is greatly increased in comparison to the effect of the desethylating factor alone. This indicates an interaction of both factors which is greater against thionates than against the corresponding phosphates. Probably, the penetration factor delays the entry of thionates more than that of the corresponding oxon compounds.

Resistance shows a maximum in the double homozygote. Increased excretion is of necessity a consequence of the increased metabolism and, as such, not a cause of resistance.

Changes in species-specific behaviour are referred to as behaviouristic resistance or the ability to avoid doses of toxic substances which would otherwise be lethal [1017]. For example, ERLENMEYER-KIMLING, HIRSCH and WEISS [267], using a normal population of *Drosophila melanogaster*, succeeded by repeated selection in obtaining two strains, one of which, when placed on perpendicular surfaces, ran only vertically upwards while the other ran only downwards. A practical example is that of mosquitoes which avoid surfaces treated with DDT, and become exophilous, exophagous or zoophagous, i.e. they change their habitat, feeding habits and source of food. These effects are relatively difficult to demonstrate in nature, but they certainly play a role [367]. (For "behaviouristic avoidance" see also [277].) This is clearly demonstrated by an interesting investigation of EBELING, WAGNER and REIERSON [256] on the acquired behaviour of cockroaches *(Blattella germanica)* which were able to avoid various insecticides according to their repellent properties and learnt to resist the strong impulse to seek dark hiding places when the light areas were not treated with an insecticide. Time-mortality curves reflect the time required to learn and the success of the learning process. An effective agent for the control of 'roaches, therefore, must possess a high initial potency, otherwise selection of the "intelligent" 'roaches occurs.

The term "behaviouristic resistance" is defined from a rather descriptive point of view; one must look for the basic mechanisms involved. In this connection, POLLES and VINSON [744] have provided important evidence which points in the same direction as do the cockroach experiments of EBELING *et al.* [256]. If third-instar larvae of *Heliothis virescens* are brought into contact with cotton plant leaves treated with *malathion* ULV droplets of a suitable size, it is seen that the larvae attempt to avoid contact with the drops, whereas larvae pre-treated with *malathion* do not show this selective behaviour. The cause is the repellent property of *malathion* – a property which other insecticides may possess. The behaviouristic resistance may thus be caused in many cases by the selection of repellent-sensitive insects. Nothing is known about the "repellency spectrum" of the organophosphates.

It is also possible that individuals of robust constitution may be selected from a normally sensitive species. This is referred to as "vigour tolerance". According to BUSVINE [154], this hypothesis is not very convincing, for all natural selection is selection for "vigour". In a few generations, selection under the pressure of insecticides should produce a result which natural selection would not achieve over many thousands of generations. Vigour resistance was postulated in order to explain the occurrence of a weak, unspecific cross-resistance. By cross-resistance we understand the fact that a strain selected by insecticide A can at the same time show resistance to an insecticide B from another class of chemical compound, although the pressure of insecticide B was never involved in the selection.

GRAYSON and COCHRAN [352] define cross-resistance as the case when "resistance to more than one insecticide occurs following exposure to only one compound" ("true cross-resistance", "uncomplicated cross-resistance").

WINTERINGHAM and HEWLETT [1029] reviewed the correlations in the resistances resulting from chlorinated hydrocarbons, carbamates and organophosphates. In the case of organophosphates the resistance maximum is generally associated with the selecting agent itself; cross-resistance occuring mainly within the class of organophosphates also occasionally against DDT and some carbamates. (This usage of the term cross-resistance is found in the literature, but it is a matter of debate whether it is possible to speak of cross-resistance within the same class of compounds.) Diffuse cross-resistance spectra might be based on vigour resistance but it suffices to explain them as the change of a single morphological property, such as the cutaneous absorption or the penetrability of the nerve sheath [517]. On the other hand, the more cross-resistance is confined to certain compounds, the more probable is it that a common enzymatic resistance mechanism can be postulated (simultaneous mechanism).

The fact that *malathion* often evokes no cross-resistance to other organophosphates is easily explained by the fact that it is metabolized in resistant strains to malathionic acid by the action of carboxylases on the carbethoxy group. This is the very reason why esters without this selectophore group cannot be metabolized by this mechanism. On the other hand, a genetically controlled detoxication enzyme with low substrate specificity can lead to a broad cross-resistance spectrum, e.g. it is to be expected with phosphatase types that – in addition to *malathion* – other similar esters containing phosphoryl radicals will fall within the cross-resistance spectrum.

Cross-resistance must be distinguished from poly- or multi-resistance which develops when a strain selected by insecticide A is then exposed to insecticide B and becomes resistant to this insecticide too, and so on (additive mechanism).

"Multi-resistance" according to GRAYSON and COCHRAN [352] occurs when a strain of insect is selected sequentially or concomitantly with two or more unrelated toxicants and because of this multiple selection it becomes resistant to more than one type of material".

The most important mechanism of resistance to the organophosphate insecticides, qualitatively, is that of a genetically controlled potentiation of enzymatic hydrolysis in the insect (physiological resistance). Taking into consideration the activation of thiono-phosphates by oxidation within the organism (see p. 227), as well as their degradation within the organism by hydrolysis, (see p. 228) an inversion of the relationship between oxidation and hydrolysis for toxicity mentioned on page 231 provides a measure of the resistance to organophosphates of the *parathion* series:

$$\frac{1}{T} \equiv R \equiv \frac{\text{Enzym. hydrol. (detoxification)}}{\text{Enzym. oxid. (activation)}}$$

This expression tells us that a resistant insect is able, by an increased formation of hydrolases, to alter the relationship between oxidation and hydrolysis to the detriment of the activating oxidation. A change in this ratio can result in "resistance" which does not depend on selection, but is encountered during certain

stages of the life cycle in an otherwise normally sensitive pest species. For example, a reduction of the fat-body fraction occurring in certain stages results also in a reduction of the activating oxidation mechanisms which are localized in the fat-body, in favour of the detoxifying hydrolysis mechanisms. The application of insecticides must, therefore, be directed against particularly susceptible stages in the development of a pest.
This does not, however, necessarily exclude other mechanisms or resistance.
BOUSH and MATSUMURA [120] reported for the first time on insecticide degradation by *Pseudomonas melophthora* (ALLEN and RIKER), an obligate extra-cellular bacterial symbiote of the apple maggot (*Rhagoletis pomonella* (WALSH)). *P. melophthora* occurs in all stages of the insect life-cycle and can readily metabolize organophosphates. *P. melophthora* provides, presumably, a favourable physical environment for the larvae, the insect *R. pomonella* transports, transmits and acts as a reservoir for the symbiotes. The metabolic pathway for its chemical degradation would appear to be predominantly hydrolysis of the ester by the very active bacterial enzyme systems. Under the special conditions of the experiment the substrates most preferred were O=P compounds. S=P derivatives were metabolized more slowly, i.e. the microorganism does not possess significant oxidation activities. As Table 12 shows, the hydrolytic effects are quite considerable but cannot be considered as the only factor responsible for the resistance. They are, however, important as supporting mechanisms and further investigations are required, for it is conceivable that phosphoric acid esters with bactericidal side-effects may delay the formation of resistance. In the first place it is necessary to determine whether there is a qualitative or a quantitative distinction between the symbiotes in sensitive and resistant insects, and finally whether there is a correlation with insect resistance.

Table 12. Degradation of various insecticides by *Pseudomonas melophthora* [120]

Labeled insecticides	Water-soluble metabolites produced (%)	Solvent-soluble metabolites produced (%)	Original insecticide remaining (%)
C^{14} dichlorovos	83.3	8.5	8.3
C^{14} diazinon	26.7	2.4	71.0
H^3 parathion	15.5	4.8	79.6
H^3 DFP	61.1	5.6	33.3
C^{14} dieldrin	5.5	4.5	90.0
H^3 carbaryl	6.4	45.5	48.1

Over the last 15 years numerous papers have provided ample evidence of a genetic relationship between resistance and an increase in certain hydrolases. A starting point was the observation of OPPENOORTH and VAN ASPEREN [695, 696, 60, 61], that the carboxyl esterase [3.1.1.1.] content of organophosphate-resistant house flies *(Musca domestica)* was significantly reduced ("aliesterases", aliphatic ester-hydrolyzing enzymes). The explanation of this was that the "aliesterases" originally present had been transformed by gene mutation into phosphatases ("modified aliesterases"), i.e. were able to function as degradation enzymes for organophosphates. At the same time MARCH [596] found that in tests *in vitro* a *malathion*-resistant strain of *M. domestica* metabolized malaoxon more rapidly than did *malathion*-susceptible individuals. HOLLINGWORTH, METCALF and FUKUTO [415] also attributed the resistance of house flies to *fenitrothion* and *parathion-methyl* to an increased activity of organophosphate metabolizing enzymes. They explained the high level of resistance as a saturation of the penetrating and activating mechanisms of the resistant strain at high insecticidal doses. The consequence of this resistance, which does not depend upon hydrolysis, is an increased accumulation of the activated products, i. e. of the P=O compounds. The degradation of both insecticides in the resistant strain at high dosage is then just as rapid as with a sensitive strain and low dosage.

In the majority of cases there is a good correlation between insecticidal pressure, rate of development of resistance and hereditary mechanisms. Resistance to organophosphates is almost exclusively associated with a single gene. If the resistance is dominant or semidominant, then the R,R homozygotes and R,s heterozygotes are able to survive the application of the insecticide. If, on the other hand, the resistance is recessive, the insect population remains susceptible until the quantitatively smaller fraction of r,r homozygotes appears. In this case resistance is considerably slower in developing. Resistance to organophosphates is usually controlled by a dominant gene [154].

In general, physiological resistance is of a complex nature despite its dependence upon a single gene. Resistant *Culex tarsalis* larvae contained only a third of the quantity of malaoxon as did sensitive Culex larvae. There are two reasons for this: firstly, an increased organophosphate hydrolase content of the R-strain and, secondly, a raised carboxyl esterase fraction, by which *malathion* itself – and not merely the activation product malaoxon – can be metabolized. In *Culex tarsalis*, unlike the house fly, it was not possible, to demonstrate a fall in carboxylases, together with an increase in the phosphoric ester-hydrolase fraction. The carboxylesterase on the other hand was genetically inseparable from the resistance phenomenon. As a typical carboxylase inhibitor EPN has a synergistic action with *malathion* in the resistant strain [603]. The carboxylesterase activity was especially marked in the mitochondria in a resistant strain of *Culex tarsalis*. It was some 13 times higher than in the sensitive strain. The carboxyl esterases were concentrated particularly in the intestine [604]. In an investigation of the hydrolysis products of *parathion* and *paraoxon* in one sensitive and two resistant strains of house fly, MATSUMURA and HOGENDIJK [605] found evidence that resistant individuals were able to metabolize *parathion* itself to diethyl phosphorothioic acid. The activity of these "thionases" against *paraoxon* was only slight (c.f. [739]).

OPPENOORTH and VAN ASPEREN [696] on the other hand considered that the "oxonases" were responsible for resistance. In addition, since these are said to be dependent upon a single genetic factor ("low aliesterase gene"), the genetics involved must be examined more closely in order to clarify whether the gene responsible determines both a "thionase" and an "oxonase".

From the midgut of lepidopterous larvae, such as *Antherea pernyi* (Saturnidae), *Lymantria dispar* (Lymantriidae) and *Mamestra brassicae* (Noctuidae) JARCZYK [442] in 1963 was able for the first time to isolate two organophosphate-hydrolyzing enzymes, of which one preferentially cleaved E 600 *(paraoxon)* and other P=O esters, while the second preferred ®E 605 *(parathion)* and other P=S esters. Later JARCZYK extended his investigations to further types of Lepidoptera, *Musca domestica* (normally sensitive and 50-fold resistant to *parathion*) and others [443].

Subsequently both enzymes were found to hydrolyze both ester types but they differed in respect to their turnover number, their rate constants and substrate specificity, their Michaelis-Menten constants and activation energies. Reduced glutathion activates, while oxidized glutathion inhibits the enzyme action *in vitro* and also the compounds $(n\text{-}C_4H_9S)_3PO$ (®DEF) and $(n\text{-}C_4H_9S_3)P$ (®Folex) [444].

For the classification of these new enzymes JARCZYK suggested the names phosphoric acid triester nitrophenol hydrolase [3.1.7.*n*] and thionophosphoric tri-ester nitrophenol hydrolase [3.1.7*n*].

In crossing and selection tests with spider mite *(Tetranychus pacificus)*, ANDRES and PROUT [17] found that *parathion*-resistant mites reacted not only with rise in LC_{50} (concentration of the active substance at which 50% of the animals in the test die), but also by division into a resistant and a sensitive population. They were able to show, by backcrossing, that a main gene is responsible for resistance and that the resistance is dominant. With a resistant strain (Blauvelt) of *Tetranychus telarius*, MATSUMURA and VOSS [607] were able to demonstrate an enhanced capacity for detoxification of *malathion*, malaoxon and *parathion*; in the case of *malathion* this is attributable to increased quantities of carboxylesterases. As an first example they also could establish that the resistant strain contained a higher phosphatase activity than the sensitive strain (Niagara). The quantity of *malathion* taken up by both resistant and sensitive strains was equal, the rate of cholinesterase inhibition *in vivo* was higher for the sensitive Niagara strain than for the resistant Blauvelt strain. In both strains it was not possible to detect either a qualitative or quantitative difference in cholinesterase. JARCZYK was able to isolate enzymes from organophosphate-resistant spider mites which, like the enzymes from the intestine of lepidopterous larvae (see p. 211), hydrolytically metabolize esters of the *parathion* series [443]. The resistant strains of spider mite contained a larger quantity of metabolizing enzymes than the sensitive strains. He was able to demonstrate the same conditions in a *tetradifon*-resistant species of *T. telarius*. A mutagenic change in the cholinesterases and hence a coupled resistance is, however, quite feasible. Evidence was provided by SMISSAERT [905] using a *demeton*-selected resistant strain of *Tetranychus telarius* from Leverkusen. This strain had been crossed with sensitive individuals and again selected with *parathion*. SMISSAERT determined the bimolecular reaction constants k_2 for both the sensitive and resistant strains. The lower k_2 value found for the resistant strain is a clear indication that it possesses a modified cholinesterase. The vitality of the resistant strain is not diminished in comparison with that of the sensitive strain, and the resistance appears to be controlled by two alleles. About the same time, MATSUMURA and VOSS [607] found, for the first time, a change in the substrate specificity of a *malathion*-metabolizing cholinesterase in *Tetranychus urticae.* Similarly BALLANTYNE and HARRISON [66] found that, in the Leverkusen R-strain and two resistant strains from New Zealand, the resistance gene determines the structure or a part of the structure of the cholinesterase. The altered enzyme possesses a decreased sensitivity to organophosphate inhibitors and provides the resistant spider mites with more time to detoxify the inhibitor. Using the previously mentioned strains of *Tetranychus telarius*, VOSS, DAUTERMANN and MATSUMURA [993] investigated the relationship between activity and structure of the phosphoric acid molecule. A Blauvelt strain with a 60-fold *malathion* resistance resulting from increased carboxyl esterase and phosphoric ester hydrolase showed remarkable resistance to carbalkoxy analogues of the *malathion* and malaoxon type. However, the resistance factor fell rapidly when a group higher than a propyloxy group was attached to the phosphorus atom. The change in resistance factor (LC_{50} Blauvelt/LC_{50} Niagara) was directly related to the toxicity of these compounds in the susceptible Niagara strain: the more toxic a substance was to the sensitive strain, the higher was the level of resistance. The resistance factor, therefore, was entirely dependent upon the susceptibility of the Niagara strain. The methyl, butyl and amyl carboxyester derivatives exhibited less interstrain-difference, either because the Niagara carboxylesterase is also able to metabolize these compounds, or because these homologues are relatively nontoxic to spider mite for other reasons than reduced power of penetration. If the resistance factor is plotted against toxicity (LC_{50} Niag.), a linear relationship is obtained which, however, is only valid for closely related compounds. The factor responsible for resistance in the Blauvelt strain is rather unspecific; it enables resistant mites to counter the attack of many chemically related acaricides. Consequently, in the synthesis of these compounds it is not possible by making minor molecular changes to overcome resistance in the spider mite, at least not with the *malathion* analogues. With cockroaches *(Blattella germanica)*, however, VAN DEN HEUVEL and COCHRAN [903] discovered something at variance with this; they found that a *malathion*-selected *Blattella* strain which, because of the resistance factor carboxylesterase, showed no cross-resistance to other organophosphates remained sensitive to a vinyl *malathion* (dimethyl 3-(O,O-dimethyl phosphoryloxy)-glutaconate).

HERNE and BROWN [402] investigated *parathion* resistance in the Niagara strain of *Tetranychus urticae* KOCH. They showed that resistance was completely dominant in reciprocal crosses with the susceptible (S) counterpart strain.

If the resistance phenomena and their various mechanisms in the spider mite are compared, then it is apparent that the American and European strains show considerable differences:

1) Niagara and Blauvelt R-strains show undiminished ChE activity. Due to a mutated ChE, the Leverkusen R-strain possesses 50% of the normal ChE activity.
2) The Leverkusen R-strain is substantially less susceptible to *malaoxon*, *paraoxon* and *diazoxon*.

HERNE and BROWN attribute this difference between strains of different origin to the fact that the resistant European strains were selected for the most part by *parathion* and *malathion*, whereas the American strains were selected by *demeton* and *oxydemeton methyl*. If non-European strains are selected with *parathion* (c.f. BALLANTYNE and HARRISON [66]), they exhibit the same reduced ChE activity to *paraoxon* as did the Leverkusen R-strain.

The organophosphorus resistance of cattle tick larvae *(Boophilus microplus)* is similar, as LEE and BATHAM found [540], to that described by SMISSAERT [905] for resistant spider mites.

The resistant strain possesses at least two different types of enzyme of which one is far less sensitive to organophosphates and carbamates than the others in the same strain or in susceptible strains. It can be fully appreciated that possession of such an enzyme combination is a great advantage to tick species that are under the pressure of insecticidal carbamates or phosphates. Zymograms of these pests that are important in livestock farming would be very helpful.

Another possible way of lowering resistance to organophosphates is by specifically inhibiting the metabolizing enzymes in both insects and spider mites ("anti-resistants"). The addition of tris-*n*-butyl phosphorotrithioite (Folex), the analogous phosphate (DEF) or triphenyl phosphate to *malathion* lowers the resistance of *Musca domestica* or *Culex tarsalis* [742]. With *Tetranychus telarius* HENNEBERRY and SMITH [401] achieved a reduction of *malathion* resistance by a factor of 3.5 by the addition of the same compounds (otherwise known as carboxylase inhibitors) as well as EPN, which MATSUMURA and BROWN [603] used against *malathion*-resistant *Culex tarsalis*. Since many of these synergists cause ataxia in the hen, i.e. are neurotoxic in action (e.g. tri-*o*-tolyl phosphate, $(RS)_3P$, $(RS)_3P{=}O$ etc.), their use, at least from a toxicological point of view, would be questionable. PLAPP and TONG [742] deny a strict correlation between neurotoxicity and synergistic action, a correlation which might at first be assumed (cf. p. 204/205). Further investigations are certainly desirable in order to settle the point.

According to the investigations of JARCZYK [445], diphenyl and triphenyl phosphate synergists behave as competitive inhibitors, and would have to be applied in the ratio of at least one mol inhibitor to one mol synergist. Also there remains the danger that insects and spider mites might once more become resistant to these displacement inhibitors (poly or multi-resistance). JARCZYK was able to find true inhibitors, i.e. those which act non-competitively *in vitro* and which are still completely active in a ratio of 1 mol inhibitor: 100 mol insecticide.

Some special problems of resistance in practical crop protection can be best expressed in the form of the following questions:

What biochemical mechanisms are involved when a resistant population continues to develop without the pressure of an insecticide (stability of resistance)?

How do resistant and sensitive individuals of a species differ biochemically and biologically from one another? Is resistance reversible?

Because of the large range of harmful arthropods, their biology and biochemistry and the multiplicity of chemical classes of insecticides and, within them, the individual compounds, there can be no conclusive answers to these questions.

For instance, THOMAS and BRAZZEL [951] investigated the biological differences between a sensitive and resistant population of an inbred strain of the cotton boll weevil *(Anthonomus grandis)*. Selection was carried out over 14 generations with *endrin*; in comparison to the sensitive control population the resistance was 75 – 100 fold. The development period of resistant individuals was prolonged by about 12.5 hours, and the fertility of the resistant *Anthonomus* females was reduced by about 22%. Both factors have an unfavourable influence on the biotic potential. Also the duration of the embryonal, larval and pupal stages was markedly different. No differences were observed in mortality rates, sex ratio, length of the period before and during oviposition or in the percentage of viable larvae. In field strains of *Anthonomus grandis* or in other insect species a completely different situation might be found. Biochemical investigations of spider mite of the species *Tetranychus urticae* showed that, in organophosphate-resistant female individuals, the activity of the alkaline phosphate monoesterases [3.1.1.8] was greatly diminished compared with that found with normally sensitive individuals. The same picture was found with *tetradifon*-resistant female individuals of *T. urticae* [443].

DITTRICH [240] worked with a Leverkusen strain of *Tetranychus urticae* which possessed a recessive factor against *oxydemeton methyl* resistance. All other resistance factors were excluded by inbreeding. Under homozygote conditions (r,r) the vitality was markedly reduced; the heterozygotes (S,r) were capable of competing with the S,S type or were slightly superior. The heterozygotes were thus able to hold their ground alongside the normal sensitive spider mite; the resistant homozygotes (r,r), on the other hand, were maintained only up to a state of equilibrium between elimination and re-formation within the population. After seven generations the homozygote strain perished.

In California *Panonychus citri* (citrus red mite) is one of the main pests in citrus growing. Field strains that had become resistant to *demeton* and *chlorfenson* were collected and reared further without insecticidal pressure. GILMORE and MUNGER [339] obtained the following results: The resistance index ($LD_{50 res.}$: $LD_{50 sen.}$) for the *demeton*-resistant strain fell from 163 to 35 within 27 months (58 generations). At this point *carbophenothion* (®Trithion) was inadvertently applied, the resistance index rose to 117 and after 41 months (89 generations) fell to 2. The *chlorfenson* resistance proved considerably more stable. After 20 months (42 generations) the resistance index fell from 131 to 60, rose to 123 after the *carbophenothion* treatment and after 33 months (71 generations) had reached 56.

KEIDING [473] described the *diazinon* resistance of Danish strains of the house fly *(Musca domestica)*. This resistance seemed to be controlled by two semi-dominant main genes. The homozygotes (R,R) showed a resistance index of 70, the hybrids one of 20. The main factor was the stability of this intermediate *diazinon* resistance, which is presumably heterozygote. The development of resistance proceeded in the following manner: presumably during the first years only the heterozygote resistance level was reached. When the pressure of *diazinon* fell, e.g. in winter, reversion of resistance took place. Reapplication of *diazinon* raised resistance to the old level or even higher. At the end of the season the population consisted of a mixture of resistant hetero- and homozygotes. Although the homozygotes disappeared in winter, re-treatment with insecticide stabilized the population as a mixture of homo- and heterozygotes at an index of 30 – 50.

When *diazinon* was withdrawn, slow reversion resulted but when other phosphate insecticides were used, the level of *diazinon* resistance was again rapidly established. Hence resistance to phosphorus insecticides is, in principle, no more readily re-established than is that to the chlorinated hydrocarbons.
After 4 years, *malathion* had attained a stable resistance index of 500–1000, and reversion was no longer observed.
KEIDING concluded that resistance to an insecticide A may fall substantially when a change is made to an insecticide B possessing a different mode of action. The stability of the resistance to insecticide A is a function of the time for which the insecticide is applied and of the degree of insecticidal pressure ("age" of the resistance). It depends upon adaptation to the environment of the resistant genotypes and also the heterozygotes. Reversion frequently leads to a heterozygote population with only a few highly resistant individuals, most maintaining the resistance level of the hybrids (often heterozygotes). It would appear to be a rare occurrence for a high percentage of resistant homozygotes to survive in nature if the pressure of the insecticide is not maintained. "Young" resistance regresses more readily and completely than "aged" resistance. If the heterozygotes are sufficiently susceptible, the partly reversed population may once more be treated with insecticide A, although it is likely that, after only one generation, the old level of resistance will again be established.
In a publication in 1967 KEIDING [474] provided ample evidence in support of the following conclusions: there is little possibility of restoring susceptibility to a pesticide once resistance has become established; a high degree of resistance may be maintained for a very long time yet diminish once the pressure of selection falls; the R-genes present after selection can survive many generations, often as heterozygotes, and provide for rapid adaptation to renewed pesticidal pressure; resistance is reestablished far more rapidly than at first selection, even when apparently complete reversion of resistance has taken place in a population.
In the main the results of KEIDING with *Musca domestica* confirm the results published by UNTERSTENHÖFER on resistance of spider mites to acaricides [982, 986].
It is desirable to have insecticides A and B with a negative correlation, i.e. insecticide B selects the genotypes resistant to insecticide A. Considering the small number of insecticides which are effective against resistant pests, in comparison to the far larger number of compounds that can be used against sensitive pests, then the chance of finding insecticides with negative correlation must be regarded as extremely small, at least with the statistically orientated methods of research usual today.
At the 6th International Crop Protection Congress in Vienna, STEINHAUSEN [922] described *formetanate* [N-methyl (3-(N′,N′-dimethyl aminomethylene-

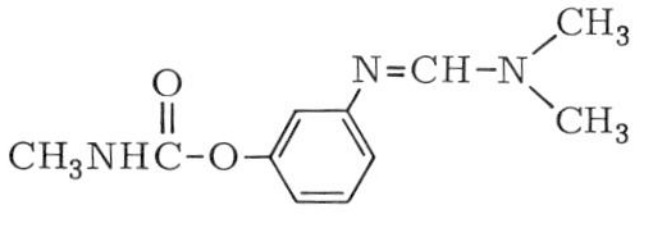

formetanate

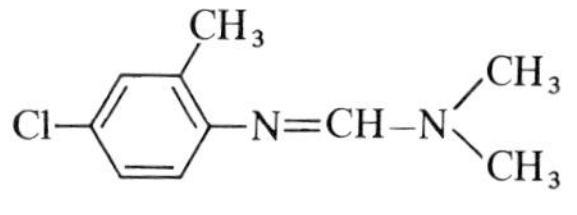

chlorphenamidine, ®Galecron

imino)-phenyl carbamate], synthesized by the firm of Schering in about 1962 [718], as an example of a compound with negative cross-resistance ("resistance-induced enhanced susceptibility").

Formetanate is said to be 4–12 times more active against various strains of organophosphate-resistant spider mites (*Tetranychus* spp.) than against normal sensitive laboratory strains. It remains to be seen whether, under field trial conditions and after several *Tetranychus* generations, these findings can be confirmed.

Similar effects have been reported for the structurally analogous *chlorphenamidine* [N,N-dimethyl formamidine N′-(2-methyl 4-chlorophenyl)], CIBA C-8514, Schering 36268, ®Galecron [241]. The LC_{50} ratios between the Leverkusen S and R strain of *Tetranychus urticae* are:

410 for *oxydemeton methyl*
52.6 for *parathion methyl*
0.0093 for *chlorphenamidine*

It can be regarded as proven that the phenomenon of negative cross-resistance is not restricted to phosphorus-free compounds. The decisive factor is that resistance-specific biochemical factors are inhibited. This is also possible within the organophosphates. Nevertheless, objective investigations are necessary into the physiological activities of the R^3O-group in the general formula (I), for it seems somewhat unlikely

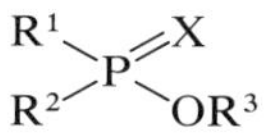

(I)

that all the compounds cited Section 3.2 and 3.3 act only as purely phosphorylating agents. As far as the organophosphates are concerned, there are no reports in the crop protection literature, in toxicological or biological papers, of the additive effects exerted by R^3OH, after the enzymic hydrolysis or inhibition; in many cases these effects may influence the specific spectrum of activity of a compound.

In practice the reverse is usually true, i.e. insecticide B lowers or stops the reversion of the resistance to insecticide A. Since, from a genetical point of view, reversion is the result of selection against the resistant genotype, and in addition of a "dilution" of the population by immigration of susceptible individuals from untreated populations, Keiding suggests that for practical crop protection, only those areas should be sprayed where it is absolutely essential. He also believes, at least in theory, that it should be possible to expose susceptible individuals at a time when the population is at a minimum, for example in winter.

There have been numerous cases of an intensification of certain pest populations after the use of insecticides. There are several possible explanations for this phenomenon. Root and Skelsey [779] attribute such outbreaks in part to the fact that the insecticides used act selectively against certain types of phytophagous insects, disturbing interspecific competition. This was confirmed by Bartlett

[73], who studied 59 different insecticides and attributed the abnormal increases of both aphids and mites in some cases to the suppression of competing pests. In other cases, however, a pest stimulation effect had to be postulated.
LUCKEY [581] named this effect "insect hormoligosis", meaning the stimulating effect of a stressor in subtoxic quantities on the growth of organisms existing under suboptimal conditions. The principle of hormoligosis is not limited to either insects or insecticides; other stressors can be climatic factors, nutritional factors, radiation, etc. A further conceivable mechanism is the elimination of the natural enemies of pests of which BARTLETT [73] gives numerous examples. If the great rapidity with which such upsets occur is taken into consideration, it may well be that we are concerned here only with a supporting effect.
It may also be taken as a certainty that the plant itself can influence the multiplication of pests. CHABOUSSOU [172] attributed the mass reproduction of sucking insects, such as mites, aphids, scales, etc., mainly to physiological and biochemical changes in the plant itself under the influence of crop protection agents.
For the dependence of a living organism upon its nutrition CHABOUSSOU suggested the term "trophobiosis". Factors which can influence the trophobiosis, for example, are the K:Ca ratios in the plant and, associated with this, the content of free amino acids and reducing sugars in the leaf. In the case of insecticidal phosphoric acid esters, another possible parameter is the phosphate content of the leaf. CHABOUSSOU also discusses the hypothesis that the resistance of many pests might depend upon the influence of trophobiosis.
It would not seem to be of much importance whether a particular effect is solely responsible for a pesticide-induced upset of pests; it is more likely that a complex of the above-mentioned mechanisms is usually involved. What would appear to be of importance is the conclusion that the use of very specific substances, both in chemical and in integrated crop protection, is not favourable. Furthermore, in the control of pests it is necessary to consider not only the target species but also the competitors and their dependence upon the host plant (trophobiosis).

I Damage to Plants

Abb. I.1

Damage to leek caused by free-living nematodes. (Photo: B. HOMEYER)

Abb. I.2

Spruce seedlings: top, plant damaged by free-living nematodes; bottom, healthy plant. (Photo: B. HOMEYER)

Abb. I.3

Carrots deformed by root-knot nematodes. (Photo: B. HOMEYER)

II Widespread Pests

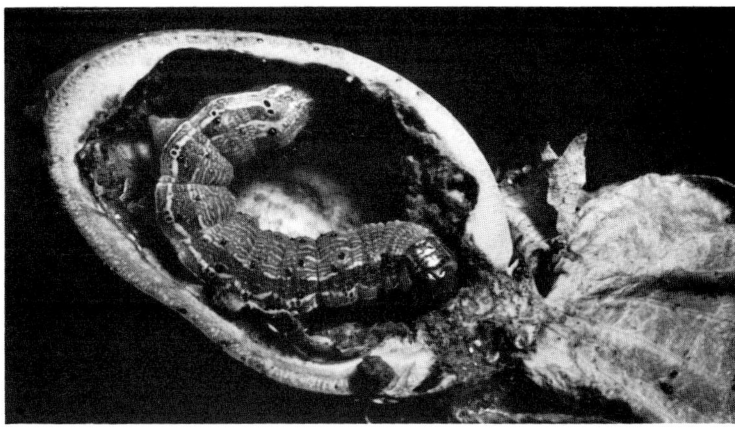

Abb. II.1 Cotton bollworm (*Heliothis zea*) in cotton boll (cut open). (Photo: v. EICKSTEDT)

Abb. II.2 Larva of Prodenia eridania on cotton plant. (Photo: v. EICKSTEDT)

II Widespread Pests (continued)

Abb. II.3 Damage due to larvae of *Laphygma frugiperda* feeding on corn plants. (Photo: v. EICKSTEDT)

Abb. II.4 Banana root damaged by nematodes. (Photo: P. KRAEMER)

II Widespread Pests (continued)

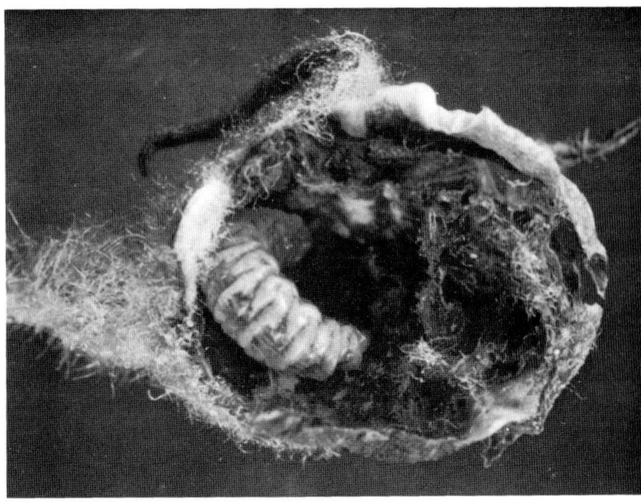

Abb. II.5 Larva of apple-blossom weevil (*Anthonomus pomorum*) in destroyed bud. (Photo: H. J. ROTH)

Abb. II.6 Parthenogenesis in aphid (*Megoura viciae*, Mexico). (Photo: H. J. ROTH)

II Widespread Pests (continued)

Abb. II.7 Cotton-boll weevil (*Anthonomus grandis*, "picudo"). (Photo: G. Müller)

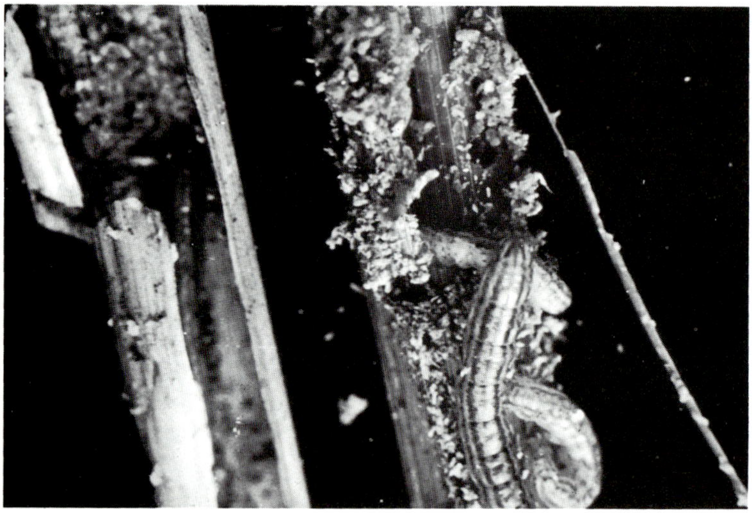

Abb. II.8 Rice-stem borer (*Chilo supp.*). (Photo: Nitokuno K. K.)

II Widespread Pests (continued)

Abb. II.9 Colorado beetle (*Leptinotarsa decemlineata*). (Photo: H. J. Roth)

Abb. II.10 Damage caused by locusts feeding (*Morocco*). (Photo: Pit Müller)

II Widespread Pests (continued)

Abb. II.11 Mediterranean fruit fly (*Ceratitis capitata*). (Photo: H. J. ROTH)

II Widespread Pests (continued)

Abb. II.12 Sigatoka disease (*Mycosphaerella musicola*) on bananas. (Photo: E. HAESKE)

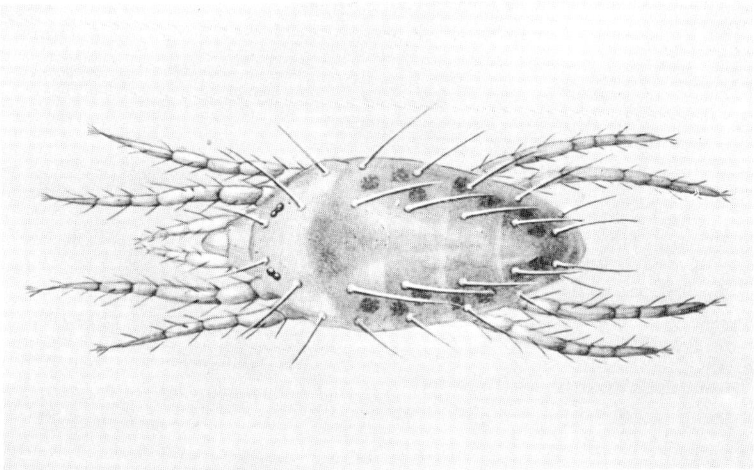

Abb. II.13 Spider mite. (Photo: Bayer-Pflanzenschutzcompendium)

4.5. Metabolism

The biochemical existence of a living organism depends upon the effectiveness of internal compensatory mechanisms which protect against external physical and chemical disturbances (the principle of homoeostasis). The object of the degradation of foreign substances is to convert them into a water-soluble, i.e. excretable form. This purpose is served by a relatively small number of basic chemical reactions, for example, oxidation which converts lipophilic thiono-compounds into the more soluble oxygen derivatives and thio-ethers into sulfoxides and sulfones. C=C double bonds yield epoxides. In the case of aromatic compounds their aliphatic side chains as well as O- and N-methyl groups are hydroxylated (e.g. tri-*o*-cresyl phosphate).
For example, ARIAS and TERRIERE [56], TERRIERE, BOOSE and ROUBAL [949] reported on the rapid hydroxylation of naphthalene and naphthole in *Musca domestica.*

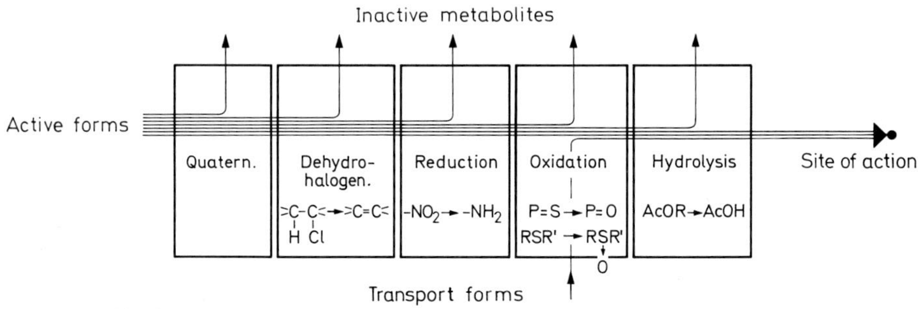

Fig. 28. Chemical reactions involved in metabolism. A molecule in its active form is detoxified to some extent on the way to the target. Other compounds (inactive transport forms) can be activated by metabolic reactions and therefore reach the site of action in an organism.

Reduction has been demonstrated with nitro-groups. It is interesting that nitro-reductases also exist in bacteria such as *Escherichia coli* in the intestinal tract of vertebrates. In addition to oxidation an important role is played by the hydrolysis of esters, special reactions being de-halogenation and de-hydrohalogenation. Finally, the excretion of degradation products can be effected after conjugation with "vehicles" such as glucuronic acid. In the following, examples of such reactions will be given only in so far as they involve the modification or degradation of organophosphates. A detailed account of the enzymatic metabolism of substances foreign to the body is given by BRODIE, GILLETTE and LADU for the years up to 1957 [133]; the period until 1963 is reviewed by SHUSTER [902].
SMITH [912] reviews the detoxication mechanisms in insects (position in 1961). JARCZYK [443] provides a comprehensive account of the literature until 1965 concerning detoxication reactions with special reference to pest control agents in insects.

a) Oxidation

Oxidation is one of the most important degradation reactions especially of the thiono-phosphates. It does not, however, result in detoxication; on the contrary, it enhances toxic action, thus having an activating effect. In vertebrates the oxidation reactions take place in the liver microsomes in the presence of $NADPH_2$ and oxygen [133], in insects in the fatbody, which according to that function may be regarded as the "liver" of the insect because the fatbody not only acts as a depot for metabolic products, but participates very actively in metabolic processes [484]. The presence of reduced NADP and oxygen is also necessary for oxidation in the fatbody of *Schistocera* sp. The oxidative activation of thiono-esters proceeds as follows:

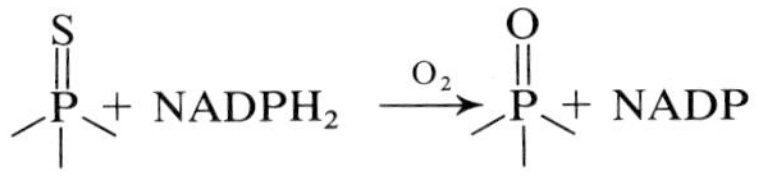
(1)

a reaction which would appear to be generally applicable to the thiono-esters. NAKATSUGAWA and DAHM [674] for example, were able to demonstrate the presence of *parathion*-activating enzymes in the fatbody microsomes of *Periplaneta americana*, which also require oxygen and $NADPH_2$ or $NADH_2$. The optimal pH for maximal activity is 7.5, the enzyme being inactivated by its own activating process. *Azinphos-methyl* is also oxidatively activated in the fatbody of *P. americana.*

NAKATSUGAWA and DAHM [673] investigated the oxidative efficiency in activating ®Gusathion *(azinphos-methyl)* to gusaoxon in cockroach tissues and found the following sequence:

malpighian tubules > fatbody > heart > nerve cord > ovary and caecum.

Diazinon seems to be converted to *diazoxon* mainly outside the nervous system of cockroaches [149].

KNAAK, STAHMANN and CASIDA [495] assume that the oxygen is converted by the peroxidase-hydrogen donor system into free perhydroxyl radicals, which form activated, rapidly-decomposing intermediates with the P=S grouping. This might offer an explanation for the otherwise inexplicable fact that, in an oxidation process, a reducing agent is required.

By oxidation of the thiono-group, the water solubility of an ester may under some circumstances be markedly increased: (for example: the solubility in water at 25°C of *parathion* is 24 mg/l, that of *paraoxon* 2400 mg/l, i.e. 1 : 100). Hydrolysis or alcoholysis are favoured and the toxicity increases. Phosphoroamidates, such as *schradan, dimefox, monocrotophos* or *dicrotophos* are also oxidatively activated, probably by direct hydroxylation of the N-methyl groups by means of $NADPH_2$ and O_2, as was demonstrated by FENWICK [278] in the oxidation of *schradan*

$$>N-CH_3 \xrightarrow{\text{Oxidation}} >N-CH_2-OH \xrightarrow{\text{Hydrolysis}} >N-H + CH_2O + H_2O \qquad (2)$$

"oxidative demethylation"
(Enzymatic reversion of the Eschweiler-Clark-reaction)

in the fatbody of the desert locust. Chelating agents reversibly inhibit this system, although the activity of the enzyme does not require the addition of metals ions. Oxidation favours the hydrolysis of *schradan*, the inhibitory action (I_{50}) increases by about 10^5 (Table 13) which can be demonstrated experimentally by oxidation with $KMnO_4$ [166]:

Table 13. Rate of hydrolysis, and inhibition of OMPA and OMPA-O

	Half-life	I_{50} (Chymotrypsin) molar
OMPA	100 years (pH 7)	$1.5 \cdot 10^{0}$
OMPA-O	18 hrs. (pH 6)	$2.4 \cdot 10^{-5}$

Not all organisms are able to perform this oxidative activation. Although the anthelmintic action of organophosphates on *Ascaris lumbricoides*, for example, may depend upon inhibition of AChE, thiono-esters are completely inactive because the microsomal peroxidases are entirely lacking. The fact that thiono-esters such as *fenchlorphos*, *coumaphos* or *famphur* can be used to control helminths in mammals must be attributed to the formation of the active anthelmintics by oxidation in the host tissue [502].
Even though the degradation reaction "oxidation" is the most important activating factor in the toxic action of the thiono-esters against vertebrates, there are nevertheless other degradation reactions with an activating effect, for example, the enzymatic dehydrohalogenation of *trichlorfon* to *dichlorvos*, the actual toxic agent. On the other hand, the removal of hydrogen chloride from DDT results in the non-toxic ethylene derivative in the sense of the homoeostasis principle. With many heterocyclic compounds protonation of basic centers may also result in an activation, as is indicated by the P-XYZ rule.

b) Hydrolysis

By far the most important detoxification reaction for the organophosphates is enzymatic hydrolysis. The basic work of MAZUR [610] in 1946 was followed by numerous investigations on this problem, primarily from a toxicological point of view (hydrolysis in the mammalian organism). The first substances to be investigated were P=O compounds, such as phosphoric and phosphonic ester fluorides, cyanides, pyrophosphates such as TEPP, or phenol phosphates like *paraoxon*. The active phosphatases were named after the substrate which they specifically degraded, for example, tabunase, sarinase, paraoxonase etc. A good review of this subject is the monograph of LOSHADKIN and SMIRNOV, which is available in the English translation of KOSOLAPOFF [578].
Hydrolases which metabolize organophosphates have not only been found in flies and mosquitos but also in many other types of insects, for example, in cockroaches, rice leaf hopper, in the intestine of lepidopterous larvae [442],

and also in the spider mite [607]. The "modified aliesterases" of resistant house flies [695, 696] should be here included.

The first investigations into enzymatic hydrolysis were mainly concerned with the demonstration and reactions of P=O ester-degrading enzymes, that is "oxonases". It was not until 1964 that attention was drawn to the "thionases", i.e. enzymes which directly hydrolyze thiono-esters [605]. They play a special role in the phenomenon of resistance.

According to "Schrader's rule" insecticidal organophosphates are in general so constructed that one of the three ester groups is more susceptible towards solvolysis than the other two. In general the enzymatic hydrolysis proceeds therefore with the removal of the toxophoric phenol, enol, alcohol or mercaptan.

Experiments by JARCZYK [443] with two specific degradation enzymes isolated from caterpillar intestines have shown that the decomposition rate of compounds of the *parathion* series in 31 cases were dependent upon:

a) the chain length and branching of the alkyl substituents in the alkoxy groups,
b) the type of ester (PS, PO, phosphate, phosphonate), and
c) the substituents in the aromatic ring.

Other mechanisms of hydrolysis are known, for example, degradation by the so-called malathionase. In the case of *malathion*, it is not the phosphoric ester but the carbalkoxy group which is hydrolytically attacked to yield the water-soluble "malathionic acids". This is the primary reason for the selective action of *malathion*, for this degradation mechanism is not normally available to insects in the same way as it is to vertebrates. On the other hand, malathionase does play a certain role in resistant insects.

In rat liver homogenates, SHISHIDO and FUKAMI [901] found a system which degrades compounds of the type of *parathion-methyl* to the corresponding demethyl compounds [304]. This enzymatic demethylation is apparently determined by the glutathion content of the tissue [305], other mercaptans are without effect. These Japanese authors assume that reduced glutathion serves directly as a methyl group acceptor in the manner described by JOHNSON [451]. ETO and OHKAWA investigated demethylation in the series of saligenin cyclic phosphorothiolates, confirming the said way of detoxification [271]. As might be expected, desethylation is of less importance. The insect species *Bombyx mori* and *Xylotropus dichotomus* as well as *Chilo suppressalis* (rice stem borer) [303] possess, in the intestine, a comparable demethylation system, but with a much reduced activity. The authors just cited consider demethylation a possible primary step in the degradation of thiono-esters. It might, therefore, also be regarded as a competitive detoxifying reaction to activating oxidation. Unfortunately, nothing is known in this respect concerning sucking insects and spider mites, but the demethylation reaction may explain why there is such a marked difference between biting and sucking insects as regards the spectrum of activity of the dimethyl esters.

The mammalian toxicity (for mice) of cis-*mevinphos*, of trans-*mevinphos*, cis and trans-Bomyl have also been related to glutathion dependent demethylation [659, 660].

The most toxic product, cis-*mevinphos*, is demethylated *in vitro* by mouse liver homogenate, while trans-*mevinphos*, cis- and trans-Bomyl are normally hydrolyzed at the phosphorus-enol ester bond, which means a more rapid detoxification than in the case of cis-*mevinphos*.

The alkylating properties of some organophosphates have been considered in connection with the mutagenic or carcinogenic activity of alkylating substances. Unfortunately, most such investigations have been carried out *in vitro*, e.g. the methylation of DNA by *dichlorvos*, which was described by LÖFROTH [553]. The results can only be expected to be conclusive, if it is possible to demonstrate that the system investigated *in vitro* can be reached by the alkylating compound also *in vivo*. For example, LÖFROTH found only 1% methylation of guanine to 7-methyl guanine after 66 hours. Under *in vivo* conditions it is impossible to exclude competitive attack of SH groups, for instance, which are much stronger acceptors than amino groups. Furthermore, such methylating compounds, being rather sensitive to hydrolysis, can by very easily detoxified by hydrolases on their way to the target. On the other hand, alkylating properties were demonstrated by means of reactions [749], which also operate with acylating compounds, as is known from other experiments. With respect to public opinion, which is rather uncritical where problems of environmental chemistry are concerned, any scientific result based on relatively limited evidence should be interpreted with great care.

In this connection it would be particularly interesting to carry out *in vitro* and *in vivo* experiments to determine whether the formation of characteristic complexes of demethyl insecticides with metal ions described by SCHRADER is significant for selectivity of detoxification and activation respectively [869]. *Schradan* also, is capable of forming complexes with a larger number of cations and these complexes, with the exception of the cobalt complex, are all more toxic in aqueous medium for the mouse (intraperitoneal) than *schradan* itself [449]. While this complex formation (reduction in electron density at the P=O oxygen, increase in electrophilicity of the phosphorus atom) enhances the acylating capacity (activating effect), it also increases the tendency to hydrolysis on the way to the site of action (detoxicating effect).

c) Combined Reactions

The difference between the enzymatic hydrolysis and inhibitory action of organophosphates consists in the fact that, during the detoxification reaction, the phosphoryl group is transferred enzymatically to water and, in the case of inhibition, to the serine alcohol of ChE respectively. Both reactions compete with each other. For a synthesis directly aimed at the production of the new insecticides, the molecule must be provided in a hydrolytically insensitive "transport form" from which the "active form" can be released in a slow reaction, e.g. oxidation. The time required for oxidation permits accumulation of the active form at the site of action. In an investigation into the relationship between systemic action and molecular structure, METCALF, REYNOLDS, FUKUTO and COLLINS [635] employed the "Principle of lethal synthesis", a conception derived by PETERS [727] in connection with the toxicity of fluoroacetamide. A lethal synthesis not only favours the systemic action (i.e. in the plant organism) in comparison to hydrolysis but also the toxic action (in the mammalian organism) or the insecticidal action (in the insect organism). JARCZYK regarded the toxicity of a compound as a relationship between enzymatic oxidation (activation) and enzymatic hydrolysis (detoxication) [443]:

$$\text{Toxicity} = \frac{\text{Enzym. oxid. (activation)}}{\text{Enzym. hydrol. (detoxication)}}$$

A very good example of this relationship is provided by the selective insecticidal action of isopropyl parathion (and n-propyl parathion) against house flies *(Musca domestica)* and honey bees *(Apis mellifera)* [633]. The low toxicity of isopropyl parathion for bees would appear, according to METCALF and FREDERICKSON [629], to depend upon a retarded oxidation *in vivo* due to the isopropoxy groups, this being 5–10 times faster in flies. Isopropyl paraoxon, the activated substance, no longer shows any selective toxicity.

For isopropyl fenthion the selectivity is less, since oxidation *in vivo* commences on the *p*-CH_3S group and, by way of the sulfoxide and sulfone steps, leads to the P=O compound, whereas one step suffices for the lethal synthesis of isopropyl paraoxon. Therefore since isopropyl fenthion is activated more slowly it is also detoxified more slowly. As might be expected the *p*-CH_3SO compound possesses a somewhat higher selectivity since further oxidation and thus hydrolytic detoxification proceeds rather more rapidly [629].

Furthermore, we are indebted to SUN [932] for the recognition of a quantitative relationship between toxicity (to insects) and rates of penetration, activation and detoxication of organophosphates. By "penetration" he meant the result of physical factors such as permeation, absorption, partition, excretion and transportation; "detoxication" represents chemical and enzymatic factors such as decomposition, metabolism and conjugation. "Activation" (oxidation) may be regarded as a special form of metabolism and for certain compounds (e.g. thionoesters, aliphatic and aromatic thioether groups) is an additional mechanism. After application of an insecticide to insects, all three effects proceed synchronously, but at different rates. The relationship of these rates to one another determines the apparent toxicity of a substance. The relationship of rate of penetration P to rate of detoxication D determines the accumulation of the substance at the target – a very important factor for toxic action. For thionophosphates or compounds such as *fenthion* or *demeton* an additional factor to consider is the rate of activation A.

If one simplifies the mechanism of action to these three stages, and if one assumes that their rates conform with first order kinetics, then it is possible to evaluate penetration, activation and detoxication by integration of their relative rates P, A and D between t_0 and t_1 or for other time intervals ($t_2, \ldots, t_n$). They obey the following equations:

$$\text{Penetration at } t_1 = \int_{t_0}^{t_1} \text{Pdt} \qquad (1)$$

$$\text{Activation at } t_1 = \int_{t_0}^{t_1} \text{Adt} \qquad (2)$$

$$\text{Detoxication at } t_1 = \int_{t_0}^{t_1} \text{Ddt} \qquad (3)$$

Eqs. (1)–(3) correspond to the hypothetical curves in Fig. 29.

Fig. 29A applies to P=S or other esters susceptible in activation: P, A and D increase rapidly with time and then decrease. The area abe (Eq. (1)) between t_0

and t_1 corresponds to total penetration, the area abd (Eq. (2)) to activation and the area abc (Eq. (3)) to detoxication. The shaded area acd represents the accumulation of the active ester (O=P compound) which also increases with time.

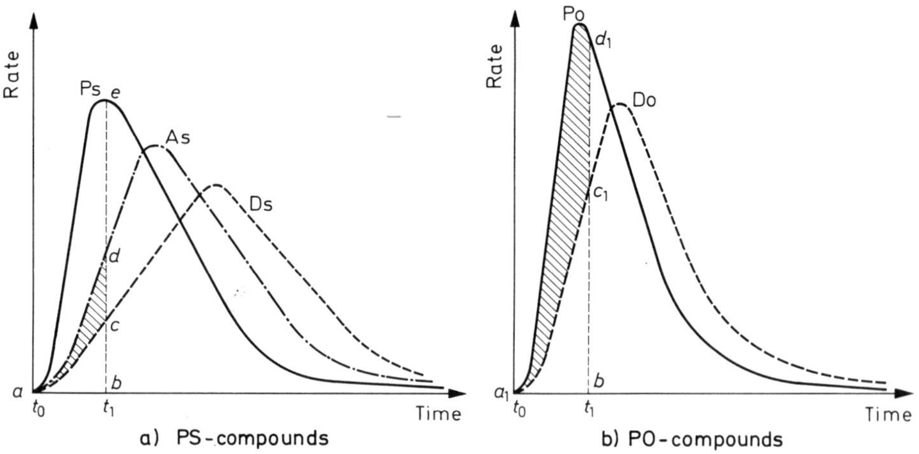

Fig. 29. Hypothetical curves for penetration, activation, and detoxication of insecticides [932]

Fig. 29B shows the penetration P_0 and detoxification D_0 of the O=P compound or of the activated ester. Between t_0 and t_1 penetration corresponds to the area $a_1 b_1 d_1$, detoxification to the area $a_1 b_1 c_1$. The accumulation of active substance at the target corresponds to the shaded area $a_1 c_1 d_1$ which is considerably larger than acd in Fig. 29A. This explains why the activated compound or O=P compound is faster-acting than the thionoesters or inactive transport forms. The actual relationships between the three factors may, however, be considerably more complex.

Using the method described by SUN in which he also suggested a graphical determination of accumulation at the target, it is possible in a first approach to evaluate differences in these three factors with respect to different insect species, susceptible or resistant strains of the same species, different properties of penetration and detoxification in different classes of chemical insecticides or their individual representatives. A disadvantage is the difficulty in determining zero time. Binding of the substances to proteins or partition into lipids give incorrect values. The advantages are that synergism, knock-down effects, rate of toxic action, species specificity or resistance can be better analyzed.

For the phosphoric acid esters given, the examples show that *parathion*, *paraoxon* and *dimethoate* penetrate rapidly into house flies. The maximal rate is very high and, after topical application, is reached rapidly. Differences in rate are determined by the specific properties of the insecticides, by the nature of the cuticle strain difference and difference in deposit. Fig. 30 shows that, in the case of house flies, *dimethoate* reaches the maximum rate of penetration in fifteen minutes (180%h); it is, however, detoxicated slowly. *Dimethoate* is, therefore, very rapid in action and is rather toxic. For a lower rate of penetration on the other hand, detoxification gains in importance.

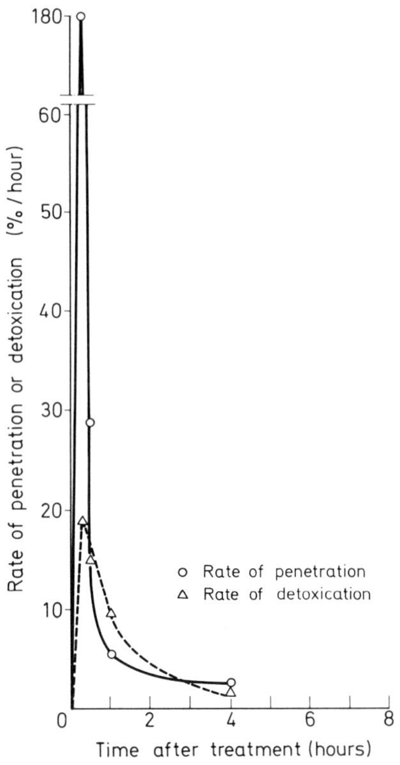

Fig. 30. Rates of penetration and detoxication of topically applied *dimethoate* in the house fly [932]

Table 14. Maximum rate of penetration of some insecticides applied topically to several species of insects [932]

Toxicants	Test insect[a]	Dose (μg/insect)	Calculated max. rate of penetration: Hours after application	Rate, %/h
paraoxon	House fly	0.004	1/12	320
parathion	House fly	0.004	1/12	290
dimethoate	House fly	3 μg/g (or 0.06 μg/fly)	0.25	180
dimethoate	Large milkweed bug	3 μg/g	0.25	120
dimethoate	Colorado potato beetle	3 μg/g	0.25	56
dimethoate	American cockroach	3 μg/g	0.25	40

Table 14 (abbrev.) illustrates a comparison of the penetration rates of various insecticides for different insects; these data were compiled by SUN from the literature (for references see original paper [932]).

d) Metabolism of Some Trade Products

The following paragraphs include a description of the degradation of certain of the more important commercial products from the series of organophosphate insecticides. It will be seen that the general rules for degradation permit the forecasting of the phosphorus-containing metabolites for an almost unlimited number of esters. Gradual differences are revealed according to the nature of the organism which performs the metabolism, i.e. plants, soil micro-organisms, insects, mites or vertebrates.

As mentioned on page 227, oxidation and demethylation begin the degradation process in compounds such as *schradan* or *dimefox*. Oxidation creates an electrophilic centre which reduces the stability of the molecule, but which increases its phosphorylating activity.

In cattle *parathion* is reduced first in the rumen [9, 899]. Another metabolic pathway suggested was:

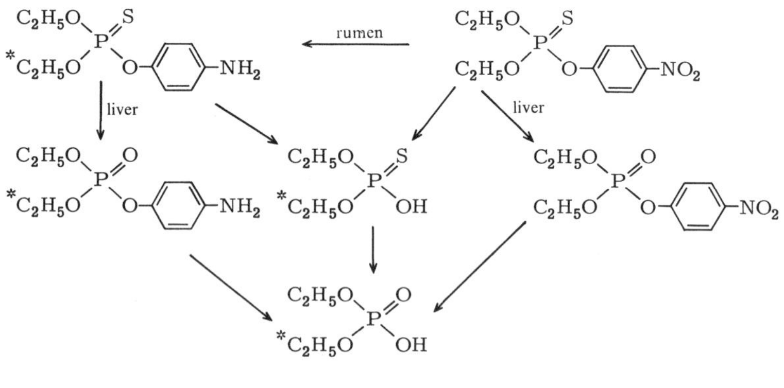

Scheme 14. *Parathion* metabolism in cattle

(The compounds marked with an asterisk are found as excretion products in the urine.)

Parathion, *paraoxon*, aminoparathion and aminoparaoxon circulate in the blood and also appear in small quantities in the milk. *p*-aminophenol is eliminated as glucuronide or sulfonic acid ester. Reduction also takes place in the case of *parathion-methyl*, EPN-O, Chlorthion etc. The *p*-amino-derivatives are much less biologically active than the nitro-derivatives, although they are more resistant to hydrolytic attack. According to KIERMEIER, KERN and WILDBRETT [482], peroxidases in milk and horseradish are, however, far more readily attacked by aminoparathion than by *parathion* itself. This is remarkable in so far as it is the hydrolases that are predominantly attacked by phosphoric acid esters. With these redox enzymes the I_{50} lies, in general, between 10^{-3} and 10^{-2} [482]. *Fenitrothion* [494, 990] is more rapidly metabolized in liver tissues than *parathion-methyl,* hydrolysis most likely taking place at the ester bond with the release of *p*-nitrophenol. This is probably one of the reasons for the selective toxicity of *fenitrothion.* Also in plants the peroxidases play a role in the *in vivo* metabolism of phosphorothionates.

®Colep is metabolized in the plant and animal organism with the removal of phenol [597]. The phenol is excreted by mammals as glucuronide, as they are apparently unable to oxidize phenol to CO_2 and water.

For *fenchlorphos,* the metabolic pathway [740] is the same both in cattle and in the rat, merely that, in the case of cattle, the metabolites are excreted more slowly.

The following scheme for the transformation and degradation has been proposed [740]:

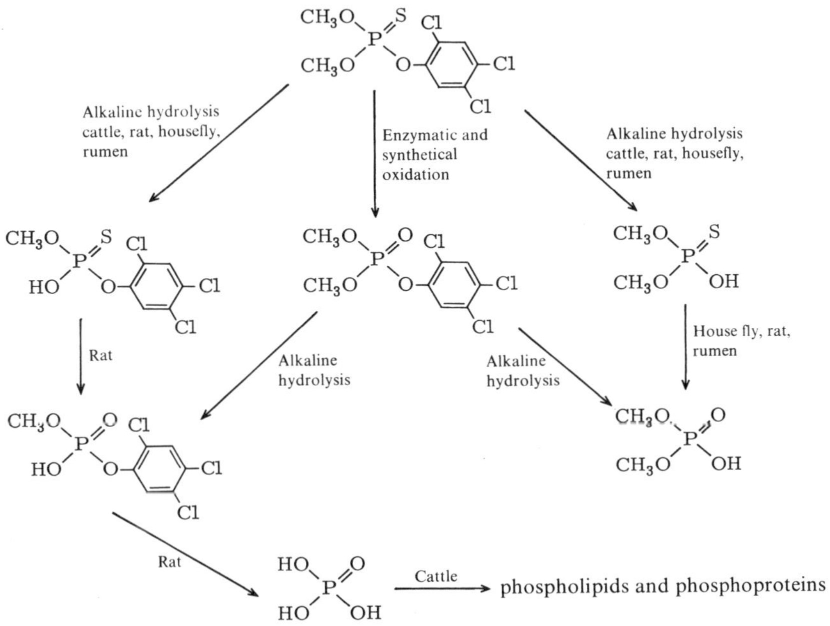

Scheme 15. Metabolism of *fenchlorphos*

In the house fly, hydrolysis takes place mainly on the phenyl group, in the rat on the methyl group. This might explain the selective toxicity for insects. On the basis of the excreted products, it is not possible to decide whether hydrolysis or oxidation takes place first. The oxygen analogue was not found among the metabolites, but its existence can be inferred from the high degree of cholinesterase inhibition in bovine blood and the brain of the fly.

Corresponding results were also obtained by STIASNI, REHBINDER and DECKERS [923]. They suggested the following scheme for the degradation of *bromophos* in the rat:

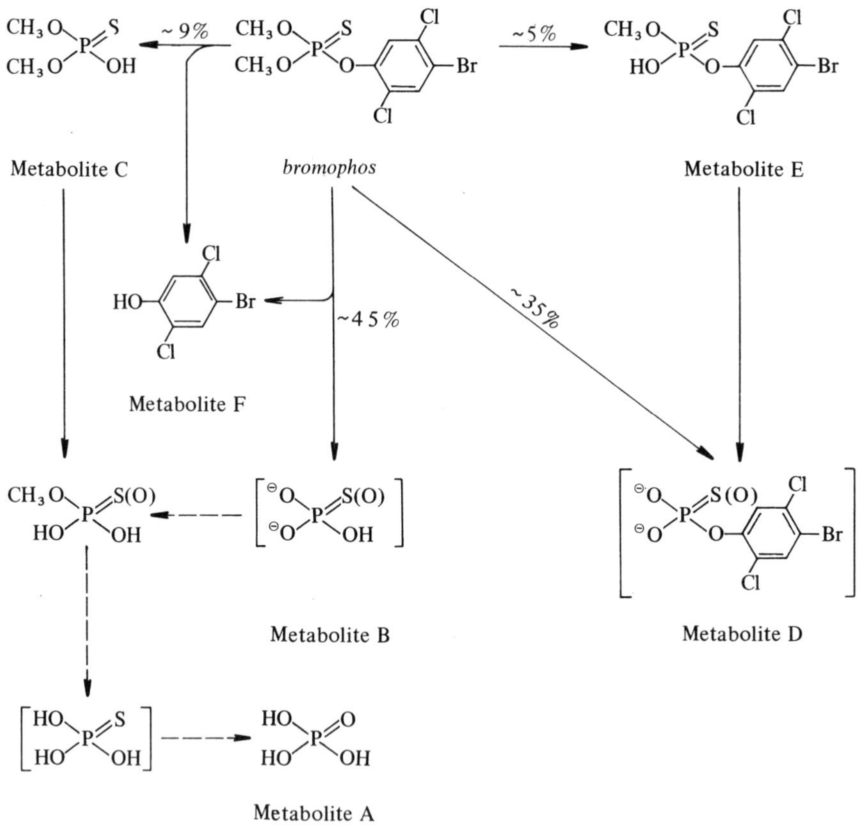

Scheme 16. Metabolic pathway as suggested for *bromophos* [923].

They were unable to demonstrate the O=P compound. It may well be therefore that the metabolism of *bromophos* is almost identical with that of *fenchlorphos*.
NIESSEN, TIETZ and FREHSE [679] investigated the behaviour of *fenthion* in the bean plant *(Phaseolus vulgaris)*. They examined the quantitative influence of light, plant enzymes and temperature on the conversion of *fenthion* and proposed the following scheme (see p. 237).
As for *demeton* [315, 319, 634, 725], *demethon methyl* [668], *disulfoton* [632], *phorate* [632] and *thiometon* [458], oxidation begins at the thioether grouping but is largely attributable to the influence of light. In contrast, oxidation of the thiono-group to the PO derivatives is effected by plant enzymes. The methyl-thio-isomers (VII)–(IX) which were also found, are obtained *in vitro* only by thermic isomerization.

The degradation of *fenthion* in insects follows in principle the same oxidation scheme. STONE [928] found that fenoxon could be demonstrated in the larvae of *Culex pipiens quinquefasciatus* only after exposure to high concentrations of *fenthion*, other metabolites found being fenthion sulfoxide, fenthion sulfone, fenoxon sulfoxide and fenoxon sulfone. Among the water-soluble products, dimethyl phosphorothioic acid predominates. Another conclusion is that, in susceptible strains of *Culex p. quinquefasciatus* larvae, *fenthion* is metabolized mainly by thionases and less frequently by oxonases.

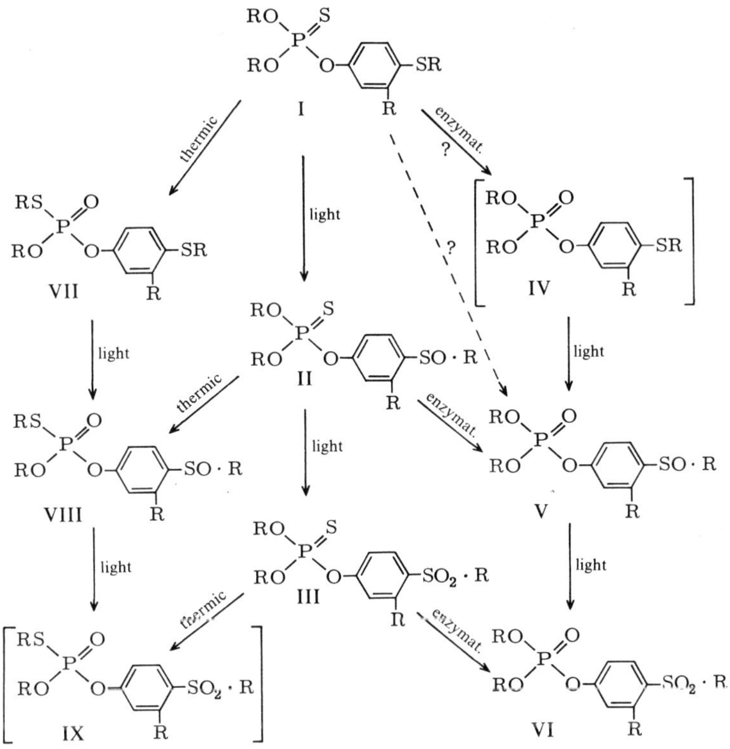

Scheme 17. Metabolism of *fenthion* in *Phaseolus vulgaris*

In blood-sucking arthropods such as *Stomoxys calcitrans* (stable fly), *fenthion* is in principle metabolized similarly, although oxidation to the sulfoxide proceeds somewhat more rapidly only when conversion to the oxygen analogue fenoxon has taken place [1040].

The metabolism of the *fenthion* analogue (I) (Bayer 9017) in calves after oral and dermal application approximates very closely to that of fenthion degradation. On primary oxidation the P=O compound would appear to be preferentially formed [1041]:

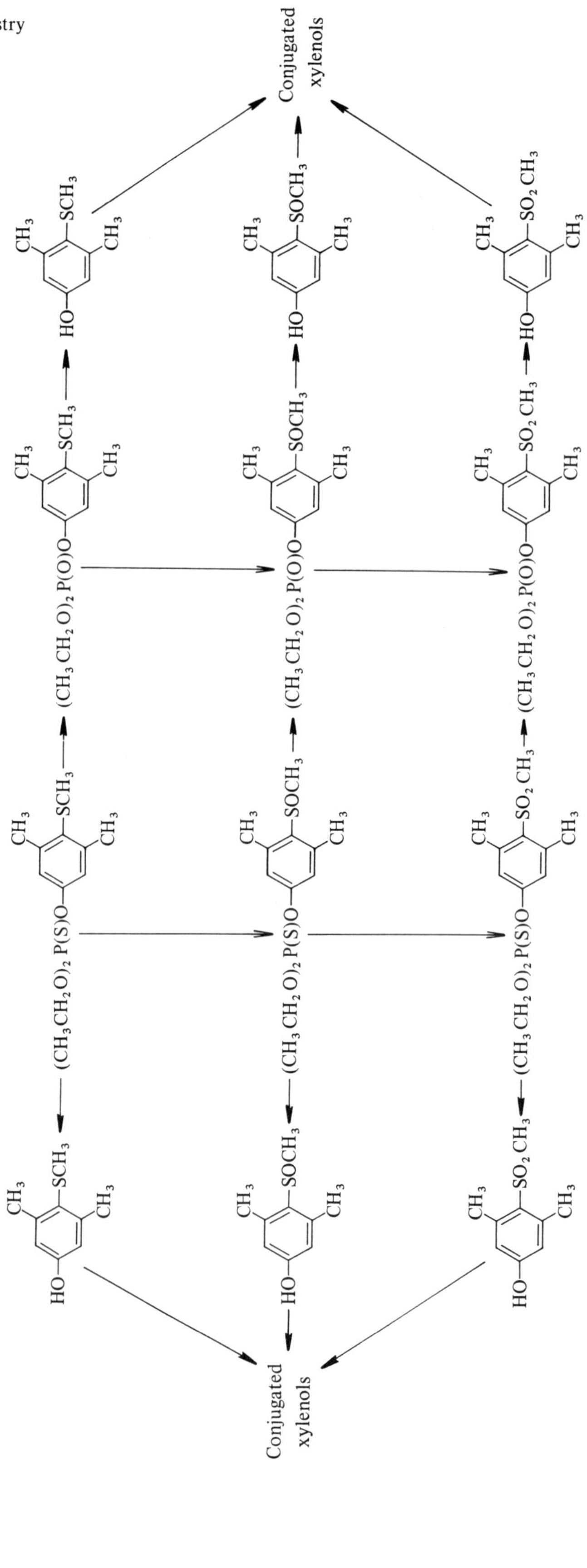

Scheme 18. Proposed pathway of metabolism of Bayer 9017 in calves [1041]

In the case of *coumaphos* the oxygen derivative which results from *in vivo* oxidation was demonstrated with certainty in tick larvae *(Boophilus microplus)* [781]:

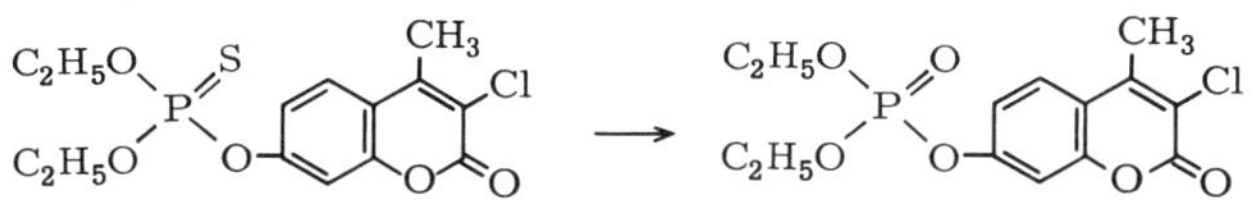

The application of sublethal doses of *coumaphos* resulted in a high proportion of water-soluble products with little P=O compound, a corresponding ratio of 1 : 1 being obtained two to three hours after application of lethal doses.
SMITH *et al.* [910] demonstrated that, in rats, ®Dursban is oxidized to the phosphate and is subsequently hydrolyzed. The following excretion products were identified:

Traces of unchanged ®Dursban
3,5,6-trichloro-pyridyl-2 phosphate
3,5,6-trichloro-2-hydroxypyridine.

Fish absorb the compound slowly, but metabolize it rapidly [910]. The metabolites then enter the water and are absorbed by plants and further metabolized. The following products are possible:

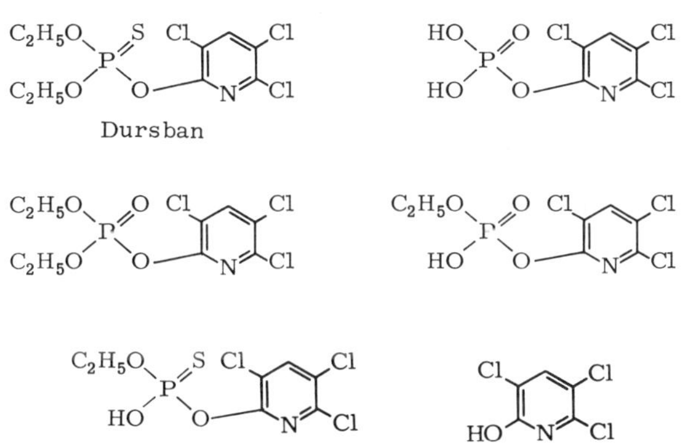

With the exception of the diethyl (3,5,6-trichloropyridyl-2) phosphate, none of the excreted products is a significant ChE inhibitor. Since, however, virtually no cholinesterase inhibition could be determined in fish, this compound can be excluded as metabolite. The main metabolite is the 3,5,6-trichloro-2-hydroxypyridine, which results from hydrolysis and is most likely metabolized further by de-halogenation and ring cleavage. In beans and corn ®Dursban is also hydrolyzed mainly to 3,5,6-trichloro-2-hydroxy-pyridine [910], with subsequent de-halogenation to di- and trioles. These can undergo ring cleavage and from CO_2. In addition to these, however, ethyl (3,5,6-trichloro-pyridyl-2) phosphate and (3,5,6-trichloro-pyridyl-2) phosphate are produced. In the soil ®Dursban is slowly hydrolyzed to the corresponding hydroxypyridine. In the alkaline range this heterocyclic phenol is rendered water soluble by salt formation and can then be taken up by the plant. It is still not known whether these metabolites result from UV or enzymatic degradation.

With *diazinon* and *thionazin* the primary degradation in the soil begins with hydrolysis at the ester bond to the heterocyclic moiety [333], the ring is then destroyed with the release of CO_2. In non-sterile soil, treated with *diazinon*, 2-isopropyl 4-methyl 6-hydroxypyrimidine could be identified with certainty as a hydrolysis product, while the hydrolysis product of *thionazin*, 2-hydroxypyrazine, was not positively identified in *thionazin*-treated soil. Instead of pyrazin, only CO_2 could be identified.

The heterocyclic hydrolysis products are apparently degraded still further by microorganisms, for these products can be demonstrated in sterile soil, from which no CO_2 release could be detected. The corresponding P=O insecticides were not found.

In the plant organism CO_2 is liberated from the heterocyclic ring. Acetoacetic acid and the corresponding amide do not appear as intermediates as might be expected from the synthesis [761]. So far the only case of conversion of a thiono-ester into the thiol ester *in vivo* has been demonstrated with *diazinon* in cereal extracts [760]. For this potent cholinesterase-inhibiting "isodiazinon" RALLS and CORTES proved the following structure:

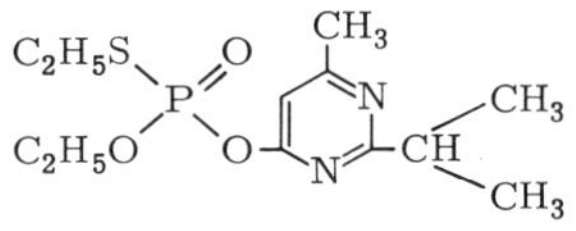

Isodiazinon

[O,S-diethyl O-(2-isopropyl 4-methylpyrimidinyl-6) phosphorothioate]. Because of its instability on storage, isodiazinon is unsuitable as an insecticide.

The metabolism of *famphur* (®Warbex) in the sheep and calf proceeds according to GATTERDAM, WOZNIAK, BULLOCK, PARKS and BOYD [327] primarily by way of O-demethyl *famphur* and *p*-(N,N-dimethyl sulfamoyl) phenol as follows (Scheme 19).

The N-demethylation could not be demonstrated in mammals but it was possible to demonstrate N-demethyl-famphur in mice and many insects [690]. The phenol component was excreted as the glucuronide [326].

In the metabolism of ®Maretin [155] it is mainly the oxygen derivative that is found in the faeces of cattle. The detoxification reactions taking place are: (1) hydrolysis of the O-ethyl bond, (2) hydrolysis of the O–N bond.

The resulting products are O-ethyl phosphoric acid and O,O-diethyl (thiono) phosphoric acid and others [122].

In the case of *phorate* degradation begins with oxidation of the thioether group to sulfoxide and sulfone [623]. Only then is the P=S group attacked with the formation of phosphate. (This same oxidation sequence was found for both *fenthion* and the *demeton* group.)

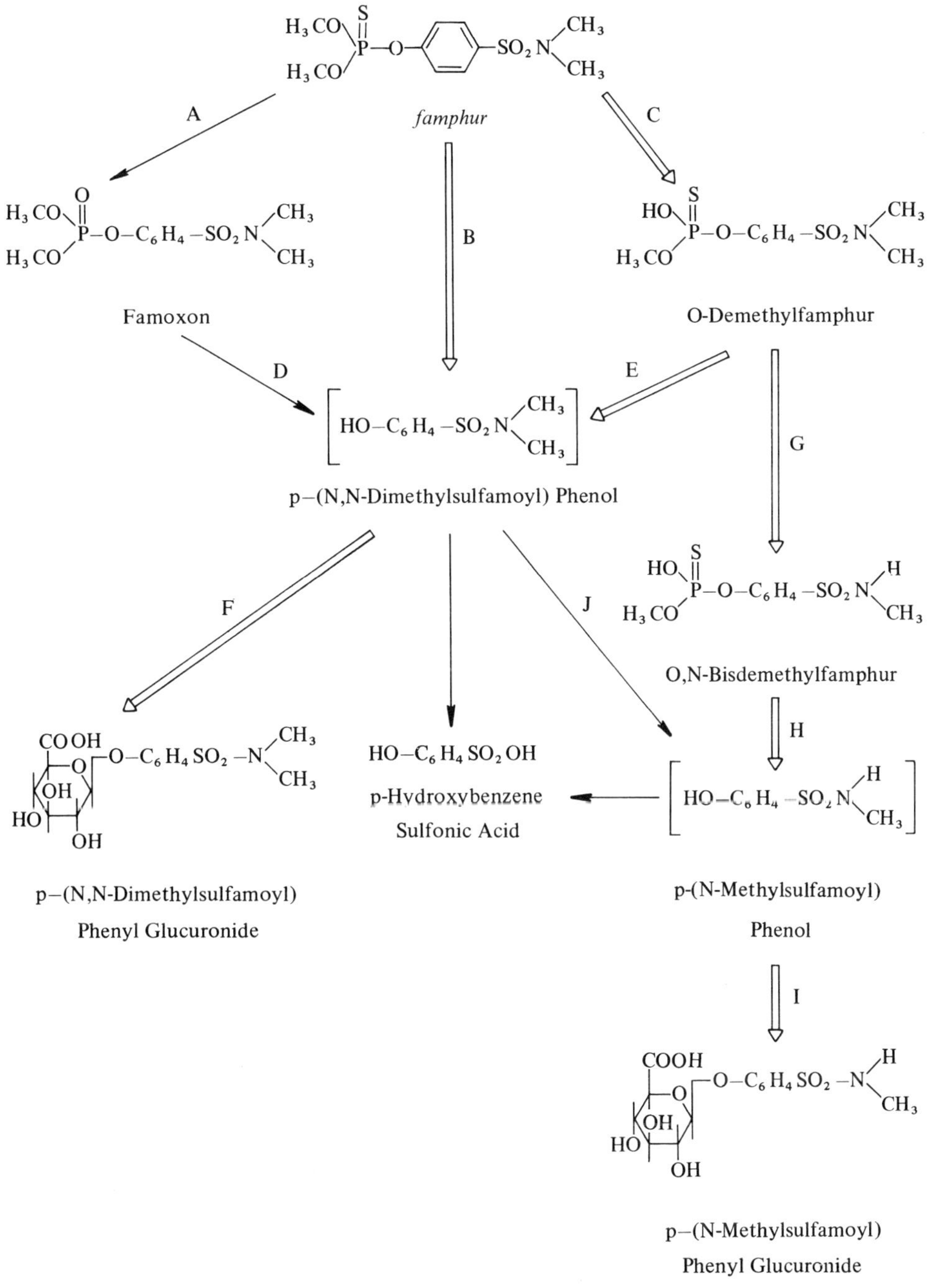

Scheme 19. Proposed metabolic scheme for *famphur* in mammals [327]

With the exception of the oxygen analogue (XIII) *phorate* itself (X)

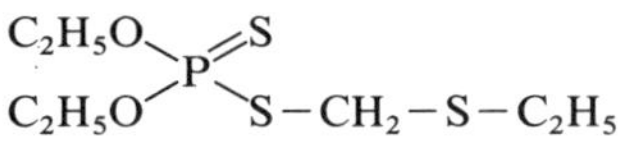
(X)

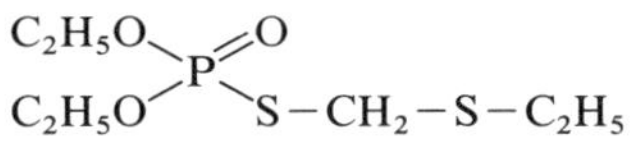
(XIII)

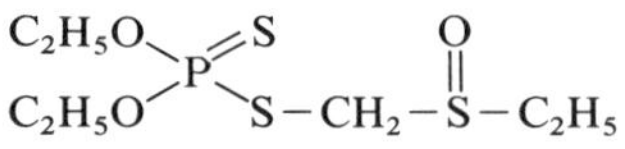
(XI)

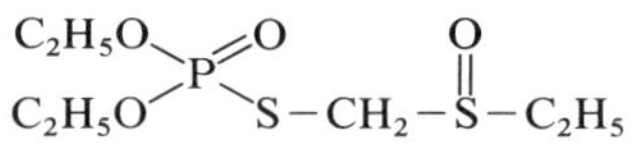
(XIV)

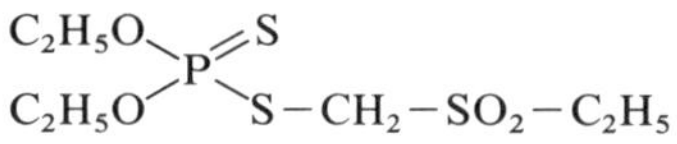
(XII)

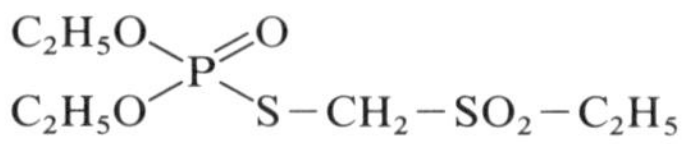
(XV)

was the most toxic compound for house flies, it exerted little ChE inhibiting action, whereas the sulfoxide of the oxygen analogue (XIV) exhibited only slight toxicity for house flies, but showed strong ChE inhibition. In the plant, sulfoxides and sulfones of *phorate* and its oxygen analogue result but not, however, the P=O compound (XIII) itself. Whereas, in plants, hydrolysis yields the dithiophosphate, and by oxidation the thiolphosphate analogue (XIV), in insects, only the former compound is found in significant quantities.
Both isomers of *demeton* are rapidly metabolized by three different mechanisms [725]. Firstly the thioether group is oxidized to sulfoxide and sulfone, secondly oxidation of thionophosphate to phosphate takes place, and thirdly hydrolysis at the P–O or P–S bond of the ethylthioethyl group to non-toxic diethyl phosphoric and thiophosphoric acids or the corresponding alcohols. These mechanisms all operate in plants, insects and mammals, except that in the mammal degradation proceeds faster. *Disulfoton* is metabolized in a similar manner. According to BULL [141] insects excrete both hydrolysis and toxic oxidation products, whereas in the rat only the hydrolysis product is eliminated slowly. In animals and plants oxidation begins on the thioether group. Oxidative metabolism in insects and plants is the same, although hydrolytic degradation in insects proceeds considerably more rapidly than in plants.
With *phosmet* [624] oxidation and hydrolysis result in a cleavage of the ring system, the compound is readily absorbed by leaves but is not distributed in the plant. The following substances are possible metabolites:

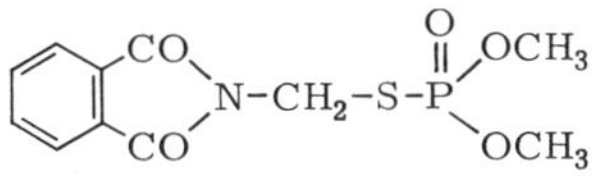

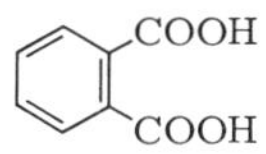

phosmet

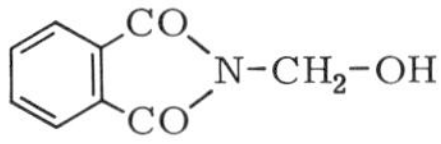

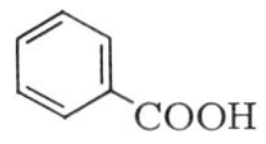

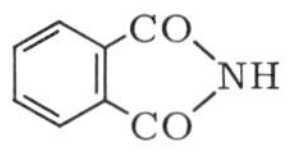

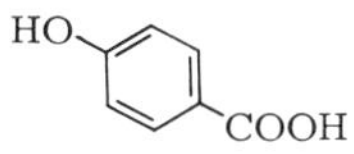

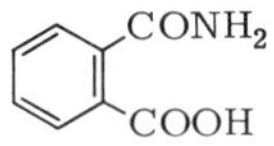

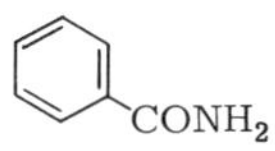

phthalimide acid

In the cotton plant hydrolysis is favoured at the expense of oxidation to the thiol analogue. The appearance of phthalimide and/or phthalic acid can be deduced from the decarboxylation product *p*-hydroxy-benzoic acid and the evolution of CO_2. Phthalimide itself cannot be detected, probably because hydrolysis leads directly to phthalic acid. The non-toxic compounds are excreted as such or as conjugated metabolites. In the soil, *phosmet* is very rapidly decomposed by hydrolysis [625]; its degradation is determined by pH and soil moisture content. If the microorganisms in the soil, are destroyed by soil sterilization, *phosmet* remains intact for longer. The hydrolysis products have not been identified but are presumably the same as in the plant.

In the rat *methidathion* is metabolized to CO_2 and three other substances which still contain the intact methoxy thiadiazolone ring and the methylene bridge. The metabolites [XVI – XVIII] were identified by synthesis.

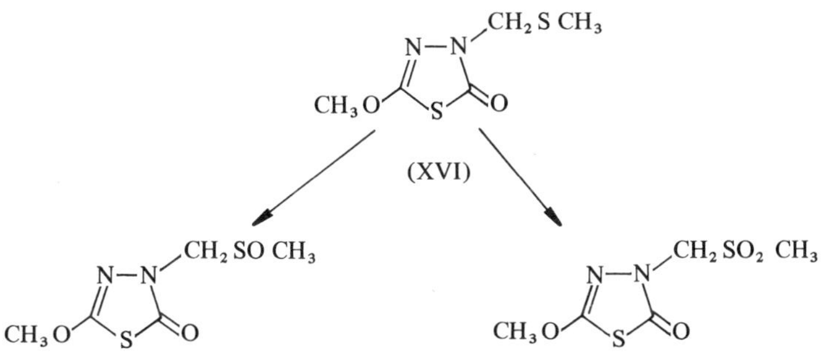

(XVI)

(XVII)

(XVIII)

It may be assumed that the thioether (XVI) is the first product, from which the metabolites isolated can readily be obtained by alkylation [783].
As might be expected, the oxygen analogue of *methidathion* is a metabolite in alfalfa, although it is not the main one for, in addition, the demethyl compound O-methyl S-[(2-methoxy-5-oxo-Δ^2-1,3,4-thiadiazolinyl-4) methyl] phosphorodithioic acid is formed. It is possible that the plant can cleave the heterocyclic ring thereby releasing CO_2, although the thiadiazole ring is rather resistant to chemical attack [169].
As UCHIDA, DAUTERMANN and O'BRIEN have reported [970], *dimethoate* is metabolized differently according to the species. In the sheep's liver, the action of amidases yields dimethoate acid (XIX), in the guinea-pig liver, dimethyl phosphorodithioic acid (XX) is produced, and in the rat and mouse liver, both compounds result. House flies and the American cockroach metabolized *dimethoate* considerably slower, which explains the selectivity of this insecticide. Groups which are responsible for the selectivity of a compound have been designated "selectophores". This term has been found applicable to *malathion* for example, which in the mammal is more rapidly metabolized at the carbethoxy group to the free acid than in the insect organism. On account of its ionic character, malathionic acid (XXI) like dimethoate acid (XIX) is weaker ChE inhibitor.

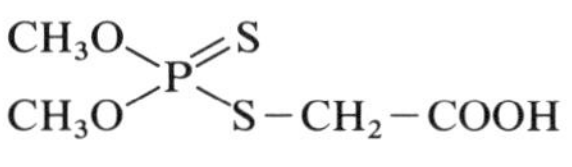
(XIX)

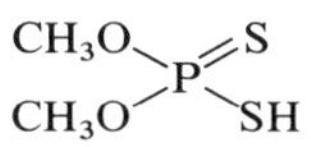
(XX)

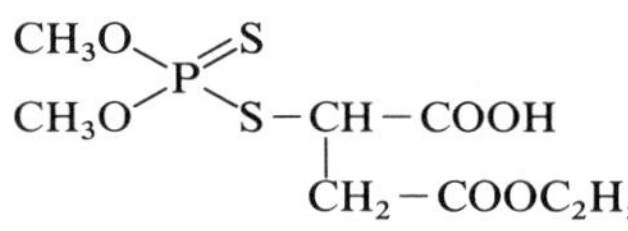
(XXI)

In many insects the degradation of *dimethoate* by amidases takes second place to the phosphatase reaction [971], an exception, for example, being the boll weevil where, as in mammalian metabolism, the first step involves an amidase reaction. The particular susceptibility of the house fly to *dimethoate* is attributable to three factors:

1) rapid penetration
2) substantial P=O production
3) the high susceptibility of the house fly ChE.

In the boll weevil, hydrolytic cleavage begins at the carbonyl-nitrogen and at the carbon-sulfur bond [145]. The oxygen analogue is also produced but is rapidly metabolized to non-toxic products. The metabolic pathway corresponds to that in mammals and plants. In addition to *dimethoate* itself the following metabolites have been identified:
Phosphoric acid, demethyl dimethoate acid (XXII), dimethyl phosphate, dimethyl dithiophosphate (XX), dimethoate acid (XIX) and the oxygen analogue (XXIII) (*omethoate*, "PO-Rogor").
Degradation in the rat occurs mainly at the nitrogen-carbon bond at high *dimethoate* concentrations (10^{-3} mol.), at small concentrations (10^{-6} mol.), at the carbon-

sulfur bond [970]. It is not possible to generalize that the amidase reaction determines the hydrolysis, for the distribution of amidases in the organism depends upon species and sex.

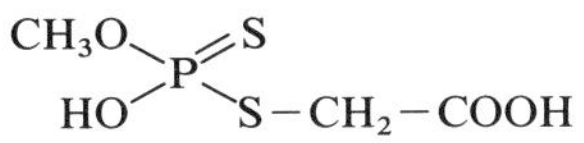

(XXII)

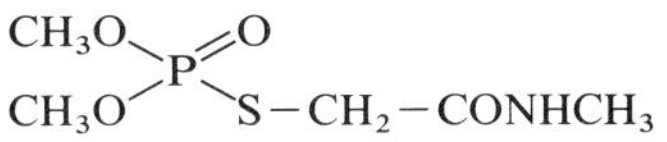

omethoate, "PO-Rogor" (XXIII)

In castrated sheep, for example, the amidase reaction is the only possible hydrolytic pathway. Here then, the amidase determines the toxic action and is the selectophore enzyme. In the sheep, cleavage takes place mainly at the carbon-nitrogen bond [174] but, in addition dealkyl derivatives of *dimethoate* result. In cattle and the rat, the metabolism appears to be similar.

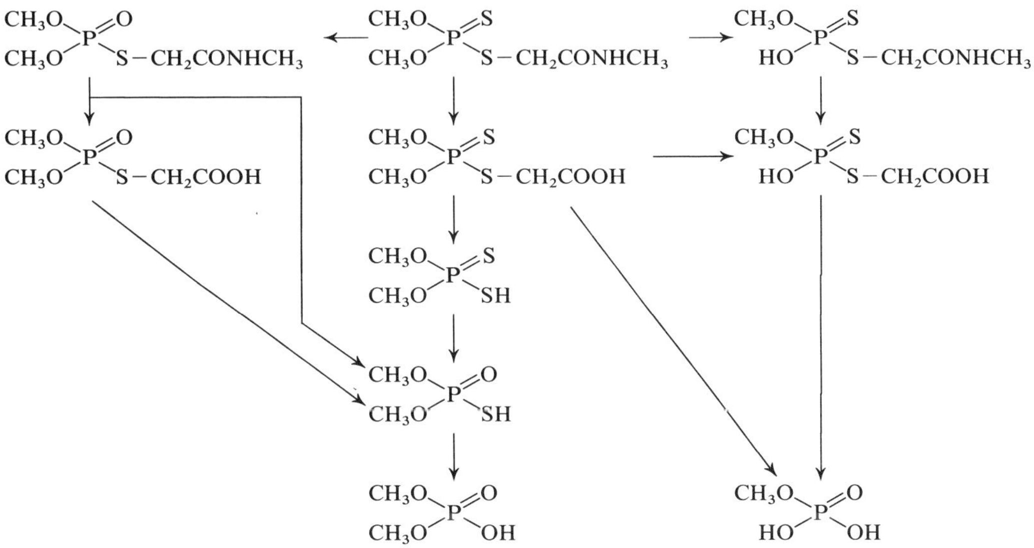

Scheme 20. Metabolism of *dimethoate*

According to BULL *et al.* [145], the action of phosphatases in plants and insects is more important for total hydrolysis than the action of carboxyesterases or of amidases. Only in the boll weevil (see above) does the carboxyamidase reaction play a comparable role. In the olive fruit, *dimethoate* undergoes oxidation to the P=O derivative [792], which is then hydrolyzed to dimethyl phosphoric acid and monomethyl phosphoric acid, i.e. non-toxic products.

MORIKAWA and SAITO [662] provided a quantitative comparison of the degradation of *dimethoate* and *vamidothion* in plants, insects and rats. The most important point was the fact that O-dealkylation is the main route in plants for both compounds. Only in the case of *dimethoate* was it possible to demonstrate carboxyl derivatives. Hydrolysis of the S−C bond of both substances is specific for rat liver homogenates. This gives the following scheme for degradation of the individual compounds:

dimethoate

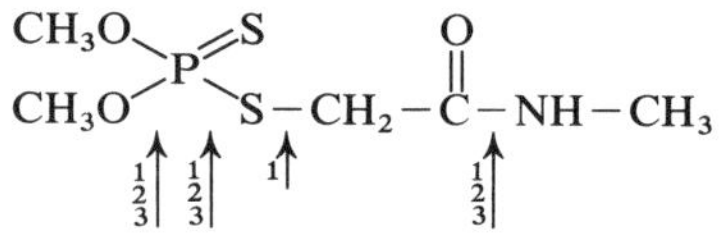

vamidothion

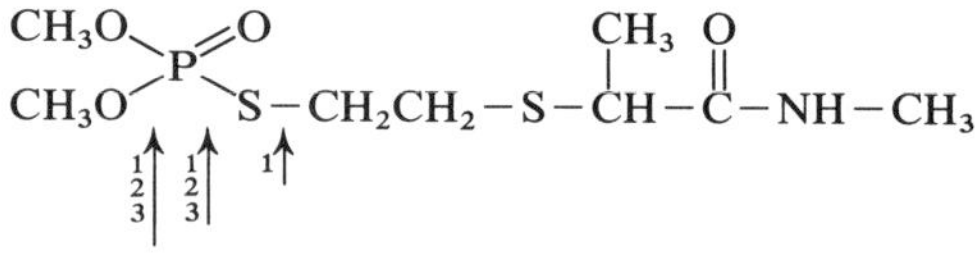

1 indicates attack by rat liver homogenates,
2 indicates attack by insects such as the house fly, American cockroach, rice stem borer, green peach aphid, green rice leaf hopper,
3 indicates attack by plants e.g. rice leaf, apple leaf, cabbage leaf.

Scheme 21 illustrates the degradation of *malathion*. Its metabolism differs very little in principle from that described for *dimethoate*. Here the selectophoric group is the carbethoxy radical and in mammals the selectophoric enzyme is malathionase present in the liver; the actual biochemical function of this enzyme is not, however, known. This point is dealt with in the section on synergism.

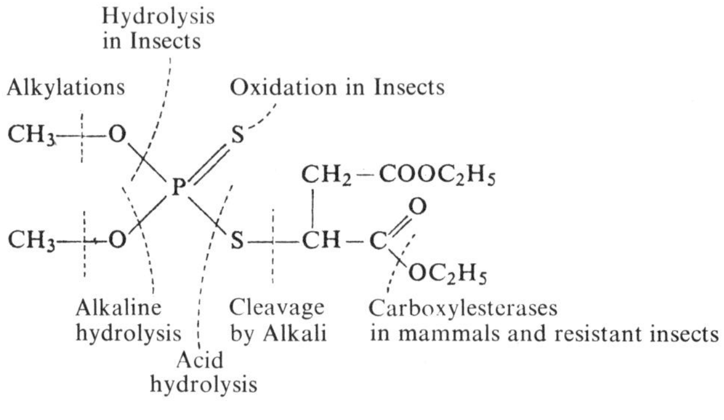

Scheme 21. Degradation of *malathion*

Dioxathion may be stored in the fatbody. In this case degradation also begins with the cleavage of the P−S and S−C bonds. Diethyl phosphorothioic and

-dithioic acid result, which may then be oxidized to the corresponding P=O compounds [738].
Trichlorfon is unstable in aqueous medium and can be metabolized by different routes [236]:

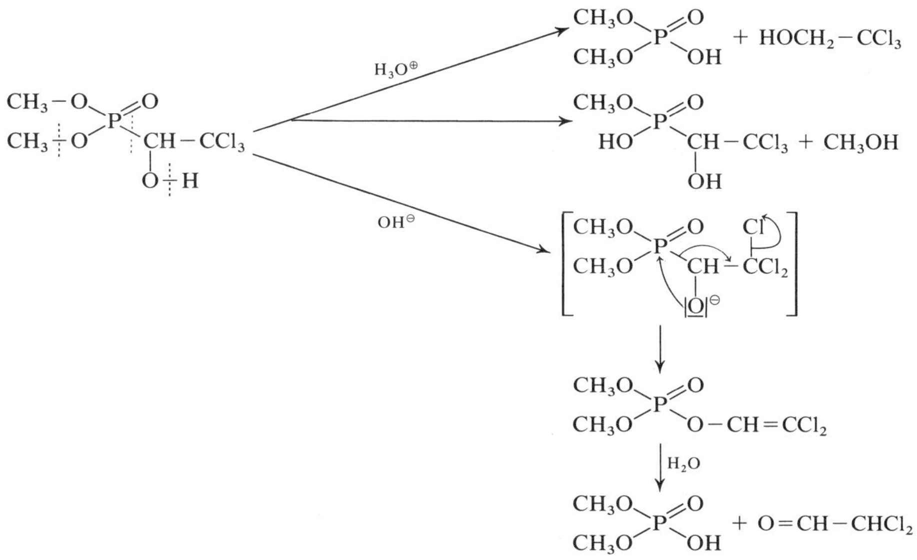

Scheme 22. Hydrolysis of *trichlorfon*

In acidic media, hydrolytic cleavage of the P–C bond and dealkylation are possible. In weakly alkaline medium, i.e. under physiological conditions, conversion to *dichlorvos* readily takes place. This activating dehydrochlorination of *trichlorfon* in the organism gives rise to the considerably more toxic *dichlorvos*. The relatively low toxicity of *trichlorfon* for mammals is attributable to the hydrolysis of the phosphonyl group and subsequent excretion of the original chloral group as trichloroethyl glucuronide [725].
In the mammal *dichlorvos* is presumably metabolized in the following manner [410] (see Scheme 23).
The resulting dichloroethyl alcohol is excreted in the urine as glucuronide. According to CASIDA [163] the C_1 fragments are eliminated in the form of unknown derivatives in the faeces and as CO_2 in the respiratory gases.
Tests with *naled* in rats and cattle have shown that hydrolytic attack begins at the $P-O-CH_3$ grouping [163], whereby additional metabolites, such as dichlorobromo-acetaldehyde result. *Naled* also reacts readily with sulfhydryl groups.
Enol phosphates containing carboxamido-groups also undergo degradation reactions such as oxidation, hydrolysis at the carboxamide group and at the enol ester bond.

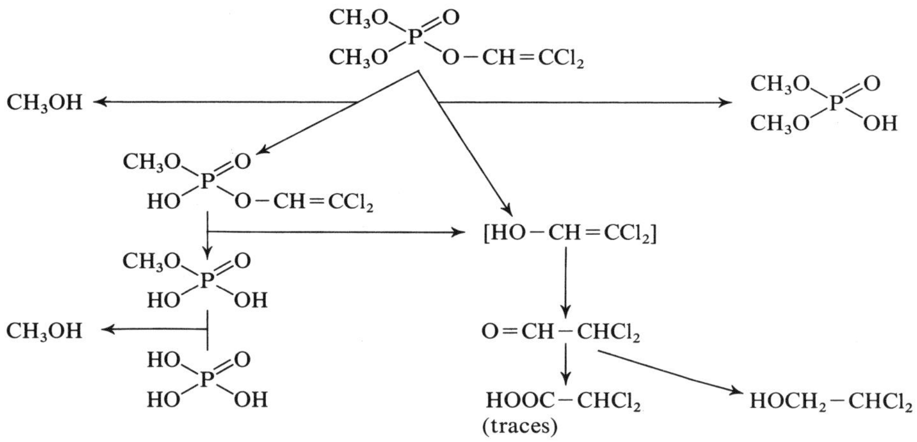

Scheme 23. *Dichlorvos* degradation

According to ANLIKER *et al.* [18], the degradation of *phosphamidon* proceeds in the bean plant as follows:

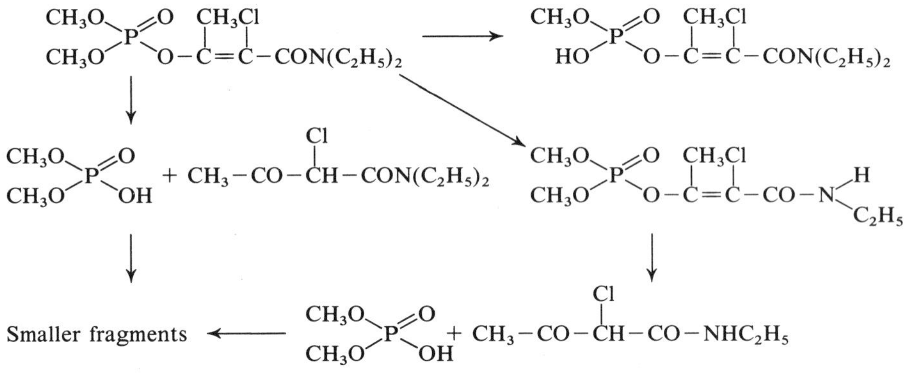

Scheme 24. Degradation of *phosphamidon* in beans

In comparison to the cis-isomer, the trans-isomer is more toxic for insects and is a stronger ChE inhibitor. Although both isomers are metabolized to similar non-toxic products, the rates of degradation differ. In plants and insects oxidative N-dealkylation of the cis-isomer is favoured [364]. The following diagram depicts the degradation of *phosphamidon* in plants and insects, the heavy arrows indicate identified products, and the light arrows unidentified products:

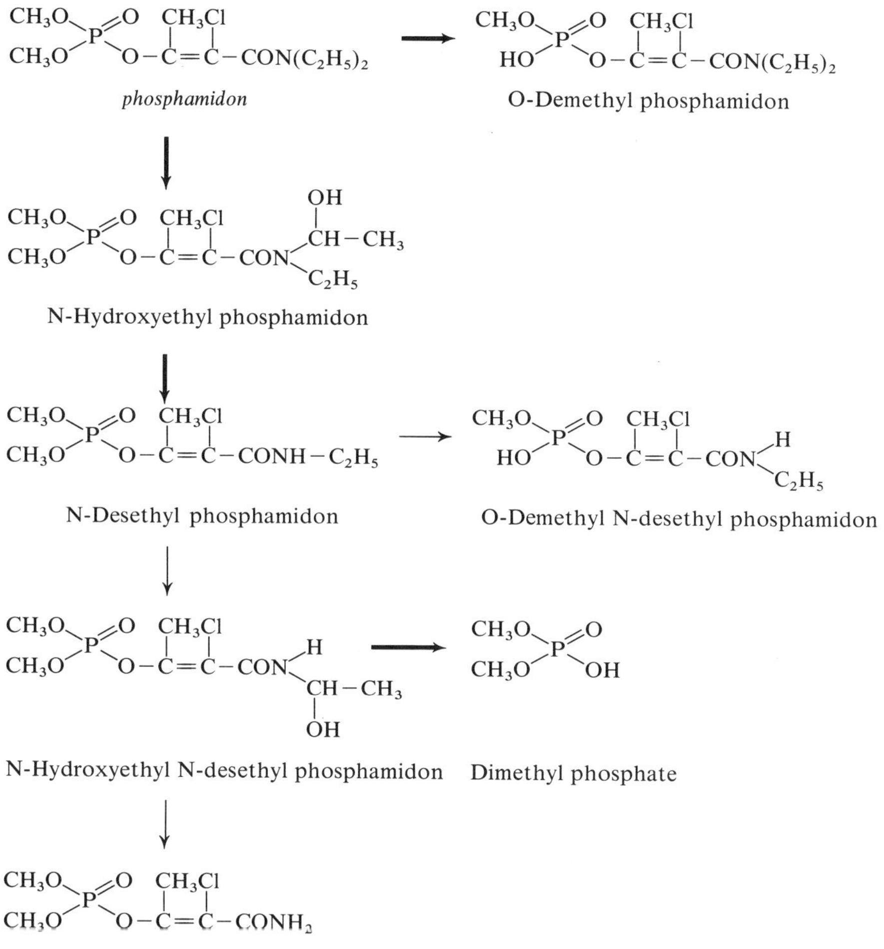

Scheme 25. Metabolism of *phosphamidon* in plants and insects [364]

The oxidation of *monocrotophos* to the N-methylol derivative is of less significance in cotton plants than in insects and mammals [551]. The main metabolic pathway is hydrolysis at the enol ester and at a methyl ester bond as follows (see Scheme 26).

Hydrolysis at the carboxamido-group does not apparently occur with *monocrotophos* but, in the case of *dicrotophos*, it plays an important role. With the latter, it can be concluded that the N-hydroxy-methyl compound is formed as an intermediate on the path to the acid. After initial oxidation, hydrolysis of *monocrotophos* proceeds rapidly in rats and insects but in cotton more slowly, hence the persistence of *monocrotophos* in this plant. Lethal synthesis also occurs (see p. 230).

According to Bull and Lindquist [142] the degradation of *dicrotophos* first of all involves the formation of the rapidly decomposing N-methylol derivative,

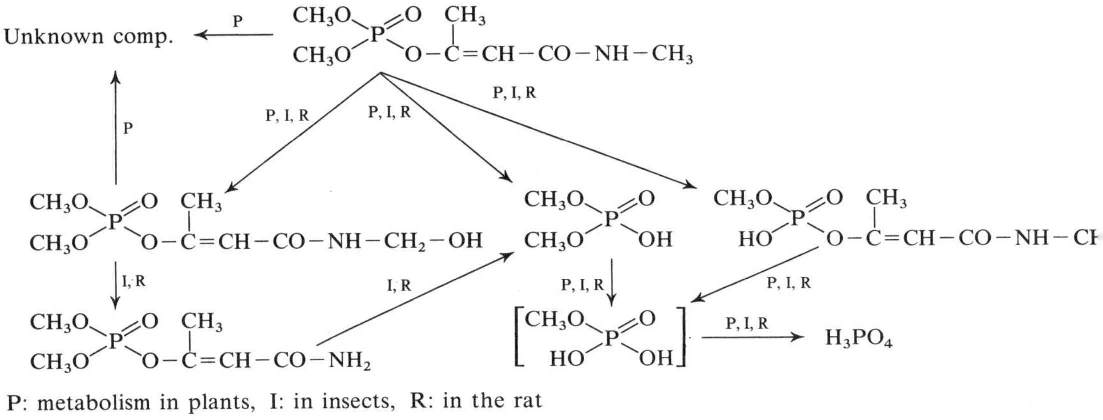

Scheme 26. *Monocrotophos* degradation [551]

produced in all biological systems by hydroxylation. Its further degradation can be pictured as follows:

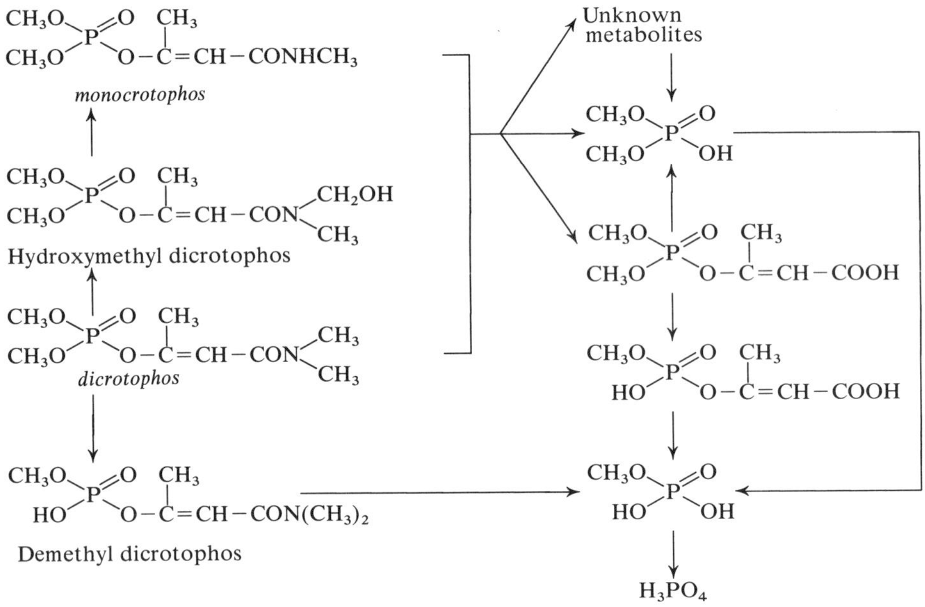

Scheme 27. *Dicrotophos* degradation [142]

In plants the hydrolysis products, dimethyl phosphate and demethyl dicrotophos are formed on the surface of the leaf. In the rat *dicrotophos* hydrolyzes at the

vinyl ester bond to dimethyl phosphate. In insects hydrolysis also takes place at the methyl ester group giving demethyl dicrotophos. The formation of N-hydroxymethyl compounds is a factor significant for the acceleration of hydrolytic degradation. This analogy to *schradan* metabolism is met with in all insecticides containing carbalkoxamido-groups.
Dicrotophos does not persist in the soil, presumably the metabolizing enzymes are the same or similar to those which have a detoxifying action in *Musca domestica* [144]. The degradation reactions are hydrolysis and oxidative dealkylation. *Sesamex*, a well-known pyrethrum synergist weakens or blocks both reactions [627]. The monomethyl derivative of *dicrotophos* is highly toxic to the boll weevil.
Dicrotophos can be transformed to *monocrotophos* by oxidative demethylation [143] (see Scheme 27, p. 250) The N-methylol compound appears as an intermediate. *Monocrotophos* may be further deaminated in a second hydroxylation process. N-methylol compounds can be demonstrated both in the rat and insect. The unsubstituted amide is not produced to any appreciable extent, since further degradation proceeds more rapidly than the formaldehyde cleavage.
According to CASIDA *et al.* [164] when *mevinphos* is fed to cattle, blood ChE is depressed, the substance being rapidly detoxified with the excretion of the known metabolites.

4.6. Toxic Action

The mechanism of action of phosphoric acid esters in the mammalian organism has already been discussed (see p. 164 f.). Attention must now be paid to the chemical aspects of some special effects of AChE inhibition from a toxicological point of view, such as the aging and reactivation of AChE.
In the first stage the phosphorylated AChE can be regenerated both *in vitro* and *in vivo*. This reactivatability decreases progressively until an irreversible stage of inhibition is reached in which substances such as 2-PAM no longer show a reactivating action. This is called the "aging" of AChE [395]. The first measurements of the time needed for the transition of an enzyme from the regeneratable into the irreversible phosphorylated form were made by WITTER and GAINES [1032]. The half-life for the aging of phosphorylated chicken brain ChE after poisoning with DDVP or *malathion*, i.e. for dimethyl phosphoryl ChE, is about 2 hours. The rate of the formation of the non-reactivatable form in terms of half-life depends upon the alkyl groups of the phosphoryl radical and increases in the order diethyl phosphate < diisopropyl phosphate < dimethyl phosphate.
BRADY and STERNBURG [127] were concerned with the problem of aging and recovery in insects (American cockroaches and house flies). They observed *in vivo* only a slow aging rate and also a slow *in vivo* recovery. This *in vivo* recovery of ChE could be attributed to the synthesis of AChE, for the rate of this process was independent of the alkyl substituents R on the inhibited cholinesterase [127]:

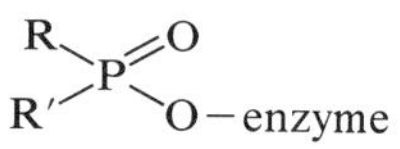

$R = CH_3-,\ C_2H_5-,\ n\text{-}C_3H_7-,\ i\text{-}C_3H_7-,\ n\text{-}C_4H_9-$

That little or no reactivation of inhibited ChE occurred *in vivo*, even before significant aging occurred as was found by BRADY and STERNBURG [127], may be explained not only by the synthesis of new ChE, but also by the above-mentioned reactivation of dealkylated ChE by means of five or six-membered rings, as well as by the reactivation of the dehydroalanine group in the aged enzyme.

The generally accepted chemical explanation for aging is that, in phosphoryl serine, one alkoxy group is hydrolyzed or dealkylated. The resulting diester must be hydrolytically inert. Thus, BENSCHOP and KEIJER [86] were able to show that the rates of aging of acetylcholinesterase [3.1.1.7] and butyryl cholinesterase [3.1.1.8] inhibited by a number of cycloalkyl and substituted benzyl methanephosphonofluoridates, correlated, with the rates of unimolecular solvolysis. Their results indicate that the rate-determining step in the aging reaction consists of the unimolecular fission of the C–O-bond in the alkoxy group.

In fly brain ChE, O,S-dimethyl phosphoroamidothioate (®Tamaron) causes substantial aging of the inhibited enzyme. The high toxicity of ®Tamaron against flies is attributed to substantial aging of the inhibited enzyme and only partial recovery in spite of low cholinesterase-inhibition [758]. ®Tamaron belongs to the group of organophosphates, the leaving group "acyl" of which is mercaptan. Presumably not only the phosphorylating properties of the molecule are of importance, but there is an additional action of the mercaptan at the target.

This finding correlates with the preparative behaviour of phosphonic acid esters and, to some extent, with that of the phosphoric acid tri- and di-esters. However, some questions remain unanswered:

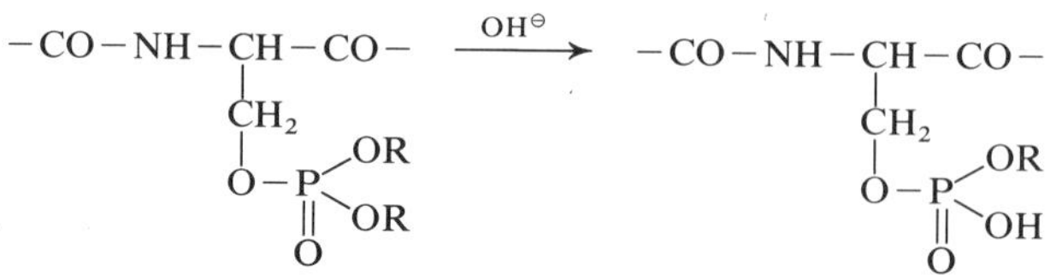

Scheme 28. Aging by dealkylation

If aging is interpreted as hydrolysis (PO or CO cleavage), then the dimethyl esters should, according to their greater susceptibility to hydrolysis, be more toxic than the diethyl esters. The latter are, in fact, less toxic and if this is explainable in terms of increased degradation on the way to the site of action, then a comparison *in vitro* should show that the irreversible form of the inhibited AChE appears more rapidly with dimethyl esters than with diethyl esters, as indeed the series by WITTER and GAINES indicates, at least for chicken brain.

Attention has already been drawn to the fact that during the depolymerization of the ribonucleic acids (see p. 32) the reactivity of phosphoric acid diesters increases when, in the course of the hydrolysis, participation of neighbouring groups makes possible the formation of five-membered rings. The spontaneous hydrolysis of phosphorylated acetylcholinesterase may be explained in a similar manner by the formation of a six-membered ring system. The consequence would be that, instead of aging, a reactivation of the inhibited enzyme could be expected, for the resulting monoester is unstable:

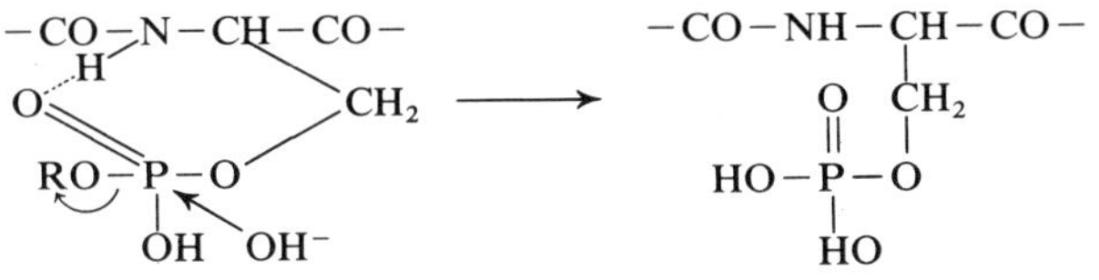

Scheme 29. "Six-ring hydrolysis" instead of aging

The paper of SHALITIN and BERNHARD [900], already mentioned, is of particular interest in connection with the problem of "aging". In their peptide model containing an imide structure they found, in addition to the normal hydrolysis of acetylated serine, up to 68 % β-elimination to give the dehydro-alanine derivative (I) (Eq. (1)):

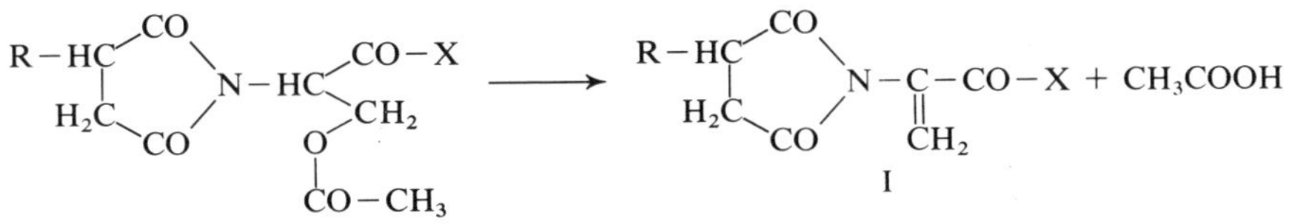

(1)

In this way they were able to confirm the work of RILEY, TURNBULL and WILSON [774], according to which diphenyl phosphoric acid esters of serine derivatives undergo rapid β-elimination with the formation of the dehydro-alanine compound:

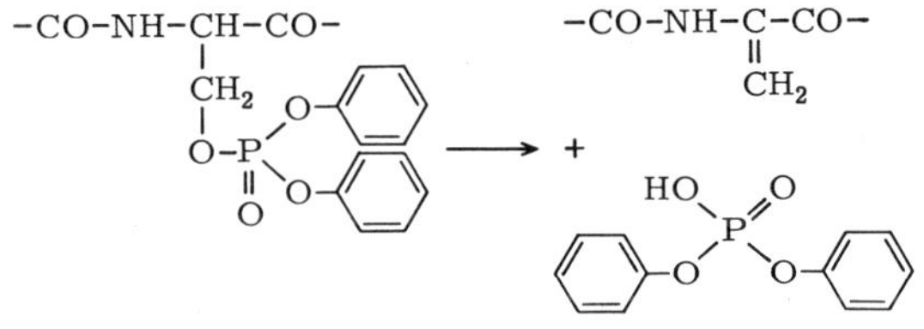

Scheme 30. Aging as β-elimination

These experiments with models of serine enzymes show that β-elimination is likely to be involved in the process of aging. This would suggest that there is

little chance of regenerating aged AChE hydrolytically, for basic catalyzing reactivators might provoke β-elimination and hence aging, instead of effecting recovery. The subsequent addition of water to the double bond of the dehydroalanine derivative is theoretically conceivable, but would have to be demonstrated experimentally ("Regrowth" of irreversibly inhibited AChE).
There now followed the development of a series of substances which were able to regenerate phosphorylated AChE at the first reversible stage. In as far as these reactivators deserve attention, they show the α-effect referred to on page 34. The hypothesis that it might be possible to regenerate phosphorylated AChE *in vivo* by hydrolytic reactions, must have given impetus to the investigation of simple oximes such as mono-isonitroso-acetone or diacetylmonoxime as reactivators [1026]. The first success *in vivo* was provided by the 2-pyridin aldoxime in quaternary form as the iodide (2-PAM) or methosulfate (P2S):

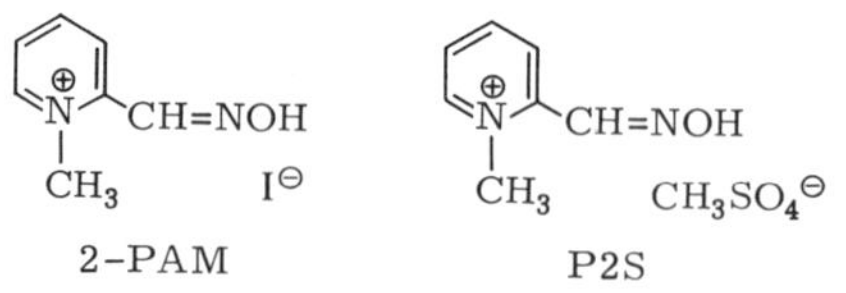

The simple oximes were effective only at concentrations which excluded their therapeutic use. The introduction of a positive charge on the molecule produced, among other things, the required adhesion at the site of action. The reversible inhibitor neostigmine served WILSON [1027] as model.

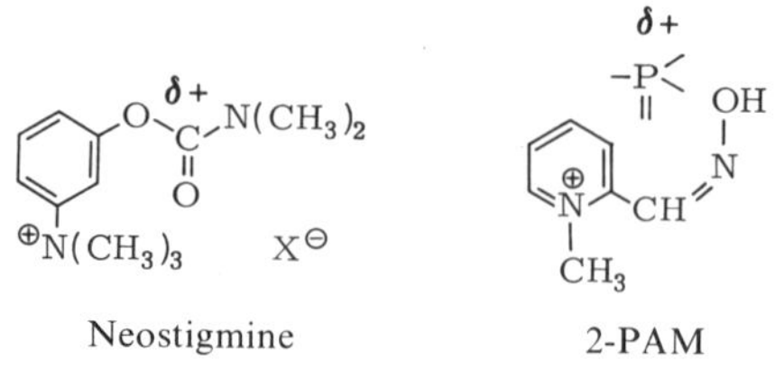

A great disadvantage of 2-PAM is that it does not penetrate the blood-brain barrier and is unsuitable to reactivate the inhibited AChE of the central nervous system.
Further study led to TMB-4 which was too toxic for therapeutic use [264]. LÜTTRINGHAUS and HAGEDORN [582] then developed the compound BH 6 (LüH 6, Toxogonin) [Oxy-bis-(4-hydroximino-methyl-1-methyl pyridinium) dichloride], in which the central methylene group of TMB-4 was replaced by an oxygen atom. ERDMANN, ENGELHARD and CLARMANN [263, 264] described this compound as less toxic than TMB-4, more active therapeutically than 2-PAM and more reactive than TMB-4. Toxogonin also has the advantage of being more active in the central nervous system [262].

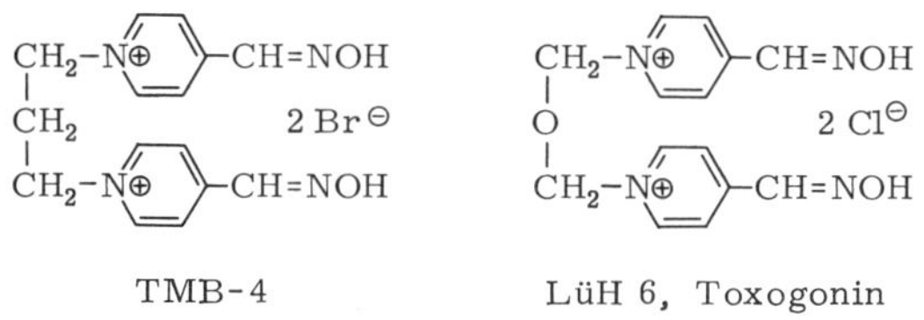

With compounds possessing two pyridine rings, it is evident that some importance can be attached to the bridge between the two rings.

In a paper by ENGELHARD, PRCHAL and NENNER [260] reviewing acetylcholinesterases, the mode of action of the reactivators is related to the enzymatic hydrolysis of acetylcholine in the enzyme-inhibitor complex. In the following kinetic scheme the Michaelis constant K_m is equated formally with the reactivation constant K_r:

$$EI + R \underset{k_{-1}}{\overset{k_1}{\rightleftharpoons}} EIR \xrightarrow{k_2} E + IR \tag{2}$$

$$K_r = \frac{[EI]\cdot[R]}{[EIR]} = \frac{k_{-1} + k_2}{k_{-1}} \tag{3}$$

where E represents the enzyme AChE, I the inhibitor (phosphoryl radical) and R the reactivator (e.g. Toxogonin).

The first step is the formation of an enzyme-inhibitor-reactivator complex (EIR) which decomposes to the free enzyme (E) and reaction products (IR).

The larger the value of K_r for a reactivator with a given blocking group, the poorer is the formation of the complex (EIR). The smaller K_r is for a reactivator, the more easily can it be regarded as a weak reversible inhibitor, which competes with cations for the anionic site in the enzyme.

Reactivation as the basic displacement of the phosphoryl radical on the alcoholic group of serine appears at first plausible, but the idea is contradictory for the first step in detoxication must lead to an oxime phosphate which has a re-acylating action [530, 531]. In so far as it is stable enough to immediate hydrolysis, the phosphorylated reactivator (IR) may under some circumstances be a more potent inhibitor than the inhibitor (I) itself. In such cases the toxicity is potentiated by the reactivator, as is found, for example, for the action of *dimethoate* on plasma ChE [266] and for mixtures of Sarin and 2-PAM or TMB-4 [531].

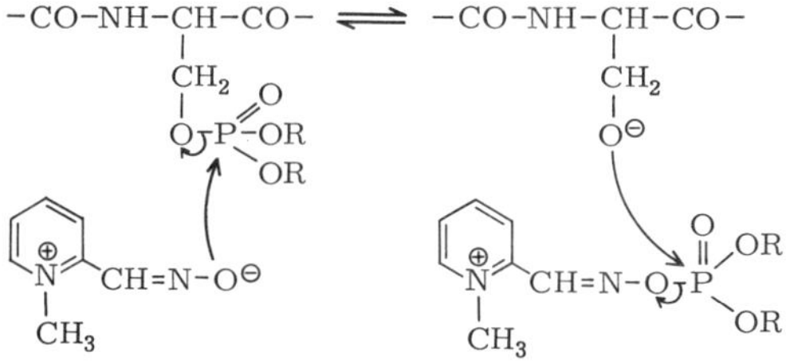

Scheme 31. Reactivation and re-acylation by PAM

Furthermore, the question remains: why is the ester bond to serine preferentially hydrolyzed? First, the bonding polarities of all three ester groups are not so very different that the preferential hydrolysis of a particular ester bond can be postulated. Second, the alkoxy groups are more readily accessible sterically than the serine ester bond. Third, for each seryloxy group there are two alkoxy groups which, taking into consideration all three factors, must present a considerable competition for the seryloxy group.

The success of this competition would involve the formation of a dealkyl ester at the site of action, which, according to accepted ideas, means the aging of the enzyme. Dimethyl esters ought to be more difficult to reactivate than diethyl esters. The only support in favour of the hydrolytic reactivation of phosphorylated cholinesterases by oximes with the formation of the dealkyl ester is the hypothesis of the six-ring mechanism discussed on page 253, which is in conflict with the concept which is at present accepted of aging via dealkylation. There are, in addition, supporting mechanisms conceivable by which anions of such oximes bring about the basic hydrolysis of a phosphoric acid ester before it reaches the target. Indications for this are to be found in a paper by KÜHN, FISCHER and LOHS [523]. If 2-PAM is reacted with readily alkylating organophosphates without catalysts by heating in acetone or alcohol, then one does not obtain, as would be expected, the phosphorylated oxime, but a salt of N-methyl pyridinium cation and the dealkyl ester. As esters, *trichlorfon*, DDVP, and *parathion-methyl* were used. The oxime structure itself remains intact. With potassium iodide itself, methyl iodide also results [869].

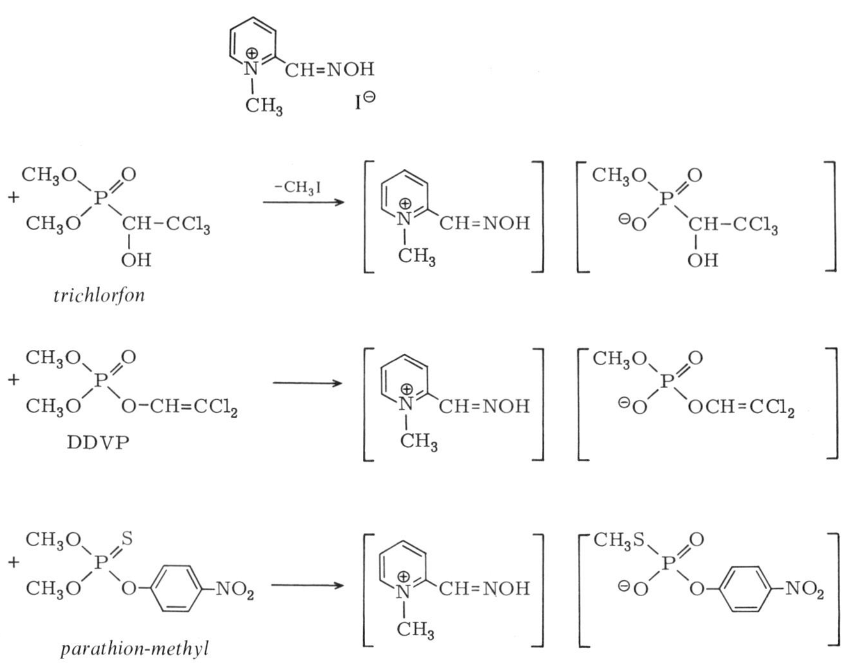

Scheme 32. Dealkylation by 2-PAM

The toxicity was measured as the LD_{50} in mg/kg for mice (intraperitoneally). For the dealkyl salts it is reduced, while for the pyridinium ion the toxicity is maintained. *Trichlorfon* is an exception which can be explained by the fact that the pyridinium ion is more toxic than *trichlorfon* itself. (*In vivo* the DDVP derivative is perhaps produced from dealkyl trichlorfon.) In the case of *parathion-methyl* the detoxifying effect is further enhanced by rearrangement to the methyl thiol compound.

Whether such mechanisms are possible *in vivo* and are associated with the iodide ion of 2-PAM, or whether other ions are also suitable, remains to be investigated. If the iodide ion is a deciding factor, then other salts, e.g. ammonium iodide substituted with long chain alkyl groups, should have detoxifying properties. Even though the dealkylation of toxic organophosphates is certainly not the main reaction of the oximes, it is, nevertheless, in principle possible. Without further intensive *in vitro* tests and model experiments it is difficult to decide thus point (cf. [265]).

Because there are certain difficulties in interpreting the reactivation of phosphorylated AChE as an aldoxime-catalyzed hydrolysis, particular consideration should be given to a paper by HAGEDORN, GÜNDEL and SCHOENE [362].

It is known that aldoximes can be converted to nitriles by reaction with diethyl phosphorochloridate in pyridine or also thermically in a suitable solvent [669]. HAGEDORN discussed the decomposition of the inhibitor-reactivator complex (IR) as the rate-determining step of the reaction. Here also the primary step would be the acceptance of the phosphoryl radical from the inhibited enzyme by the pyridine aldoximes which, under physiological conditions, are partially transformed into resonance-stabilized betaines:

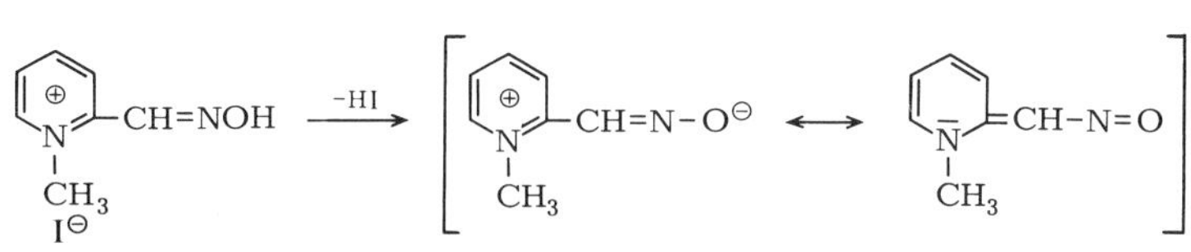

As secondary step, however, a β-cis-elimination to nitrile and diester acid is proposed.

The total reactivation might then be described by the following Scheme 33.

These mechanisms, in which reactivation is seen as a function of nitrile formation, are supported by the following findings:

1) Ketoximes, which after phosphorylation undergo only a pH- and time-dependent hydrolytic cleavage and do not form a nitrile, are virtually inactive as reactivators.
2) If UV spectra are taken of Toxogonin or TMB-4 and excess triethyl phosphate in buffer solution (pH 7.4–7.8), after a few hours curves are obtained which are identical with those of the corresponding authentic bis-(4-cyanopyridinium) salts.

3) As NENNER [676] found in an investigation into the kinetics of phosphorylation and reactivation of acetylcholinesterase, the pK_1-values of the more active reactivators are lower than the pK value of 4-PAM (pyridine-4-aldoxime-N-methyl iodide):

Compound	pK_1-values (37°C; 0.1 m KCl)
4-PAM	8.30 (8.34)
LüH 40	(8.12)
TMB-4	7.85 (7.78)
2-PAM	7.79 (7.68)
Toxogonin	7.50 (7.54) (Data in brackets according to STARK, see p. 259)

4-PAM no longer exhibits a nitrile spectrum, it is without reactivating action.

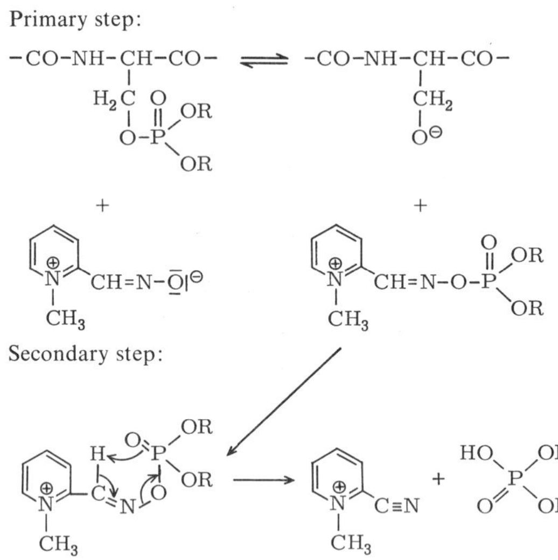

Scheme 33. Reactivation as β-elimination

LüH 40 [pyridinium-(4-aldoxime N-methoxymethyl) chloride], on the other hand, is transformed into the nitrile under the same conditions.

In summary, it may be concluded that, although the effect of electron withdrawal on the pyridine system is to increase the acidity of the oxime group in the 4-(even more in the 2-) position, this increase in acidity is of little consequence. What might, however, be decisive for the reactivating action could be that the same electron withdrawal on the pyridinium system causes acidification, i.e. loosening, of the aldehyde hydrogen, thus favouring a synchronous, cyclic elimination mechanism. This concept, if followed up, would provide a mechanistic parallel to β-elimination as aging. The role of pyridine aldoxime would consist in transferring the aging of the inhibited AChE to an "aging" of the oxime phosphate. Just as a dehydroalanine derivative results from the phosphorylated AChE by an irreversible step, so would the inhibitor-reactivator complex decompose irreversibly into cyanide and acid phosphate.

The hypotheses were clearly confirmed experimentally by Stark *. Using a potentiometric method, STARK redetermined the acidities of numerous AChE-oxime reactivators or was able to correct them. It was found that the reactivating action of the oximes of various pyridinium-aldehydes falls within a very narrow *p*H range. If, for example, the *p*K values of the 4,4′-oxime of bis-pyridiniumaldehyde lie outside the range 7.78 – 8.10, then the reactivating activity rapidly diminishes. In the case of the 4-mono-oximes there is a range of only 0.05 units between *p*K 7.83 and 7.78. The reason for this at *p*K values below 7.6 is that the oxime anion is no longer able to attack the phosphoric acid triester nucleophilically, i.e. the oxime is unable to accept a phosphoryl radical as was shown as the primary step in Scheme 33. At higher *p*K values the acidity of the methine protons (i.e. of the original aldehyde hydrogen atom) is too low to permit nitrile formation (second step in Scheme 33). The very small *p*K range in which reactivating activity can be expected, may therefore be explained by the interaction of two parameters each inversely proportional to the other:

1) adequate nucleophilicity of the oxime anion for the primary step.
2) adequate methine acidity for the secondary step according to the Scheme 33.

As will be seen from Fig. 17 on page 165, both the myoneural junction of the motor system and the ganglia of the sympathetic system of the cholinergic system are stimulated by nicotine, whereas the end-plates of the parasympathetic system are stimulated by muscarine.

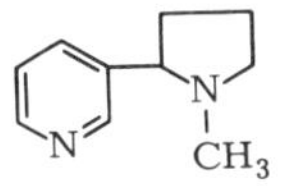

Nicotine

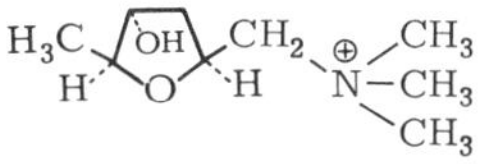

L(+)-Muscarine

Accordingly, in cases of poisoning by organophosphates (more correctly: by the endogenous acetylcholine toxification evoked by the organophosphates) various effects are to be observed [491]:

1) *Muscarine-like effects:*
   Sweating, tears and salivary secretion, often miosis (constriction of the pupils), malaise, vomiting, diarrhoea, increased secretion in the respiratory tract, bronchospasm and an asthma-like condition, cyanosis and oxygen deficiency.

2) *Nicotine-like effects:*
   Tremor, restlessness, muscle cramp and weakness, facial fibrillary twitching to epileptiform tonic-clonic convulsions.

* I. STARK: Reactivity of phosphorylated acetylcholinesterase with quaternary pyridine aldoximes: Determination of a relationship between oxime acidity and capacity of reactivation. Thesis, Freiburg 1971.

3) *Effects on the central nervous system:*
Loss of sensitivity to light and spatial orientation, insensitivity to pain, absence of reflexes, loss of muscle tone and unconsciousness. The heart and circulation respond with bradycardia changing to tachycardia, with hypotension followed by a rise in blood pressure and collapse.

At very high doses, death results from paralysis of the striated respiratory muscles, from paralysis of the respiratory center, cardiac arrest or pulmonary edema. Clinical manifestations of poisoning are generally observable when the ChE activity in the blood has fallen to below 30% (70–75% inhibition). The first acute symptoms may appear after only a few minutes (after inhalation), after 30–60 min (oral) or not until after several hours (cutaneous absorption). These periods are, however, very much dependent upon the structure of the organophosphate, upon the amount and the type of formulation and upon secondary factors such as stomach contents, etc.
The syndrome indicates the type of therapy, since the antagonists at the appropriate myoneural junctions and ganglia are known:
The antagonist for nicotine in the motor system is curare (in the pure form *d*-tubocurarine), antagonists for nor-adrenaline at the sympathetic endplates are the sympathicolytic agents such as 2-halogen-ethylamine; finally, the antagonist for muscarine is atropine, the active principle of deadly nightshade *(Atropa belladonna)*.

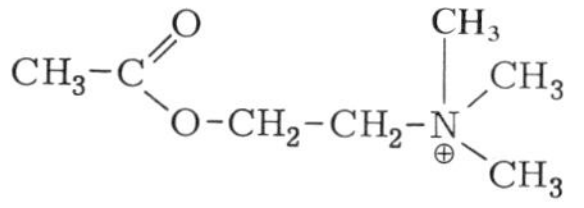

Acetylcholine

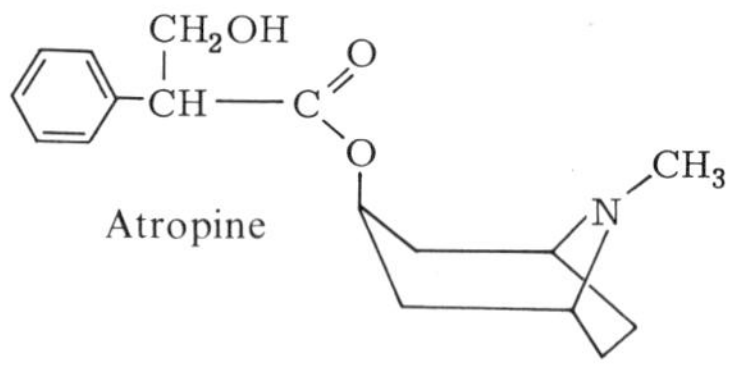

Atropine

Atropine attacks the receptor and has proved the most effective agent for the treatment of organophosphate poisoning. It also passes the blood-brain barrier into the central nervous system. The structural similarity to the natural agonist acetylcholine is very apparent. Strictly speaking, atropine is not a direct antagonist of organic phosphorus compounds but rather a competitive reversible inhibitor of acetylcholine at the receptor sites. For practical application, atropine possesses a very welcome property, to which its therapeutic superiority over other antagonists is partly attributable, i.e. its action on the diameter of the pupils by which its dosage may readily be controlled. By intravenous or intramuscular injections of atropine sulfate, it is possible to "titrate" toxic esters in the organism. Should constriction of the pupils and flow of saliva re-occur, then atropinization is continued (24–48 h) until the organophosphate liberated by desorption processes no longer results in renewed bouts of toxicity.
Coleman, Patton and Bannard [201] reported very promising properties in the treatment of organophosphorus poisoning for Parpanit (Geigy 2747,

Caramiphen, Pentaphen), a cholinolytic drug, used to treat Parkinson's syndrome. Parpanit is the hydrochloride of the diethyl amino-ethyl ester of 1-phenylcyclopentane carboxylic acid (II).

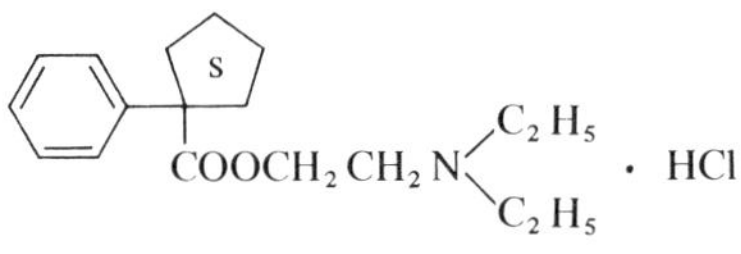

Parpanit (II)

The practical directions for the first-aid treatment of organophosphate poisoning are as follows [491]:

1) Call a doctor immediately.
2) Remove mask and protective clothing.
3) Wash the body with soap and water and *not* with alcohol.
4) Ensure fresh air and keep the patient warm.
5) Administer an aqueous slurry of medicinal charcoal.

On no account give milk or alcohol, for both are ideal vehicles for promoting absorption and toxication by organophosphates. Any further treatment must be given by the doctor. This begins with the intravenous administration of 2 mg atropine sulfate, in cases of severe poisoning 3–5 mg may be given as an initial dose. If there is respiratory distress, oxygen should be administered before treatment with atropine. A further 2 mg atropine is injected every 10–15 min until there is a definite improvement.

Atropine treatment must be given in all cases; in addition a reactivator such as 2-PAM (500–1000 mg) or toxogonin (250 mg) may be given by slow intravenous injection; the reactivator must be discontinued unless a definite response is seen after the first injection.

4.7. Neurotoxic Action

In 1930 during the prohibition era in America about 15,000 persons suffered severe paralysis of the legs but less of the arms, 10–14 days after drinking illicitly manufactured "Jamaica Rum". The motor nerves of the brain stem were seldom involved. One of the ingredients of this rum was Jamaica ginger. SMITH, ELVOVE and FRAZIER [913] found the toxic factor to be tri-o-cresyl phosphate (TOCP), which was added to falsify the viscosity and colour. Hence such cases became known as "ginger-paralysis". Industrially TOCP was much used as a plasticizer and as a lubricant by the motor oil industry. A second case of mass poisoning arose from the latter one. In 1959 in Morocco about 10,000 persons suffered severe paralysis, attributed at first to a virus infection. The real cause was, however, soon discovered to be salad oil to which some dealers had added American motor oil containing TOCP [911].

These incidents stimulated an intensive investigation of the mechanism of the neurotoxic action of TOCP and other organophosphates (e.g. [70]). Tests were carried out in the hen, which is particularly sensitive to neurotoxic compounds. The emphasis was first placed on the neurophysiological and neuropathological aspects [341, 342].
The biochemical mechanism has not yet been clarified.
Recently JOHNSON [452] was able to show that neurotoxic organophosphates are covalently bound *in vivo* to a protein in the brain and spinal cord of hens. *In vitro*, the reactivity of this target to DFP in brain tissue that was dosed with non-neurotoxic esters is of the same order as in normal brain tissue; the specific binding activity on the other hand is very small when the brain tissue of hens has been pretreated with neurotoxic esters. This indicates that inhibition of brain cholinesterases is not the only factor involved in the neurotoxic action of these compounds.
ALDRIDGE and BARNES [12] discussed the existing findings and theories. It would appear to be fairly certain that the primary attack of a neurotoxic organophosphate takes place at the distal end of an axon, in the spinal column at the upper end of the ascending long tracts and at the lower end of the descending tracts. The damage to the myelin sheath and to the Swann cells is a secondary occurrence ("dying back"). The primary damage to certain neurons does not immediately lead to destruction of the neuron but it is no longer able to maintain the distal end of the neuron completely functional. It is only then that its function is lost and structural damage ensues. The primary action of the neurotoxic substance is, therefore, to produce deficiencies in the neuron, the symptoms of which resemble many avitaminoses, although doses of thiamine or other vitamins are neither preventive nor do they effect a cure. A successful hypothesis must take into consideration the fact that the ataxia does not appear until 10–14 days after the poisoning, although the blocking of the cholinesterases may already be alleviated. Neurotoxic substances are usually not stable in the organism for such a long time. Furthermore, according to WITTER and GAINES [1031], the paralytic syndromes caused by DFP, TOCP, etc., when compared with those of DDVP and *trichlorfon* as non-neurotoxic substances, are in no way correlated with the inhibitory action on the brain or plasma cholinesterases.
Although very little is known about the mechanism of the neurotoxic action, it is possible to form a rough picture of the inhibition reaction.

1) GLEES [340] found that a degree of antidotal action was provided by cortisone acetate after small doses of TOCP, an indication that, for neurotoxic substances, this ester acts as a competitive inhibitor on some esterases. It is possible, that, with cortisone acetate, the optimum activity corresponding to the optimum structure has not yet been reached; perhaps immunochemical processes are involved.
2) POULSEN and ALDRIDGE [746] selected SM-1 as model and argued that the natural substrates of neuron esterases must be both aliphatic compounds and structural analogues of SM-1. They synthesized the phenol esters of ω-phenyl carboxylic acid and found two esterases which were able to degrade the acetic acid derivative (PPA) and the proprionic acid derivative (PPP)

at the same rate that acetylcholine is inactivated by acetylcholinesterase. Both esterases are strongly inhibited by DFP and SM-1. PPA and PPP may well indicate that the corresponding enzymes do not possess an anionic site:

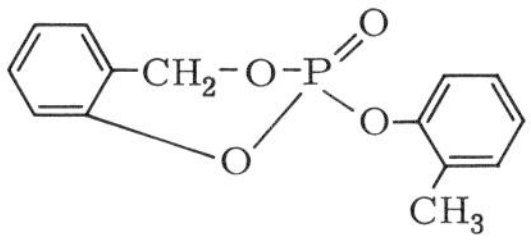

"SM 1"

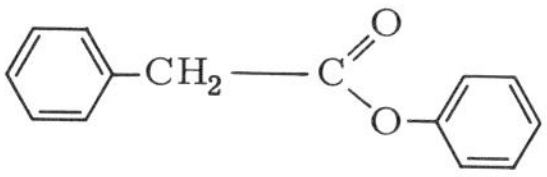

"PPA"

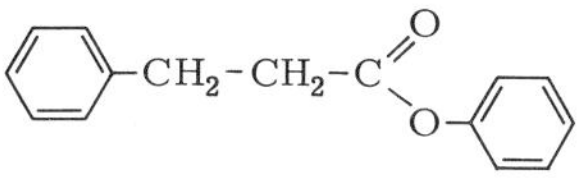

"PPP"

3) The special properties of the phosphoryl fluorides are note-worthy: they are specific inhibitors of esteratic sites, possess special penetration and distribution properties and are very favourable sterically (molar phosphorylating potential). The neurotoxic properties of the phosphoryl fluorides also suggest that esterase inhibition is involved. It is understandable that the size of the alkyl radical up to C_5 and cyclohexyl should be of limited influence (see Table 16), for the space for the alkoxy groups has an upper limit: it is determined by the bulkiness of the phenyl groups in the molecules of SM-1 or PPA and PPP. Phosphorofluoridates up to a certain molecular size might, therefore, slip into the target unhampered sterically.

Table 15. Structure and neurotoxic action according to ALDRIDGE and BARNES [12]

Neurotoxic	Non-neurotoxic	Lit.
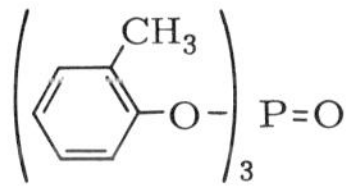	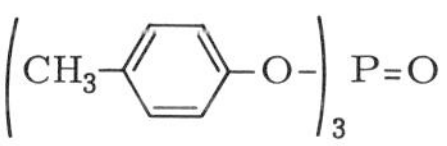	[10]
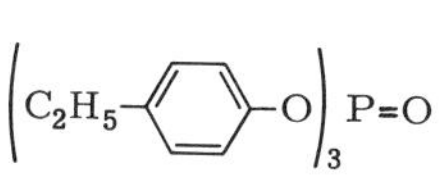	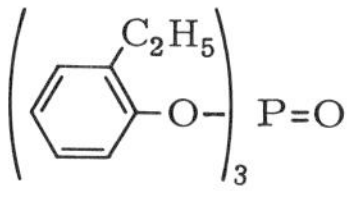	[10]
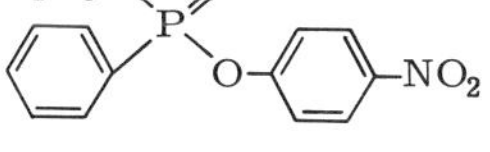	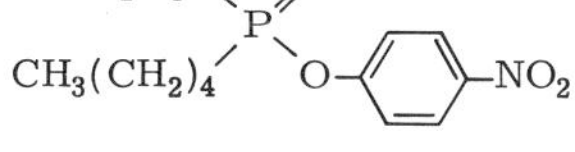	[11]
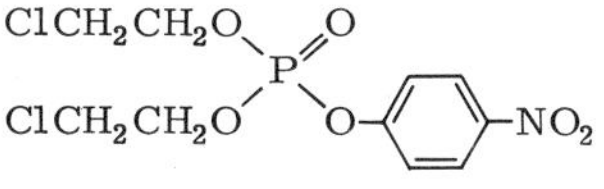	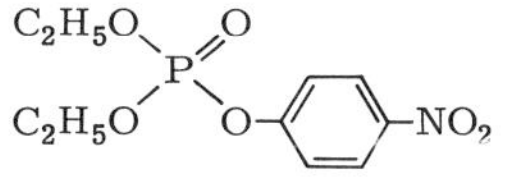	[11]

Table 15 (Continued)

Neurotoxic	Non-neurotoxic	Lit.
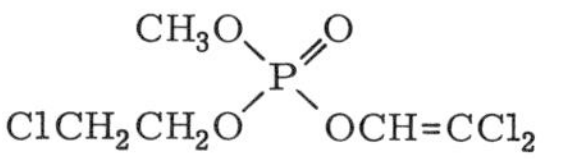	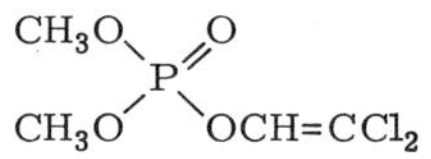	[11]
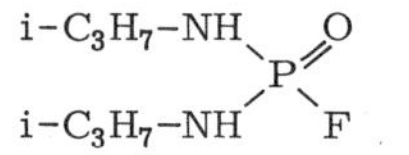	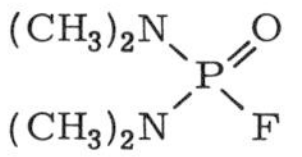	[69]
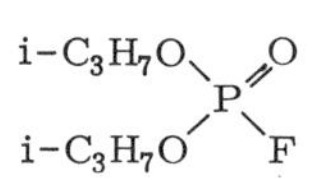	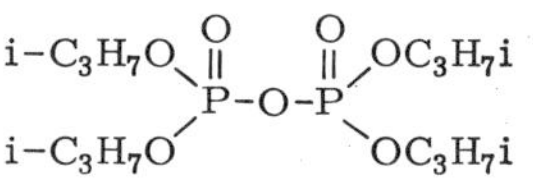	[234]

These results all indicate that certain esterases are blocked directly in the neuron; they are those that play an important role in the metabolism of the neuron and are protected by a lipophilic barrier which, like a sieve, separates the neurotoxic from the non-neurotoxic compounds. Subsequent biochemical damage is then of a secondary nature. It is a consequence of lethal interruptions of the metabolism of the nerve which leads to death of the axon from the distal end to the neuron. CAVANAGH [171] described this process as "dying back" which would also explain the demyelinization. Inhibition of the neuroesterases proceeds rapidly, but the subsequent physiological reactions like dying back are time-dependent. Reactivators must, therefore, as in the case of AChE inhibition, be administered very soon after poisoning; should suitable compounds be found they would be given prophylactically.

Table 15 reveals that it is possible to distinguish between two larger groups of neurotoxic organophosphates. The phenol esters present an irregular picture and it is difficult to find a common denominator that will give a clue to the structural conditions required for neurotoxic action. Recognition of the fact that TOCP, for example, does not exert a direct neurotoxic action but acts rather by way of the metabolite SM-1, as was suspected by ALDRIDGE and structurally clarified by ETO, CASIDA and ETO [268], shows that in a discussion on structure and neurotoxic action, considerable attention must be paid to lethal syntheses. As has already been mentioned, only a few reaction types are possible here, but these can in each case give rise to a large number of metabolites.

Very recently, ETO, ABE and TAKAHARA published important papers*,** on the metabolism of tri-*p*-ethyl phenyl phosphate (TPEP, Formula I in Scheme 34) and the neurotoxicity of its metabolites. The lethal synthesis is, as with TOCP, an hydroxylation of the alkyl side-chain mainly to O,O-bis-(4-ethyl phenyl) O-(4-α-hydroxyethyl phenyl) phosphate (II) and the analogous bis- and tri-hydroxy derivatives (II, IV). By the action of a soluble dehydrogenase these compounds are transformed to the corresponding α-oxo-phenylphosphates (V – VIII), all of them damaging the sciatic nerve and causing ataxia in hens.

---

* ETO, M., ABE, M., TAKAHARA, H.: Agr. Biol. Chem. (Tokyo) **35**, No. 6, 929 (1971).

** ETO, M., ABE, M.: Biochem. Pharmacol. **20**, 967 (1971).

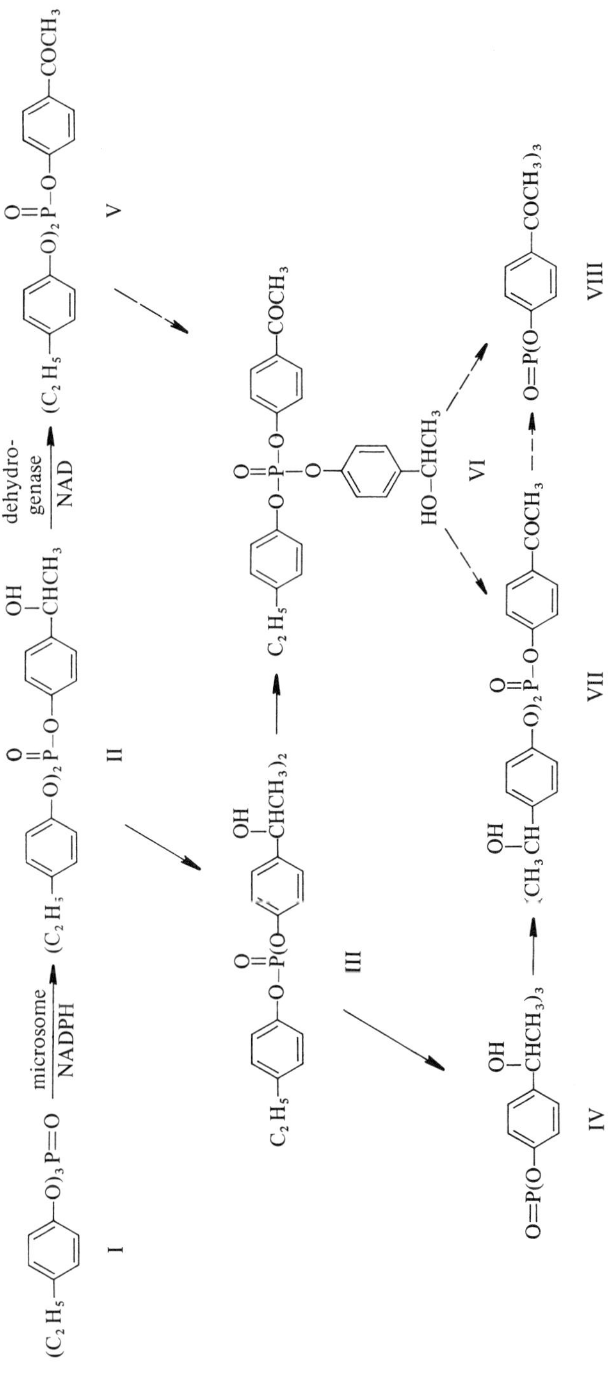

Scheme 34. Metabolic pathway of tri-*p*-ethylphenyl phosphate [ref. see page 264]

On the basis of these results it is possible to explain the opposing properties of the alkylphenyl phosphates in Table 15:
o-Methyl phenyl phosphates are transformed to cyclic Saligenin phosphates which are neurotoxic, the *p*-methyl phenyl phosphates being oxidized to phosphates of *p*-hydroxybenzaldehyde, the reactivity of which may be high enough for detoxification by secondary reactions.
The *o*-ethyl phenyl phosphate is unable to form cyclic Saligenin phospate types after hydroxylation, the *p*-ethylphenyl phosphates being further dehydrogenated to the rather stable and neurotoxically active phosphates of *p*-hydroxyacetophenone.
m-Alkyl phenyl phosphates do react to similar hydroxy- or oxo-metabolites but are not neurotoxic. In this case, the same reaction – hydroxylation followed by dehydrogenation – does not relate to neurotoxic activation, but to a normal degradation.

Nevertheless, in general the metabolism of neurotoxic substances and their chemical reactions at the site of action are little understood, even though ALDRIDGE and BARNES [11] were able to forecast the neurotoxic action of certain compounds of type (IX, X) on the basis of their structural relation to PPA and PPP. Furthermore, it is notable that a neurotoxic action was found among chloroethyl esters [11].

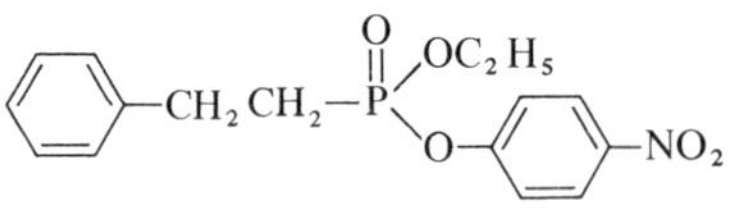

(IX)

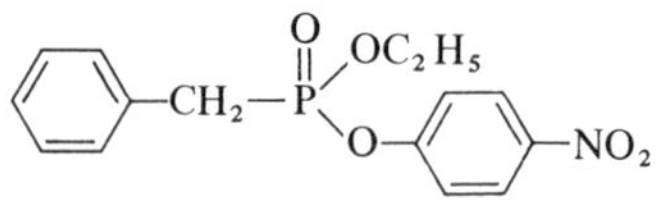

(X)

In the second group, which includes the fluorophosphates (see Table 16) and the fluorophosphonates (see Table 17), neurotoxic compounds are relatively numerous. Dimethyl phosphorofluoridate is somewhat of an exception for relatively high doses are required. This is fairly simply explained by increased degradation on the way to the site of action [233].

Table 16. Neurotoxicity of compounds of the type

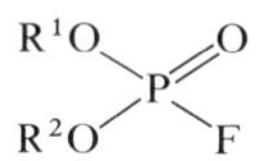

in the hen [12]

R^1	R^2	Minimum dose for ataxia (mg/kg)
CH_3	CH_3	30
C_2H_5	C_2H_5	0.75
C_3H_7	C_3H_7	0.25
i-C_3H_7	i-C_3H_7	0.3
C_4H_9	C_4H_9	0.5
i-C_4H_9	i-C_4H_9	1.5
s-C_4H_9	s-C_4H_9	1.5
C_5H_{11}	C_5H_{11}	2.5
C_3H_7, CH_3 >CH	C_3H_7, CH_3 >CH	2.5
Cyclohexyl	Cyclohexyl	2.5
C_2H_5	C_2H_7	1.0

Table 17. Neurotoxicity of compounds of the type

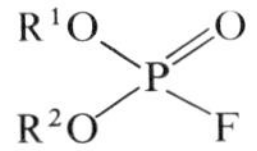

in the hen [12]

R^1	R^2	Minimum dose for ataxia (mg/kg)
i-C_3H_7	CH_3	1.0
i-C_3H_7	C_2H_5	1.0
CH_3	i-C_3H_7	5.0
C_2H_5	CH_3	2.9
i-C_4H_9	C_2H_5	2.9

At least one alkoxy group on the phosphorus atom would appear to be a prerequisite for neurotoxic action, the general conditions for which are given by structure (XI):

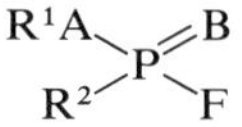

(XI)

R^1 Alkyl
R^2 Alkyl, alkoxy and alkyl amino groups
A An oxygen or nitrogen atom
B An oxygen or sulfur atom

The fact that a fluorine atom is a common feature of many neurotoxic phosphoric acid esters, led to a correlation of this property with the fluoride ion liberated by hydrolysis and inhibition. Phosphorofluoridates are first and foremost normal acid halides having the advantage of being hydrolytically more stable in certain *p*H ranges and in addition being more lipophilic than the chlorides or bromides. For these reasons they possess superior penetration and partition properties. A further advantage is the small size of the fluorine atom, whereby steric limitations in reactions at the target are of less importance than, for example, with the large phenolic or heterocyclic radicals which also obey the Schrader rule. All these properties enhance both accumulation at the site of action and penetration into the central nervous system. Phosphorofluoridates possess no group in the molecule which would be suitable for interaction with the anionic site of AChE. They should, therefore, be regarded as ideal acylating inhibitors, i.e. as selective inhibitors of esteratic sites in the enzyme. The possible number of enzymes which can be inhibited is, therefore greater than with the structurally specialized triesters. Thus, all serine enzymes are inhibited and also the "neuron esterases", so far of unknown structure. Above all, the fluorides are suitable for the inhibition of acylcholine acylhydrolase [3.1.1.8]. The increase in activity resulting from a transition from phosphoro- to phosphonofluoridates is in agreement with the rules for hydrolysis and phosphorylation. When, as in the case of the O-alkyl esters of alkanephosphonic acid the alkoxy group is branched, the hydrolytic side reactions on the alkoxy group are suppressed and the true phosphorylating reaction "enforced" leading to an increase in activity. With regard to the toxicity of fluoride ions, there is an informative paper by HEILBRONN-WIKSTRÖM [396] according to which fluoride ions exert a reactivating action on phosphorylated cholinesterases. The rate of reactivation is proportional to the fluoride ion concentration, it is also *p*H-dependent and greater in slightly acid medium than in a basic medium where oxime reactivation is optimal. We have, therefore, the following reactivation scheme:

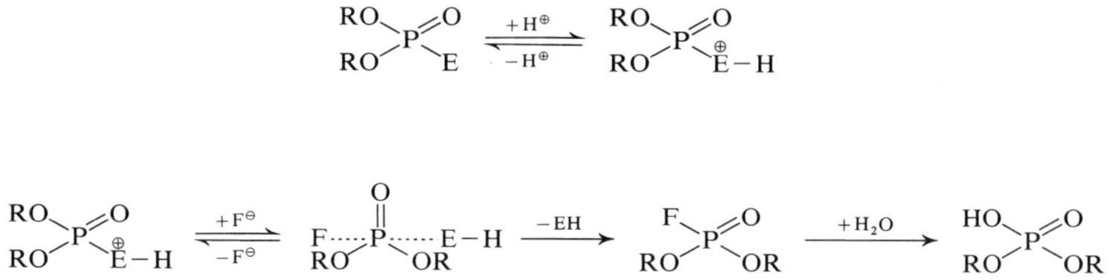

Scheme 35. Reactivation by fluoride ion [396]

In itself, this effect is not surprising, for the fluoride ion is the strongest basic ion in the periodic system; it has a higher charge density than the hydroxyl ion. The fluoride ion is able, therefore, to displace the hydroxyl ion on the central phosphorus atom, even in aqueous solution where it is not so much the nucleophilicity but rather the basicity of the entering group that is important. Furthermore, one is reminded that hydrogen fluoride, like the water molecule, is able to form very stable associates. This might mean that fluoride ions, by forming dipoles in the organism, interfere with hydrogen bridges. These considerations are supported by the fact that fluoride ions are weak reversible inhibitors of cholinesterase and serve as fairly effective antidotes for compounds such as Sarin or Tabun. (It would be interesting to learn how in this respect long chained substituted ammonium fluorides behave and whether there is a correlation with the analogously constructed iodides.) Hence it would seem reasonably certain that the toxicity of the fluoride ions themselves must depend on an attack on other stages of the metabolism.

A distinction must be made between this role of fluorine and that in compounds such as fluoroacetic acid. In this case the metabolism which normally begins with oxidation or hydroxylation is hindered, i.e. the toxicity depends upon the opposite behaviour of fluorine, its "non-cleavability".

A third group of neurotoxic substances fits less well into this scheme. These are some trithiol phosphites and phosphates:

The spectrum of activity of this class of substance is, in general, shifted from insecticidal towards herbicidal-fungicidal activity. As with other neurotoxic compounds, synergistic action is found with *malathion* [162, 1031]. On hydrolysis, mercaptan is evolved. Although the di- and tri-thiol esters may be regarded as acylating agents, a special action of the mercaptan released solvolytically on enzyme inhibition is to be anticipated. Little is known in this field and further experiments, particularly with respect to the practical application of these substances in crop protection, is required. Even if ALDRIDGE and BARNES [12] consider the neurotoxic action of a substance to be of only academic interest, it is, nevertheless, of great practical significance. A pesticide with neurotoxic properties is unacceptable. When such properties are observed in the toxicological investigations, the substance is no longer followed up, even if it has exceptional insecticidal properties.

5. Appendix

5.1. Bibliography

a) Chemistry of Phosphorus Compounds

HOUBEN-WEYL, Methoden der organischen Chemie (ed. E. MÜLLER und K. SASSE): Organische Phosphorverbindungen, Band XII/1 (1963), Band XII/2 (1964). Stuttgart: Georg Thieme.

HUDSON, R. F.: Structure and mechanism in organo-phosphorus chemistry. London and New York: Academic Press 1965.

IUPAC. Organic Chemistry Division and Gesellschaft Deutscher Chemiker Symposium on Organo-Phosphorus Compounds held in Heidelberg, Germany, 20–22 May 1964. London: Butterworths 1964.

KOSOLAPOFF, G. M.: Organophosphorus compounds. New York: John Wiley & Sons, Inc. and London: Chapman & Hall, Ltd. 1950.

PURDELA, D., VÎLCEANU, R.: Chimia compuşilor organici al fosforului şi al acizilor lui. Editura Academici Republicii Socialiste România 1965.

VAN WAZER, J. R.: Phosphorus and its compounds. Vol. I: Chemistry. New York and London: Interscience Publishers, Inc. 1958; – Vol. II: Technology, biological functions and applications. New York and London: Interscience Publishers, Inc. 1961.

b) Insecticides (Activity, Biology, Metabolism)

Bayer-Pflanzenschutz, Compendium. Leverkusen: Farbenfabriken Bayer AG 1962.

EICHLER, W.: Handbuch der Insektizidkunde. Berlin: VEB Verlag Volk und Gesundheit 1965.

FREAR, D. E. H.: Pesticide index. 4th edition. College Science Publishers, State College Pennsylvania 1969.

FREAR, D. E. H., FRIEDMAN, S., DEMARTINO, TH.: Pesticide handbook-entoma. 22nd edition. College Science Publishers, State College Pennsylvania 1970.

HEATH, D. F.: Organophosphorus poisons. Anticholinesterases and related compounds. Oxford-London-New York-Paris: Pergamon Press 1961.

KHORANA, H. G.: Some recent developments in the chemistry of phosphate esters of biological interest. New York: John Wiley & Sons 1961.

LOHS, K. H.: Synthetische Gifte. Berlin: Deutscher Militärverlag 1967.

LOSHADKIN, N. A., SMIRNOV, V. V.: A review of modern literature on the chemistry and toxicology of organophosphorus inhibitors of cholinesterases. Glen Ridge, New Jersey: Associated Technical Services, Inc. 1962.

MARTIN, H.: The scientific principles of crop protection. 5th edition. London: Edward Arnold (Publishers Ltd.) 1964.

MARTIN, H.: Pesticide manual. 1st edition. British Crop Protection Council 1968.

METCALF, R. L.: Advances in pest control research, Vol. I (1957) – VII (1966). New York and London: Interscience Publishers, Inc.

METCALF, R. L.: Organic insecticides. New York and London: Interscience Publishers, Inc. 1955.

O'BRIEN, R. D.: Insecticides. Action and metabolism. London and New York: Academic Press 1967.

O'BRIEN, R. D.: Toxic phosphorus esters. Chemistry, metabolism and biological effects. London and New York: Academic Press 1960.

O'BRIEN, R. D., YAMAMOTO, I. (eds.): Biochemical toxicology of insecticides. Proceedings of the 5th U.S.-Japan Cooperative Science Program held in Tokyo, June 1969. New York: Academic Press 1970.

PERKOW, W.: Wirksubstanzen der Pflanzenschutz- und Schädlingsbekämpfungsmittel. Berlin und Hamburg: Verlag Paul Parey 1971.

ROSIVAL, L., VRBOVSKÝ, L., SELECKÝ, FR. V.: Toxikologie und Pharmakodynamik der Phosphororganischen Verbindungen. Pressburg 1959.
SAUNDERS, B. C.: Some aspects of the chemistry and toxic action of organic compounds containing phosphorus and fluorine. Cambridge, Great Britain: Cambridge University Press, London, New York 1957.
SCHRADER, G.: Die Entwicklung neuer insektizider Phosphorsäure-Ester. 3. Auflage. Weinheim: Verlag Chemie 1963.
SPENCER, E. Y.: Guide to the chemicals used in crop protection. 5th edition. Research Branch, Canada Department of Agriculture.
WEGLER, R. (ed.): Chemie der Pflanzenschutz- und Schädlingsbekämpfungsmittel. 2 Volumes. Berlin-Heidelberg-New York: Springer 1970.
WEST, T. F., HARDY, J. E.: Chemical control of insects. 2nd edition. London: Chapman & Hall, Ltd. 1961.

c) Biochemistry and Enzyme Inhibition

BAKER, B. R.: Design of active-site-directed irreversible enzyme inhibitors. The organic chemistry of the enzymic active-site. New York-London-Sidney: John Wiley & Sons, Inc. 1967.
BURGER, A.: Drugs affecting the peripheral nervous system, vol. I. New York: Marcel Dekker, Inc. 1967.
KARLSON, P.: Kurzes Lehrbuch der Biochemie für Mediziner und Naturwissenschaftler. 6. Auflage. Stuttgart: Georg Thieme 1967.
KOELLE, G. B.: Cholinesterases and anticholinesterase agents. In: Handbuch der experimentellen Pharmakologie, Ergänzungswerk, Band 15 (eds. O. EICHLER and A. FARAH). Berlin-Göttingen-Heidelberg: Springer 1963.
NACHMANSOHN, D.: Chemical and molecular basis of nerve activity. New York and London: Academic Press 1959.
TRIGGLE, D. J.: Chemical aspects of the autonomic nervous system. London: Academic Press 1965.
WEBB, J. L.: Enzyme and metabolic inhibitors, vol. I: General principles of inhibition. New York and London: Academic Press 1963; – Vol. II: Malonate, analogs, dehydroacetate, sulfhydryl reagents, o-iodoso benzoate, mercurials. New York and London: Academic Press 1966.

d) Biochemistry of Insects

GILMOUR D.: The biochemistry of insects. New York and London: Academic Press 1961.
GOODWIN, T. W. (ed.): Aspects of insect biochemistry. Biochemical Society Symposium No. 25, London, 1. April 1965. New York and London: Academic Press 1965.

e) Statistical Analysis

CAVALLI-SFORZA, L.: Grundbegriffe der Biometrie, insbes. der statistischen Methoden bei der Wertbemessung biologisch wirksamer Substanzen. Stuttgart: Fischer 1964.
FINNEY, D. J.: Statistical methods in biological assay. 2nd edition. London: Charles Griffin Comp. 1964.
WEBER, E.: Grundriß der biologischen Statistik. 6. Auflage. Stuttgart: Fischer 1967.

5.2. Trade Names and Common Names

The nomenclature of the compounds used in practice or mentioned in the literature is complex and difficult to survey. Different firms use different trade names for the same product; one firm may use separate trade names for the same active substance for different uses. These trade names are protected by law and cannot be freely used in the literature without an indication that they are trade names, for example, the raised ®.

For this reason "common names" had been evolved which are freely usable, unprotected trivial names serving as abbreviations for the scientific name. These names are proposed by national institutes such as the American National Standards Institute, the Interdepartmental Committee of Pest Control (USA), the British Standards Institution (BSI), the Canadian Standards Association, as well as by international institutes, such as the International Organization for Standardization in Geneva (ISO). The common names appear in the publications of these institutes and of state authorities, for example, in the Crop Protection Indices of the Biological Federal Institute for Agriculture and Forestry in Brunswick, or in technical literature such as the World Review of Pest Control [50], Journal of Economic Entomology, Pesticidal Science, and the Bulletin of the Entomological Society of America.

There are also code designations, often originating from the development period of a compound and indicating that it is a trial product.

Example: L 13/59 (internal designation by firm)
= Bayer 15922 (internal designation by firm)
= Bayer 4824 (internal designation by firm)
= ®Dipterex (for agriculture use)
= ®Dylox (for agriculture use, USA)
= ®Tugon (for hygiene use)
= ®Neguvon (for use in vet. med. sector)
= *trichlorphon* (D, B common name)
= *trichlorfon* (I, C, USA common name)
= Chlorofos (USSR)
= Phoschlor (P)

Among code designations of national or international registration authorities are the ENT numbers of the Entomological Society of America or the OMS numbers (Organisation mondial de Santé = World Health Organization).

In East European countries many of the proposed common names have been accepted and, in addition, special designations are known [150].

In the following tables the more important active substances are arranged: a) alphabetically according to the common name and b) according to their trade names. When possible the source is indicated: D = German, I = ISO, B = British, F = French, C = Canadian, USA = American, USSR = Russian, P = Polish. The arrow indicates the most commonly used names. In many cases it is doubtful whether a trade name or a common name is involved: it is then included under the trade name without ®.
The marking of the trade names mentioned in this publication by the sign ® as Registered Trade Marks does not claim to be complete. No responsibility is taken for checking that trade names which are not marked by the sign ® are not registered as Trade Marks.

Table 18. Trade names and common names

Page	Trade names	Other names	Scientific names
102	®Abate		4,4′-Bis-(O,O-dimethyl thionophosphoryloxy) diphenyl sulfide
91	®Accothion	→®Folithion	
	®Acetoxon	Azetophos	O,O-Diethyl S-(carbethoxymethyl) phosphorothioate
142	®Afidamon	→ *phosphamidon* (D, I, B, C, USA)	
135	®Aflix	→ *formothion*	
152	®Afugan	→ HOE 2873 *pyrazophos*	O,O-Diethyl O-[5-methyl 6-carbethoxy pyrazolo-(1,5a)-pyrimidyl-2] phosphorothioate
105, 192	®Agridip	→®Asuntol	
94, 191	®Agritox	®Phytosol *trichloronate*	O-Ethyl O-(2,4,5-trichlorophenyl) ethanephosphonothioate
139	®Akton	SD 9098	O,O-Diethyl O-1-(2,5-dichlorophenyl) 2-chloro-vinyl phosphorothioate
103	®Alamos	®Slam C	O,O-Dimethyl O-[4-(4′-chlorophenylazo) phenyl] phosphorothioate
88	®Alkron	→ *parathion*	
135	®Amidithion	→®Thiocron	
	®Amiphos	DAEP	O,O-Dimethyl S-(N-acetyl) aminoethyl phosphorodithioate
135	®Anthio	→ *formothion*	
88	®Armin		O-Ethyl O-p-nitrophenyl ethanephosphonate
86	®Aspon	→ NPD	
105, 192	®Asuntol	®Agridip ®Co-Ral ®Muscatox ®Resitox → *coumaphos* (I, B, C, USA)	O,O-Diethyl O-(3-chloro-4-methyl-2-oxo-2H-1-benzopyranyl-7) phosphorothioate

Table 18 (continued)

Page	Trade names	Other names	Scientific names
	®Awenin		O,O-Dimethyl N-carboxypropyl phosphoroamidate
127	®Azidithion	→ *menazon*	
141	®Azodrin	®Nuvacron *monocrotophos*	O,O-Dimethyl O-(1-methyl 2-N-methylcarbamoyl) vinyl phosphate
116	®Azothion		O,O-Diethyl S-(N,N-dimethyl dithiocarbamyl) methyl phosphorodithioate
106	®Bacdip	*oxinothiophos* *quintiofos* (WHO)	O-Ethyl O-quinolyl-8 benzenephosphonothioate
107	®Basudin	→ *diazinon*	
	®Basudin-R	Exudin-R *(diazinon + phenkapton)*	
108	®Bayrusil	®Ekalux *diethquinalphione* (B) *diethchinalphion* (D)	O,O-Diethyl O-quinoxalyl-2 phosphorothioate
96	®Baytex	→ *fenthion*	
113	®Baythion	→ *phoxim*	
156	®Betasan	®Disan ®Prefar *bensulide* (USA)	O,O-Diisopropyl S-[2-(N-benzenesulfonyl) aminoethyl] phosphorodithioate
141	®Bidrin	®Carbicron *dicrotophos* (B)	O,O-Dimethyl O-[1-methyl 2-(N,N-dimethyl) carbamoyl] vinyl phosphate
	®Bilacil	®Dipterex → *metriphonate*	
138	®Birlane	→ *chlorfenvinphos*	
84	®Bladan	→ TEPP	
86	®Bladafum	→ *sulfotepp*	
142	®Bomyl		O,O-Dimethyl O-1,3-di(carbomethoxy) 1-propen-2-yl phosphate
131	®Bopardoil	→ *dimethoate*	
	®Bromthion		O,O-Dimethyl O-(2-bromo-4-nitrophenyl) phosphorothioate
155	®Butyl-DEF	®DEF-Defoliant TBTP 5	S,S,S-Tri-*n*-butyl phosphorotrithioate
141	®Carbicron	→®Bidrin	
147, 200	®Cerezin		O-Methyl O-cyclohexyl S-(4-chlorophenyl) phosphorothioate
91	®Chlorthion	*chlorthion* (D, F, USA)	O,O-Dimethyl O-(3-chloro-4-nitrophenyl) phosphorothioate
130	®Cidial	®Erucin ®Erusan ®Papthion *dimephentoate* *phentoate*	O,O-Dimethyl S-(ω-carbethoxy) benzyl phosphorodithioate
141	®Ciodrin	®Simax *crotoxyphos*	O,O-Dimethyl O-[1-methyl 2-carbo-(α-methyl) benzyloxy] vinyl phosphate
117	®Citram	→ *amiton*	

Table 18 (continued)

Page	Trade names	Other names	Scientific names
91	®Colep	Monsanto CP 40 294	O-Phenyl O-p-nitrophenyl methane- phosphonothioate
147, 200	®Conen		O-*n*-Butyl S-ethyl S-benzyl phosphoro- dithioate
105, 192	®Co-Ral	→®Asuntol → *coumaphos*	
	®Co-thion	®Gusathion ME + *parathion*	
127	®Cotnion	→ *azinphos* (D, I, B, C)	
95	®Cyanox	→ *cyanophos* (B)	
169	®Cygon	→ *dimethoate*	
108	®Cynem	®Nemafos ®Nemaphos *thionazin* (D, B) ®Zinophos	O,O-Diethyl O-(pyrazinyl-2) phosphorothioate
111	®Cyolane	Amer. Cyanamid 43064	2-[N-(O,O-Diethyl phosphoryl)] imino-1,3-dithiolane
111	®Cytrolane		2-[N-(O,O-Diethyl phosphoryl)] imino-4-methyl 1,3-dithiolane
89	®Dalf	→ *parathion-methyl*	
91	®Danathion	→ *fenitrothion*	
99	®Dasanit	→ *fensulfothion*	
138	®Dedevap	→ DDVP	
155	®DEF	®Butyl-DEF ®DEF-Defoliant TBTP 5	S,S,S-Tris-*n*-butyl phosphorotrithioate
136	®Delcar	→ *delnav*	
136	®Delnav	®Delcar ®Navadel *delnav* (D) *dioxathion* (I, B, C, USA)	1,4-Dioxane-2,3-(bis-O,O-diethyl thionophosphoryl)-dithiol
138	®Dermaton	®Birlane → *chlorfenvinphos*	
140	®Dibrom	*naled* (D, C, USA)	O,O-Dimethyl O-(1,2-dibromo- 2,2-dichloroethyl) phosphate
	®Dicontal	®Chlorthion + ®Dipterex	
142	®Dimecron	→ *phosphamidon*	
108	®Diothyl	→ *pirimiphos ethyl* (B)	
137	®Dipterex	®Dylox ®Neguvon ®Tugon *trichlorphon* (D, B) *trichlorfon* (I, C, USA) Chlorofos (USSR) *metriphonate* Phoschlor (P)	O,O-Dimethyl (1-hydroxy 2,2,2-tri- chloro)-ethanephosphonate
157	®Disan	→®Betasan	
124	®Disyston	®Solvirex ®Teration II	O,O-Diethyl S-(2-ethylthioethyl) phosphorodithioate

Table 18 (continued)

Page	Trade names	Other names	Scientific names
	®Disyston	→ *disulfoton* (D, I, B, C) M 74 (USSR) → *thiodemeton* (D)	
125	®Disyston S	®Disystonsulfoxid	O,O-Diethyl S-(2-ethylsulfinylethyl) phosphorodithioate
121	®Dithiometasystox	→®Ekatin → *thiometon*	
	®Dithionate	→ *sophamide* (I, B, F)	
86	®Dithione	→ *sulfotepp*	
	®Dition	→ *coumithoate* (I, B)	
100	®Dowco 109	®Narlene	O-Methyl O-(2-chloro-4-tert. butyl phenyl) N-methyl phosphoroamidothioate
33, 155, 191	®Dowco 118	→®Zytron	
100	®Dowco 132	→®Ruelene	
107	®Dowco 179	→®Dursban	
112, 149, 200	®Dowco 199	DOW 49	O,O-Diethyl phthalimidophosphorothioate
107	®Dursban	®Dowco 179	O,O-Diethyl O-(3,5,6-trichloropyridyl-2) phosphorothioate
102	®Dyfonate	→ *fonophos* (B)	
137	®Dylox	→®Dipterex	
87	®E 600	→ *paraoxon*	
88	®E 605	→ *parathion*	
	®E 605-Combi	®Metasystox R + ®E 605	
121	®Ekatin	®Dithiometasystox ®Intrathion *thiometon* (I, B, C) M 81 (USSR)	O,O-Dimethyl S-(2-ethylthio ethyl) phosphorodithioate
134	®Ekatin F	→ *morphothion*	
134	®Ekatin M	→ *morphothion*	
88	®Ekatox	→ *parathion* (B, I, C, USA)	
108	®Elimix	→ *pirimiphos ethyl*	
127	®Endocide	→ *endothion*	
96	®Entex	→ *fenthion*	
90	®EPN-300	→ EPN	
131	®Erucin	→®Cidial	
131	®Erusan	→®Cidial	
119	®Estox	→®Metasystox-S	
158	®Ethrel	*ethephon*	2-Chloroethanephosphonic acid
127	®Ethylgusathion	→®Gusathion	
127	®Ethyl Guthion	→®Gusathion A	
93	®Etrolene	→ *ronnel*	
107	®Exodin	→ *diazinon* (D, I, B, F, C, USA)	
134	®Fac 20	*prothoate* (I, B)	O,O-Diethyl S-(N-isopropyl carbamoylmethyl) phosphorodithioate

Table 18 (continued)

Page	Trade names	Other names	Scientific names
154	®Falone	3 Y 9 2,4-DEP (USA)	Tris-(2,4-dichlorophenoxy-ethyl) phosphite
99	®Famophos	®Warbex *famphur*	O,O-Dimethyl O-4-(dimethyl sulfamoyl phenyl) phosphorothioate
	Famoxon		O,O-Dimethyl O-4-(dimethyl sulfamoyl phenyl) phosphate
	®Fitios	→ *ethoate-methyl* (I, B)	
195	®Fluorthion		O,O-Dimethyl O-(3-trifluoromethyl 4-nitrophenyl) phosphorothioate
154	®Folex	*merphos*	S,S,S-Tri-n-butyl phosphorotrithioite
88, 89	®Folidol	→ *parathion*, also *parathion-methyl*	
89	®Folidol M	→ *parathion-methyl*	
	®Folidol-Öl	*parathion* + mineral oil	
133	®Folimat	*(P=O)-dimethoate* *omethoate*	O,O-Dimethyl S-(N-methylcarbamoyl) methyl phosphorothioate
91	®Folithion	®Accothion ®Danathion ®Sumithion *fenitrothion* (D, I, B) Metilnitrofos (USSR) Metathion (CSR)	O,O-Dimethyl O-(3-methyl 4-nitrophenyl) phosphorothioate
	®Formocarbam	→ *sophamide* (I, B, F)	
138	Forstenon	*fosfinon*	O,O-Diethyl O-[1-(2-chloroethoxy) 2,2-dichlorovinyl] phosphate
	®Fostion EP		O,O-Dimethyl S-(N-methylcarbamoyl) isopropyl phosphorodithioate
131	®Fostion MM	→ *dimethoate*	
124	®Frumin AL	→ *thiodemeton*	
101	®Fujithion		O,O-Dimethyl S-(4-chlorophenyl) phosphorothioate
139	®Gardona	*tetrachlorvinphos*	O,O-Dimethyl O-1-(2,4,5-trichlorophenyl) 2-chlorovinyl phosphate
115	®Garrathion	→ *carbophenothion*	
	®Gearphos	*parathion-methyl* + *ethyl*	
101	®Gophacide		O,O-Bis-(4-chlorophenyl) N-acetimidoyl phosphoramidothioate
127	®Gusathion	®Guthion ®Methylgusathion *azinphos-methyl*(D, I, B, C)	O,O-Dimethyl S-4-oxo-1,2,3-benzotriazin-3 (4H)-ylmethyl phosphorodithioate
127	®Gusathion A	®Ethylgusathion ®Ethyl Guthion *azinphos-ethyl* (D, I, B, C)	O,O-Diethyl S-4-oxo-1,2,3-benzotriazin-3 (4H)-ylmethyl phosphorodithioate
	®Gusathion ME	®Gusathion + ®Gusathion A	
127	®Guthion	→®Gusathion	

Table 18 (continued)

Page	Trade names	Other names	Scientific names
105	®Haloxon		O,O-Bis-2-chloroethyl O-(3-chloro-4-methyl 2-oxo-2H-1-benzopyran-7-yl) phosphorothioate
78	®Hanane	→ *dimefox*	
146, 200	®Hinosan	*edifenphos* (B)	O-Ethyl S,S-diphenyl phosphorodithioate
128	®Imidan	®Prolate *phosmet* (B)	O,O-Dimethyl S-(N-phthalimidomethyl) phosphorodithioate
147, 200	®Inejin F 254	®Inegin	O-Methyl S-benzyl benzenephosphonothioate
121	®Intration	→®Ekatin	
118	®Isometasystox	→®Metasystox (i)	
118	®Isomethylsystox	→®Metasystox (i)	
119	®Isomethylsystox-sulfoxid	→®Metasystox R	
79	®Isopestox	→ *mipafox*	
122	®Isosystox	→®Systox	
	®Ketothion		O,O-Diethyl S-acetonyl phosphorodithioate
125	®Kilval	*vamidothion* (I, B)	O,O-Dimethyl S-[2-(1-methylcarbamoyl) ethylthio]-ethyl phosphorothioate
126, 151, 200	®Kitazin		O,O-Diethyl S-benzyl phosphorothioate
152, 200	®Kitazin P		O,O-Diisopropyl S-benzyl phosphorothioate
93	®Korlan	→ *ronnel*	
96	®Lebaycid	→ *fenthion*	
97	®Lucijet	®Lujet DMP S 1751	O,O-Diethyl O-(3-methyl 4-methylthiophenyl) phosphorothioate
97	®Lujet	→®Lucijet	
138	®Mafu	→ DDVP	
	®Maitometo		O-Methyl O-(2-chloro-4-methylthiophenyl) N-ethyl phosphoroamidothioate
112	®Maretin	®Rametin	O,O-Diethyl O-naphthaloximidophosphate
96	®Mercaptophos	→ *fenthion* (D, I, B, C, F, USA)	
142	®Merkon	→ *phosphamidon* (D, I, B, C, USA)	
154	®Merphos	→®Folex	
89	®Metacide	→ *parathion-methyl*	
118	®Metaisosystox	→®Metasystox (i)	
119	®Metaisosystox-sulfoxid	→®Metasystox R	
120	®Metasystox	PO: *demeton-S-methyl* (D, I, B)	O,O-Dimethyl O-(2-ethylthioethyl) phosphorothionate

Table 18 (continued)

Page	Trade names	Other names	Scientific names
	®Metasystox	PS: *demeton-O-methyl* (I, B) and PO + PS: *demeton-methyl* (I, B, D) *methyl demeton* (USA) ®Metasystemox Methylmercaptophos (USSR)	O,O-Dimethyl S-(2-ethylthioethyl) phosphorothiolate
118	®Metasystox (i)	*demeton-S-methyl* (D, I, B) ®Isometasystox ®Isomethylsystox ®Metaisosystox	O,O-Dimethyl S-(2-ethylthioethyl) phosphorothiolate
119	®Metasystox R	*demeton-S-methyl-sulfoxid* (D) *oxydemeton-methyl* (I, B) ®Isomethylsystox-Sulfoxid ®Metaisosystox-Sulfoxid	O,O-Dimethyl S-(2-ethylsulfinylethyl) phosphorothiolate
119	®Metasystox S	®Estox	O,O-Dimethyl S-(1-methyl 2-ethyl-sulfinylethyl) phosphorothiolate
	®Methyl-äthyl-disyston	→®Teration	
115	®Methyltrithion		O,O-Dimethyl S-(4-chlorophenylthio-methyl) phosphorodithioate
88	®Mintacol	→ *paraoxon*	
	®Mocap		O-Ethyl S,S-dipropyl phosphoro-dithioate
113	®Monitor	®Tamaron, *methamidophos*	O,S-Dimethyl phosphoroamidothioate
134	®Morphotox	→ *morphotion* (iso)	
105	®Muscatox	→®Asuntol → *coumaphos*	
135	®Murfotox	→ *mecarbam*	
93	®Nankor	→ *ronnel*	
109	®Narlene	→ Dowco 109	
136	®Navadel	→ *delnav*	
137	®Neguvon	→®Dipterex	
101	®Nellite		O-Phenyl N,N-dimethyl phosphoro-diamidate
92	®Nemacide	→®VC-13 Nemacide	
98	®Nemacur P	*metaphenamiphos*	O-Ethyl O-(3-methyl 4-methylthio-phenyl) N-isopropyl phosphoroamidate
108	®Nemafos	→®Cynem	
	®Neosar		O,O-Dimethyl S-benzenesulfonyl phosphorodithioate
138	®Nerkol	→ DDVP	
94	®Nexagan	→ *bromophos* (D, I)	
94	®Nexion	→ *bromophos*	
116	®Nialate	®Rhodocide *ethion* (D, I, B, C, USA) *diethion* (F)	O,O,O′,O′-Tetraethyl S,S′-methylene bis-phosphorodithioate

Table 18 (continued)

Page	Trade names	Other names	Scientific names
	®Nichlorphos	*isochlorthion*	O,O-Dimethyl O-(3-nitro-4-chlorophenyl) phosphorothioate
89	®Nitrox	→ *parathion-methyl*	
138	®Nogos	→ DDVP	
86	®NPD		O,O,O′,O′-Tetrapropyl pyrophosphorodithionate
141	®Nuvacron	→®Azodrin	
138	®Nuvan	→ DDVP	
94	®Nuvanol N	*iodofenphos*	O,O-Dimethyl O-(2,5-dichloro-4-iodophenyl) phosphorothioate
138	®Oko	→ DDVP	
104	®Optunal		O-Methyl O-(2-carboisopropoxyphenyl) phosphoroamidothioate
130	®Papthion	→®Cidial	
88	®Parafos	→ *parathion* (I, B, F, C, USA)	
135	®Pencathion	→ *malathion* (D, I, B, F, C, USA)	
131	®Perfekthion	→ *dimethoate*	
87	®Pestox III	→ *schradan*	
78	®Pestox XIV	→ *dimefox*	
79	®Pestox XV	→ *mipafox*	
140	®Phosdrin	*mevinphos* (D, I, B, C, USA)	O,O-Dimethyl O-(1-methyl 2-carbomethoxy) vinyl phosphate
131	®Phosphamid	→ *dimethoate*	
152	®Phosphonomycin		1,2-Epoxypropanephosphonic acid
142	®Phostex		Bis-(O,O-dialkyl thionophosphoryl) disulfides [75% ethyl, 25% isopropyl]
95	®Phosvel	VCS-506	O-Methyl O-(2,5-dichloro-4-bromophenyl) benzenephosphonothioate
94	®Phytosol	→ *trichloronate* (D)	
	®Pirazinon		O,O-Diethyl O-(2-n-propyl 4-methyl pyrimidyl-6) phosphorothioate
104, 192	®Potasan	E 838	O,O-Diethyl O-(4-methyl-2-oxo-2H-1-benzopyranyl-7) phosphorothioate
156	®Prefar	→ *bensulide*	
	®Proban	*cythioate*	O,O-Dimethyl O-(4-sulfamoyl phenyl) phosphorothioate
128	®Prolate	→®Imidan	
110	®Pyrazothion		O,O-Diethyl O-(3-methyl pyrazolyl-5) phosphorothioate
110	®Pyrazoxon		O,O-Diethyl O-(3-methyl pyrazolyl-5) phosphate
139	®Rabon	*tetrachlorvinphos*	
112	®Rametin	→®Maretin	
	®Ra Vap	®Rabon + DDVP	
115	®Remadion	→ *carbophenothion*	
105	®Resitox	→ *coumaphos*	

Table 18 (continued)

Page	Trade names	Other names	Scientific names
88	®Rhodiatox	→ *parathion*	
116	®Rhodocide	→ *ethion*	
131	®Rogor	®Cygon ®Daphene ®Fostion MM ®Phosphamid ®Roxion *dimethoate* (I, B, C, USA) *dimethoat* (D)	O,O-Dimethyl S-(N-methylcarbamoyl) methyl phosphorodithioate
131	®Roxion	→®Rogor	
100	®Ruelene		O-Methyl O-(4-tert. butyl 2-chlorophenyl) N-methyl phosphoroamidate
103	®Salioxon		2-Methoxy-4H-1,3,2-benzodioxaphosphoran-2-one
103	®Salithion		2-Methoxy-4H-1,3,2-benzodioxaphosphoran-2-thione
139	®Sapecron	→ *chlorfenvinphos* (B)	
127	®Saphos	→ *menazon*	
127	®Saphizon	→ *menazon*	
107	®Sarolex	→ *diazinon*	
103	®Slam C	→®Alamos	
110, 151	®Septin	→ *triamiphos*	
141	®Simax	→®Ciodrin	
	®Sisvar		O,O-Dimethyl S-5-(N-methyl 2,2-dimethyl 3-thiavaleroylamido)-phosphorodithioate
137	®Soldep	→®Dipterex	
124	®Solvirex	→ *thiodemeton*	
93, 191	®S-Seven		O-Ethyl O-(2,4-dichlorophenyl) benzenephosphonothioate
91	®Sumithion	→ *fenitrothion*	
139	®Supona	→ *chlorfenvinphos*	
129	®Supracide	®Ultracide	O,O-Dimethyl S-[(2-methoxy 5-oxo-Δ^2-1,3,4-thiadiazolinyl-4) methyl] phosphorodithioate
95	®Surecide		O-Ethyl O-(4-cyanophenyl) benzenephosphonothioate
122	®Systox	(P=O): *demeton-S* (I, B) ®Isosystox (P=S): *demeton-O* (I, B) (P=O + P=S)-mixture: *demeton* (I, B, USA)	O,O-Diethyl S-(2-ethylthioethyl) phosphorothiolate and O,O-Diethyl O-(2-ethylthioethyl) phosphorothionate
86	®Sytam	→ *schradan*	
113	®Tamaron	®Monitor, *methamidophos*	O,S-Dimethyl phosphoroamidothioate
134	®Tartan	*cyanthoate*	O,O-Diethyl S-[N-(1-cyano-1-methyl) ethyl carbamoyl] methyl phosphorodithioate

Table 18 (continued)

Page	Trade names	Other names	Scientific names
	®Teration	®Teration III ®Methyl-äthyl-disyston	O-Methyl O-ethyl S-(2-ethylthioethyl) phosphorodithioate
124	®Teration II	→®Disyston	
	®Teration III	→®Teration	
99	®Terracur P	*fensulfothion* (D)	O,O-Diethyl O-(4-methylsulfinyl phenyl) phosphorothioate
117	®Tetram	→ *amiton*	
84	®Tetron	→ TEPP	Tetraethyl pyrophosphate
114	®Thimet	*phorate* (I, B, C, USA)	O,O-Diethyl S-2-ethylthiomethyl phosphorodithioate
134	®Thiocron	®Amidithion *medithionate*	O,O-Dimethyl S-(N-methoxyethyl carbamoyl) methyl phosphorodithioate
88	®Thiophos	→ *parathion*	
96	®Tiguvon	→ *fenthion*	
	®Tinox	*methyldemeton-methyl*	O,O-Dimethyl S-(methylthioethyl) phosphorothioate
	®Tokunol	NTN 5006	O-Ethyl O-(2-nitro-4-methyl phenyl) N-isopropyl phosphoroamidothioate
138	®Tribuphon	→ *butonate* (I, B, C, USA)	
115	®Trithion	®Garrathion ®Remadion *carbophenothion* (I, B, C, USA)	O,O-Diethyl S-(4-chlorophenylthio methyl) phosphorodithioate
93	®Trolene	→ *ronnel*	
137	®Tugon	→®Dipterex	
129	®Ultracide	→®Supracide	
113	®Valexon	→ *phoxim*	
	®Vamidoate		O,O-Dimethyl S-[2-(1-methyl carbamoylethylthio)-ethyl] phosphoro-dithioate
138	®Vapona	→ DDVP	
125	®Vation	→ *vamidothion*	
92	®VC-13-Nemacide	*dichlofenthion*	O,O-Diethyl O-2,4-dichlorophenyl phosphorothioate
99	®Warbex	→®Famophos	
110, 151	®Wepsyn 155	®Septin *triamiphos* (B) *triamiphos* (D)	5-Amino-1-[bis-(dimethylamino)-phosphoryl]-3-phenyl-1,2,4-triazole
89	®Wofatox	→ *parathion-methyl* (D, I, B, C, USA)	
108	®Zinophos	*thionazin* (D, B) ®Cynem ®Nemafos	O,O-Diethyl O-pyrazinyl-2 phosphorothioate
129	®Zolone	*phosalone* (B)	O,O-Diethyl S-(2-oxo-6-chloro-2H-benzoxazolinyl-3) methyl phosphorodithioate
93, 155	®Zytron	DMPA (USA)	O-Methyl O-(2,4-dichlorophenyl) N-isopropyl phosphoroamidothioate

Table 19. Common names and trade names

Page	Common names Code numbers	Other names	Scientific names
130	*acethion*	*azethion*	O,O-Diethyl S-carbethoxymethyl phosphorodithioate
	acetoxon	Azetofos	O,O-Diethyl S-carbethoxymethyl phosphorothioate
117	*amiton*	®Citram ®Tetram (oxalate) Inferno	O,O-Diethyl S-(2-diethylaminoethyl) phosphorothioate
130	*azethion*	→ *acethion*	
127	*azinphos-ethyl* (D, I, B, C)	®Gusathion A ®Äthylgusathion ®Ethyl guthion	O,O-Diethyl S-4-oxo-benzotriazin-3 (4H)-ylmethyl phosphorodithioate
127	*azinphos-methyl* (D, I, B, C)	®Gusathion ®Guthion ®Methylgusathion	O,O-Dimethyl S-4-oxo-benzotriazin-3 (4H)-ylmethyl phosphorodithioate
	azothoate (I)		O,O-Dimethyl O-(4-chlorophenyl-azophenyl) phosphorothioate
112	Bayer 22408	®Maretin ®Rametin S 125	O,O-Diethyl O-naphthaloximido-phosphate
156	*bensulide*	®Betasan ®Disan ®Prefar	O,O-Diisopropyl S-[2-(N-benzene-sulfonyl) aminoethyl] phosphoro-dithioate
94	*bromophos* (D, I)	®Nexion	O,O-Dimethyl O-(4-bromo-2,5-dichlorophenyl) phosphorothioate
	bromophos-ethyl	*ethylbromophos*	O,O-Diethyl O-(4-bromo-2,5-dichlorophenyl) phosphorothioate
138	*butonate* (I, B, C, USA)		O,O-Dimethyl (2,2,2-trichloro-1-n-butyryloxy)-ethanephosphonate
115	*carbophenothion* (I, B, C, USA)	®Garrathion ®Remadion ®Trithion	O,O-Diethyl S-(4-chlorophenyl-thiomethyl) phosphorodithioate
138	*chlorfenvinphos*	®Birlane ®Dermaton ®Supona	O,O-Diethyl O-[1-(2,4-dichloro-phenyl) 2-chlorovinyl] phosphate
137	Chlorofos (USSR)	→®Dipterex	
91	*chlorthion* (D, F, USA)	®Chlorthion	O,O-Dimethyl O-(3-chloro-4-nitro-phenyl) phosphorothioate
105, 192	*coumaphos* (I, B, C, USA)	®Agridip ®Asuntol } Vet. med. ®Co-Ral ®Muscatox } Hygiene ®Resitox	O,O-Diethyl O-(3-chloro-4-methyl-2-oxo-2H-1-benzopyranyl-7) phosphorothioate
	coumithoate (I, B)	®Dition	O,O-Diethyl O-(3,4-tetramethylene-umbelliferone) phosphorothioate
141	*crotoxyphos* (B)	→®Ciodrin	
95	*cyanophos* (B)	®Cyanox	O,O-Dimethyl O-(4-cyanophenyl) phosphorothioate

Table 19 (continued)

Page	Common names Code numbers	Other names	Scientific names
134	*cyanthoate* (B)	®Tartan	O,O-Diethyl S-[N-(1-cyano-1-methyl) ethyl carbamoyl] methyl phosphorothioate
	cythioate	®Proban	O,O-Dimethyl O-(4-sulfamoylphenyl) phosphorothioate
	DAEP	→®Amiphos	
138	DDVP (USA)	*dichlorphos* (D) *dichlorvos* (I, B, C) ®Dedevap ®Mafu ®Nerkol ®Nuvan ®Oko ®Vapona	O,O-Dimethyl O-(2,2-dichlorovinyl) phosphate
136	*delnav* (D)	→®dioxathion	
122	*demeton* (I, B, USA)	Mercaptophos (USSR) ®Systox	Mixture of *demeton-O* and *demeton-S*
122	*demeton-O* (I, B)	P=S isomer	O,O-Diethyl O-(2-ethylthioethyl) phosphorothioate
122	*demeton-S* (I, B)	P=O isomer ®Isosystox	O,O-Diethyl S-(2-ethylthioethyl) phosphorothioate
120	*demeton-methyl* (I, B, D)	*methyl demeton* (USA) Methyl-mercaptophos (USSR) ®Metasystox ®Metasystemox	Mixture of demeton-O-methyl and demeton-S-methyl
120	*demeton-O-methyl* (I, B, D)	P=S isomer (®Metasystox)	O,O-Dimethyl O-(2-ethylthioethyl) phosphorothionate
118	*demeton-S-methyl* (I, B, D)	P=O isomer ®Isometasystox ®Isomethylsystox ®Metaisosystox ®Metasystox (i) ®P=O Metasystox	O,O-Dimethyl S-(2-ethylthioethyl) phosphorothiolate
119	*demeton-S-methyl-sulfoxid* (D)	*oxydemeton-methyl* (I, B) ®Isomethylsystox-sulfoxid ®Metaisosystox-sulfoxid ®Metasystox R	O,O-Dimethyl S-(2-ethylsulfinyl ethyl) phosphorothiolate
154	2,4-DEP	→®Falone	
79	DFP	(Diisopropyl fluorophosphate)	O,O-Diisopropyl phosphorofluoridate
107	*diazinon* (D, I, B, F, C, USA)	®Basudin ®Exudin ®Sarolex	O,O-Diethyl O-(2-isopropyl 4-methyl pyrimidyl-6) phosphorothioate
107	*diazoxon*		O,O-Diethyl O-(2-isopropyl 4-methyl pyrimidyl-6) phosphate
92	*dicapthon* (C, USA)		O,O-Dimethyl O-(2-chloro-4-nitrophenyl) phosphorothioate

Table 19 (continued)

Page	Common names Code numbers	Other names	Scientific names
92	*dichlofenthion*	→®VC-13-nemacide	
138	*dichlorphos* (D)	→ DDVP	
138	*dichlorvos* (I, B, C)	→ DDVP	
141	*dicrotophos* (B)	→®Bidrin	
116	*diethion*	→ *ethion*	
108	*diethquinalphion*	®Bayrusil ®Ekalux	O,O-Diethyl O-quinoxalyl-2 phosphorothioate
78	*dimefox* (D, I, B, F, C, US	®Hanane ®Pestox XIV ®Terrasytam	N,N,N′,N′-Tetramethyl phosphorodiamidofluoridate
130	*dimephenthoate*	*phenthoate* ®Cidial ®Erucin ®Papthion	O,O-Dimethyl S-(ω-carbethoxy) benzyl phosphorodithioate
131	*dimethoate* (I, B, C, D, USA)	®Bopardoil ®Cygon ®Daphene ®Fostion MM ®Perfekthion ®Phosphamid ®Rogor ®Roxion	O,O-Dimethyl S-(N-methylcarbamoylmethyl) phosphorodithioate
133	*(P=O)-dimethoate*	*omethoate* ®Folimat	
136	*dioxathion* (I, B, C, USA)	*delnav* (D) ®Delnav ®Delcar ®Navadel	1,4-Dioxane-2,3-(bis-O,O-diethyl thionophosphoryl)-dithiol
124	*disulfoton* (D, I, B, C)	*thiodemeton* (D) M 74 (USSR) ®Disyston ®Frumin AL ®Solvirex ®Teration II	O,O-Diethyl S-(2-ethylthioethyl) phosphorodithioate
86	Ditio (USSR)	→ *sulfotepp*	
86	Ditiofos (USSR)	→ *sulfotepp*	
97	DMP	→®Lucijet	
93, 155	DMPA (USA)	®Zytron	O-Methyl O-(2,4-dichlorophenyl) N-isopropyl phosphoroamidothioate
99	DMSP (USA)	→ *fensulfothion*	
96	DMTP	→ *fenthion*	
112	DOW 49		O,O-Diethyl phthalimidophosphorothioate
112	DOW 50		O,O-Diethyl O-[7-oxabicyclo-(2,2,1)-hept-5-ene 2,3-dicarboximido]-phosphorothioate
146, 200	EDDP	→®Hinosan	
146, 200	*edifenphos* (B)	®Hinosan	O-Ethyl S,S-diphenyl phosphorodithioate

Table 19 (continued)

Page	Common names Code numbers	Other names	Scientific names
127	*endothion* (D, I, B, C, USA)	®Endocide	O,O-Dimethyl S-(5-methoxy-4-oxo-4H-pyranyl-2) methyl phosphorothioate
	endoxan		2-[Bis-(2-chloroethyl)-amino]-tetrahydro-1,3,2-oxazaphosphorin-2-oxide
90	EPN	®EPN-300	O-Ethyl O-(4-nitrophenyl) benzenephosphonothioate
158	*ethephon*	→®Ethrel	
116	*ethion* (D, I, B, C, USA)	*diethion* (F) ®Nialate ®Rhodocide	O,O,O′,O′-Tetraethyl S,S′-methylene bis-phosphorodithioate
	ethoate-methyl (I, B)	®Fitios	O,O-Dimethyl S-(N-ethylcarbamoyl) methyl phosphorodithioate
99	*famphur*	®Famophos ®Warbex	O,O-Dimethyl O-[4-(dimethylsulfamoyl) phenyl] phosphorothioate
93	*fenchlorphos* (B)	→ *ronnel*	
91	*fenitrothion* (D, I, B)	Metilnitrofos (USSR) ®Accothion ®Danathion ®Folithion ®Sumithion	O,O-Dimethyl O-(3-methyl 4-nitrophenyl) phosphorothioate
99	*fensulfothion* (D)	®Dasanit ®Terracur P	O,O-Diethyl O-(4-methylsulfinyl phenyl) phosphorothioate
96	*fenthion* (D, I, B, C, USA)	®Baytex (Hygiene) ®Entex (Hygiene) ®Lebaycid (Agricult.) ®Tiguvon (Vet. med.)	O,O-Dimethyl O-(3-methyl 4-methylthiophenyl) phosphorothioate
102	*fonofos* (B)	®Dyfonate	O-Ethyl S-phenyl ethanephosphonodithioate
134	*formothion* (D, I)	®Aflix ®Anthio	O,O-Dimethyl S-(N-methyl N-formyl carbamoyl) methyl phosphorodithioate
138	*fosfinon*	→ Forstenon	
152	HOE 2873	®Afugan	O,O-Diethyl O-[5-methyl 6-carbethoxy pyrazolo-(1,5a)-pyrimidyl-2] phosphorothioate
94	*iodofenphos*	®Nuvanol N	O,O-Dimethyl O-(2,5-dichloro-4-iodophenyl) phosphorothioate
	isochlorthion	®Nichlorphos	O,O-Dimethyl O-(3-nitro-4-chlorophenyl) phosphorothioate
135	Karbofos (USSR)	→ *malathion*	
124	M 74 (USSR)	→ *thiodemeton*	
121	M 81 (USSR)	→ *thiometon*	
135	Malaoxon		O,O-Dimethyl S-(1,2-dicarbethoxy)-ethyl phosphorothioate
135	*malathion* (D, I, B, F, C, USA)	Karbofos (USSR)	O,O-Dimethyl S-(1,2-dicarbethoxy-ethyl) phosphorodithioate
135	*mecarbam* (I, B, C, USA)	®Murfotox	O,O-Diethyl S-[(N-methyl N-carbethoxy) carbamoyl] methyl phosphoro-thioate

Table 19 (continued)

Page	Common names Code numbers	Other names	Scientific names
135	*medithionate*	→®Thiocron	
127	*menazon* (I, B, USA)	®Azidithion ®Saphizon ®Sayphos	O,O-Dimethyl S-(4,6-diamino-1,3,5-triazinyl-2) methyl phosphorodithioate
122	Mercaptofos (USSR)	→®Systox →*demeton*	
154	*merphos*	®Folex	
	Metafos (USSR)	Mixture of *parathion* and *parathion-methyl*	
91	Metathion (CSR)	→®Folithion	
98	*metaphenamiphos*	®Nemacur P	O-Ethyl O-(3-methyl 4-methylthiophenyl) N-isopropyl phosphoroamidate
113	*methamidophos*	→®Tamaron	
129	*methidathion* (B, D)	®Supracide	O,O-Dimethyl S-[(2-methoxy-5-oxo-Δ^2-1,3,4-thiadiazolinyl-4) methyl] phosphorodithioate
	methocrotophos (B)		O,O-Dimethyl O-[2-(N-methoxy N-methyl carbamoyl) 1-methyl] vinyl phosphate (Cis isomer)
	methyl-ethyl-thiometon	→®Teration	
120	*methyl demeton* (USA)	→®Metasystox	
	methyldemeton-methyl	→®Tinox	
118	*methyl-O-demeton*	→®Metasystox (i)	
89	*methyl parathion*	→ *parathion-methyl*	
120	Metilmercaptophos (USSR)	→®Metasystox	
91	Metilnitrofos (USSR)	→ fenitrothion (D, I, B)	
137	*metriphonate*	→ *trichlorfon* →®Dipterex →®Bilacil	
140	*mevinphos* (D, I, B, C, USA)	®Phosdrin	O,O-Dimethyl O-(1-methyl 2-carbomethoxy) vinyl phosphate
79	*mipafox* (I, B, C)	®Isopestox ®Pestox XV	N,N-Diisopropyl phosphorodiamidofluoridate
141	*monocrotophos* (B)	→®Azodrin	
134	*morphothion* (I, B, C, USA)	®Ekatin F ®Ekatin M ®Morphotox	O,O-Dimethyl S-(morpholinocarbonyl)-methyl phosphorodithioate
140	*naled* (D, C, USA)	®Dibrom	O,O-Dimethyl O-(1,2-dibromo-2,2-dichloroethyl) phosphate
87	Oktametil (USSR)	→ *schradan*	
	Oleo-Diazinon	*diazinon* + mineral oil	
	Oleo-Malathion	*malathion* + mineral oil	
	Oleo-Parathion	*parathion* + mineral oil	
133	*omethoate* (B)	→®Folimat	

Table 19 (continued)

Page	Common names Code numbers	Other names	Scientific names
87	OMPA	→*schradan*	
114	Ortho 12420		O,S-Dimethyl N-acetyl phosphoroamidothioate
106	*oxinothiophos*	®Bacdip	O-Ethyl O-quinolinyl-8 benzenephosphonothioate
119	*oxydemeton-methyl*	→®Metasystox R	
87	*paraoxon*	®E 600	O,O-Diethyl O-4-nitrophenyl phosphate
88	*parathion* (I, B, F, C, USA)	*äthylparathion* *parathion-ethyl* Tiofos (USSR) ®Alkron ®E 605 ®Folidol ®Niran ®Rhodiatox	O,O-Diethyl O-4-nitrophenyl phosphorothioate
89	*parathion-methyl* (D, I, B, C, USA)	*methylparathion* ®Folidol M ®Metacide	O,O-Dimethyl O-4-nitrophenyl phosphorothioate
93	*phenchlorphos*	→ *ronnel*	
115	*phenkapton* (D, I, B, C)		O,O-Diethyl S-(2,5-dichlorophenylthiomethyl) phosphorodithioate
130	*phenthoate*	*dimephenthoate* ®Cidial ®Erucin ®Papthion	O,O-Dimethyl S-(ω-carbethoxy) benzyl phosphorodithioate
114	*phorate* (I, B, C, USA)	®Thimet	O,O-Diethyl S-2-ethylthiomethyl phosphorodithioate
129	*phosalone* (B)	®Zolone	O,O-Diethyl S-(2-oxo-6-chloro-2H-benzoxazolinyl-3) methyl phosphorodithioate
137	*phoschlor* (P)	→®Dipterex	
128	*phosmet* (B)	→®Imidan	
142	*phosphamidon* (D, I, B, C, USA)	®Dimecron	O,O-Dimethyl O-(1-methyl 2-chloro-2-N,N-diethylcarbamoyl) vinyl phosphate
113	*phoxim*	®Baythion ®Valexon	O-(O,O-Diethyl thionophosphoryl) α-phenyl α-hydroximino-acetonitrile
108	*pirimiphos ethyl* (B)	®Diothyl ®Elimix	O,O-Diethyl O-(2-dimethylamino-4-methyl-pyrimidyl-6) phosphorothioate
	prothidathion (I, F, USSR)		O,O-Diethyl S-(2-oxo-5-isopropoxy 1,3,4-thiadiazolinyl-3) methyl phosphorodithioate
134	*prothoate* (I, B)	®Fac ®Fostion	O,O-Diethyl S-(N-isopropyl carbamoylmethyl) phosphorodithioate
152	*pyrazophos*	→®Afugan	
106	*quintiofos* (WHO)	→®Bacdip	

Table 19 (continued)

Page	Common names Code numbers	Other names	Scientific names
93	*ronnel* (C, USA)	*fenchlorphos* (B) *phenchlorphos* ®Etrolene ®Korlan ®Nankor ®Ronnel ®Trolene ®Viozene	O,O-Dimethyl O-(2,4,5-trichloro-phenyl) phosphorothioate
112	S 125	→®Maretin	
99	S 767	→®Terracur P	
97	S 1751	→®Lucijet	
96	S 1752	→ *fenthion* (D, I, B, C, USA)	
94	S 1942	→ *bromophos* (D, I)	
94, 101	S 4400	→®Agritox	
91	S 5660	→®Folithion	
133	S 6876	→®Folimat	
87	*schradan* (I, B, C, USA)	OMPA ®Pestox III ®Sytam	Octamethyl pyrophosphoroamidate
141	SD 3562	→®Bidrin	
141	SD 4294	→®Ciodrin	
141	SD 9129	→®Azodrin	
140	Shell OS 2046	→®Phosdrin	
	sophamide (I, B, F)	®Dithionate ®Formocarbam	O,O-Dimethyl S-[N-(methoxymethyl) carbamoylmethyl] phosphorodithioate
86	*sulfotepp* (D, USA)	*sulfotep* (I, B, C) Ditio (USSR) ®Bladafum ®Dithione	O,O,O′,O′-Tetraethyl pyrophosphoro-dithionate
	Sumioxon		O,O-Dimethyl O-(3-methyl 4-nitro-phenyl) phosphate
84	TEPP (I, B, C, USA)	®Tetron	Tetraethyl pyrophosphate
139	*tetrachlorvinphos* (B)	®Gardona	O,O-Dimethyl O-1-(2,4,5-trichloro-phenyl) 2-chlorovinyl phosphate (cis isomer)
148	TH 184-F		N,N,N′,N′-Tetramethyl O-pentachloro-phenyl phosphorodiamidate
124	*thiodemeton* (D)	*disulfoton* (D, I, B, C) M 74 (USSR) ®Disyston ®Frumin AL ®Solvirex ®Teration II	O,O-Diethyl S-(2-ethylthioethyl) phosphorodithioate
88	Tiofos (USSR)	→ *parathion* (I, B, F, C)	
121	*thiometon* (I, B, C)	M 81 (USSR) →®Ekatin	

Table 19 (continued)

Page	Common names Code numbers	Other names	Scientific names
108	*thionazin* (D, B)	®Zinophos ®Cynem ®Nemafos ®Nemaphos	O,O-Diethyl O-pyrazinyl-2 phosphorothioate
110, 151	*triamphos* (D)	→®Wepsyn 155	
110, 151	*triamiphos* (B)	→®Wepsyn 155	
137	*trichlorfon* (I, C, USA)	Chlorofos (USSR) Phoschlor (P) *metriphonate* *trichlorphon* (D, B) ®Dipterex ®Dylox ®Neguvon	O,O-Dimethyl (1-hydroxy-2,2,2-trichloro)-ethanephosphonate
94, 191	*trichloronate* (D)	®Agritox ®Phytosol	O-Ethyl O-(2,4,5-trichlorophenyl) ethanephosphonothioate
137	*trichlorphon* (D, B)	→ *trichlorfon*	
125	*vamidothion* (I, B)	®Kilval ®Vation	O,O-Dimethyl S-[2-(1-methylcarbamoyl) ethylthio]-ethyl phosphorothioate

5.3. References

1. Abt, W.: J. Prakt. Chem. **39**, 140 (1889).
2. Addor, R. W. (American Cyanamid Co.): A. P. 3.197.481 (1962/1965).
3. Addor, R. W.: J. Org. Chem. **29**, 738 (1964).
4. Addor, R. W.: J. Heterocycl. Chem. **7**, 381 (1970).
5. Addor, R. W., Lovell, J. B. (Amer. Cyanamid Co.): E. P. 974.138 (1962/1964), F. P. 1.327.386 (1962/1963).
6. Adolphi, H.: Z. Angew. Zool. **52** (2), 133 (1965).
7. Adrian, E. D., Feldberg, W., Kilby, B. A.: Brit. J. Pharmacol. **2**, 56 (1947).
8. Aharoni, A. H., O'Brien, R. D.: Biochemistry **7**, 1538 (1968).
9. Ahmed, M. K., Casida, J. E., Nichols, R. E.: J. Agr. Food Chem. **6**, 740 (1958).
10. Aldridge, W. N., Barnes, J. M.: Biochem. Pharmacol. **6**, 177 (1961).
11. Aldridge, W. N., Barnes, J. M.: Biochem. J. **15**, 541 (1966).
12. Aldridge, W. N., Barnes, J. M.: Neurotoxicity of drugs. In: Proceedings of the european society for the study of drug toxicity, vol. VIII, p. 162. Intern. Congr. Ser. No. 118. Amsterdam: Excerpta Medica Foundation 1967.
13. Aldridge, W. N., Davison, A. N.: Biochem. J. **51**, 62 (1952).
14. Allen, J. F.: J. Am. Chem. Soc. **79**, 3071 (1957).
15. Allen, J. F., Johnson, O. H.: J. Am. Chem. Soc. **77**, 2871 (1955).
16. Almasi, L., Paskucz, L.: Angew. Chem. **79**, 859 (1967).
17. Andres, L. A., Prout, T.: J. Econ. Entomol. **53**, 626 (1960).
18. Anliker, R., Beriger, E., Geiger, M., Schmid, K.: Helv. Chim. Acta, **44**, 1635 (1961).
19. Anonym: Agr. Res. (Washington) **15**, (1), 7 (1966).
20. Anonym (Amchem. Prod. Inc.): NE 6.802.633 (Appl. No.) (1967/1968).
21. Anonym: Ann. Chem. Pharmaz. **91**, 375 (1854).
22. Anonym (Aktiengesellschaft für Anilinfabrikation): D. R. P. 349.794 (1914/1922).
23. Anonym: Bayer-Pflanzenschutz-Kurier **12**, (3), (1967).
24. Anonym: Bayer-Pflanzenschutz „E 605" (firm's publication), Leverkusen 1953.
25. Anonym (C. H. Boehringer, Sohn): F. P. 1.133.785 (1955/1957).
26. Anonym (C. H. Boehringer, Sohn): F. P. 1.234.879 (1959/1960).
27. Anonym: Bull. World Health Organ. **29**, Suppl. (1963).
28. Anonym (132 ACS National Meeting): Chem. Eng. News **35**, 87 (1957)
29. Anonym: Chem. Labor Betrieb **12**, 559 (1967).
30. Anonym (Ciba AG): Swiss P. 310.395 (1951/1955).
31. Anonym (Degussa): F. P. 1.365.650 (1963/1964).
32. Anonym (Dow Chemical Co.): Belg.P. 579.237 (1959/1959).
33. Anonym (Farbwerke Hoechst AG): Belg. P. 676.802 (1966/1966), F. P. 1.481.094 (1966/1967).
34. Anonym (Food Machinery and Chem. Corp.): E. P. 783.697 (1953/1957).
35. Anonym (Food Machinery and Chem. Corp.): E. P. 784.985 (1953/1957).
36. Anonym (Food Machinery and Chem. Corp.): E. P. 784.986 (1953/1957).
37. Anonym (Geigy): F. P. 1.335.755 (1962/1963), E. P. 1.008.451 (1962/1965).
38. Anonym (Imperial Chemical Industry): F. P. 1.219.238 (1958/1960).
39. Anonym: J. Chem. Soc. **1952**, 5122.
40. Anonym: Med. Monatsschr. **19**, (10), 433 (1965).
41. Anonym (Montecatini): Derwent Commonwealth Pat.Rep. Issue No. 94, 22 (1956/1957).
42. Anonym (Philips' Gloeilampenfabrieken N. V.): Belg. P. 637.397 (1963/1964).
43. Anonym (Philips' Gloeilampenfabrieken N. V.): Belg. P. 637.399 (1962/1964).
44. Anonym (Sandoz Ltd.): Belg. P. 562.144 (1957/1957), E. P. 814.587 (1957/1959).

45. Anonym (Shell Chemical Co.): Agricultural Chem. **18**, 100 (April 1963).
46. Anonym (Standard Oil Development Co.): A. P. 2.553.770 (1950/1954).
47. Anonym (Sumitomo Chemical Co. Ltd.): Belg. P. 594.669 (1959/1961).
48. Anonym (Sumitomo Chemical Co. Ltd.): Belg. P. 624.058 (1961/1962).
49. Anonym (Sumitomo Chemical Co. Ltd.): NE. P. 6.717.383 (1967/1968).
50. Anonym: World Rev. Pest Contr. **6**, 80 (1967).
51. ANSCHÜTZ, R., EMERY, W. O.: Liebigs Ann. Chem. **253**, 105 (1889).
52. ARBUSOV, A. E.: ж **38**, 687 (1906) (Dissertation).
53. ARBUSOV, B. A.: Organo-phosphorus compounds, p. 307. London: Butterworths 1964.
54. ARBUSOV, A. E., ALIMOV, P. I.: Izvest. Akad. Nauk SSSR **1951**, 530.
55. ARBUSOV, A. E., RAZUMOV, A.: Zh. Obshch. Khim. **4**, 834 (1934).
56. ARIAS, R., TERRIERE, L. C.: J. Econ. Entomol. **55**, 925 (1962).
57. ARNDT, F.: Ber. **46**, 3522 (1913).
58. ARNDT, F., ROSENAU, B.: Ber. **50**, 1248 (1917).
59. ASCHER, K. R. S.: World Rev. Pest Contr. **3** (1), 7 (1964).
60. VAN ASPEREN, K., OPPENOORTH, F. J.: Entomol. Exp. Appl. **2**, 48 (1959).
61. VAN ASPEREN, K., OPPENOORTH, F. J.: Entomol. Exp. Appl. **3**, 68 (1960).
62. AUGUSTINSSON, K. B., JONSSON, G.: Acta Chem. Scand. **11**, 275 (1957).
63. BALAREFF, D.: Z. Anorg. Chem. **88**, 133 (1914).
64. BALAREFF, D.: Z. Anorg. Chem. **99**, 187 (1917).
65. BALAREFF, D.: Z. Anorg. Chem. **101**, 225 (1917).
66. BALLANTYNE, G. H., HARRISON, R. A.: Entomol. Exp. Appl. **10**, 231 (1967).
67. BARNARD, P. W. C., BUNTON, C. A., LLEWELLYN, D. R., VERNON, C. A., WELCH, J. A.: J. Chem. Soc. **1961**, 2670.
68. BARNES, J. M.: Trans. Roy. Soc. Trop. Med. Hyg. **58**, 327 (1964).
69. BARNES, J. M., DENZ, F. A.: J. Pathol. Bacteriol. **65**, 597 (1953).
70. BARON, R. L., CASIDA, J. E.: Biochem. Pharmacol. **11**, 1129 (1962).
71. BARTHEL, W. F., ALEXANDER, B. H., GIANG, P. A., HALL, S. A.: J. Am. Chem. Soc. **77**, 2424 (1955).
72. BARTHEL, W. F., GIANG, P. A., HALL, S. A.: J. Am. Chem. Soc. **76**, 4186 (1954).
73. BARTLETT, B. R.: J. Econ. Entomol. **61**, 297 (1968).
74. BARTLETT, J. B., RUDEL, H. W., CYPHERS, E. B. (Esso Research and Engineering Co.): A. P. 2.705.694 (1952/1955).
75. BAUMHOVER, A. H., GRAHAM, A. J., BITTER, B. A., HOPKINS, D. E., NEW, W. D., DUDLEY, F. H., BUSHLAND, R. C.: J. Econ. Entomol. **48**, 462 (1955).
76. BEACH, L. K. (Esso): E. P. 872.408 (1959/1961).
77. BECK, T. M., KLEIN, G. J. (Victor Chemical Works): A. P. 3.071.608 (1958/1963).
78. BECKER, E. L.: Biochim. Biophys. Acta **147**, 289 (1967).
79. BEHRENZ, W.: Vet. Med. Nachr. **1962**, 34.
80. BEHRENZ, W., FEDERMANN, M., BOLLE, W. R.: Vet. Med. Nachr. (Farbenfabriken Bayer AG) H. 4 (1959).
81. BELL, J. D., BUSVINE, J. R.: Entomol. Exp. Appl. **10**, 263 (1967).
82. BELLEAU, B., LAVOIE, J. L.: Can. J. Biochem. **46**, 1397 (1968).
83. BENDER, M. L.: Chem. Rev. **60**, 53 (1960).
84. BENOITON, L., HANSON, R. W., RYDON, H. N.: J. Chem. Soc. **1964**, 824.
85. BENSCHOP, H. P., KEIJER, J. H.: Biochim. Biophys. Acta **110**, 217 (1965).
86. BENSCHOP, H. P., KEIJER, J. H.: Biochim. Biophys. Acta **128**, 586 (1966).
87. BERGMANN, E., BONDI, A.: Ber. **63**, 1158 (1930).
88. BERGMANN, E., BONDI, A.: Ber. **64**, 1455 (1931).
89. BERGMANN, E., BONDI, A.: Ber. **66**, 278, 286 (1933).
90. BERIGER, E. (Ciba AG): D. A. S. 1.138.977 (1961/1962), F. P. 1.285.498 (1961/1962).
91. BERIGER, E. (Ciba AG): Offenlegungsschrift 1.493.569 (1965/1969).
92. BERIGER, E., SALLMANN, R. (Ciba AG): D. A. S. 1.098.939 (1956/1961), Belg. P. 552.284 (1955/1955).
93. BERKELHAMMER, G. (American Cyanamid Co.): A. P. 3.005.004 (1959/1961), A. P. 3.005.002 (1961/1961).
94. BEROZA, M., BOŘKOVEC, A. B.: J. Med. Pharm. Chem. **7**, 44 (1964).
95. BESTIAN, H.: Liebigs Ann. Chem. **566**, 210 (1950).

96. BIDSTRUP, P. L.: Brit. Med. J. **1950** (II), 548.
97. BIDSTRUP, P. L., BONNEL, J. A., BECKETT, A. G.: Brit. Med. J. **1953**, 1068.
98. BIRUM, G. H. (Monsanto Chemical Co.): A. P. 2.861.876 (1955/1958).
99. BISSEY, J. E.: J. Chem. Educ. **44**, 95 (1967).
100. BLAIR, E. H., KAUER, K. C., KENAGA, E. E.: J. Agr. Food Chem. **11**, 237 (1963); J. Agr. Food Chem. **14**, 298 (1966).
101. BLAIR, E. H., WASCO, I. L., KENAGA, E. E.: J. Agr. Food Chem. **13**, 383 (1965).
102. BLISS, C. I.: Science **79**, 38, 409 (1934).
103. BLISS, C. I.: Ann. Appl. Biol. **26**, 585 (1939).
104. BLUMENTHAL, E., HERBERT, J. B. M.: Trans. Faraday Soc. **41**, 611 (1945).
105. BOGATYREV, I. A., KUŠELEV, I. V., MELNIK, P. V.: Ref. Ž. Chim. 11 N, 502 (1966).
106. BOON, W. R.: Agric. Pesticide Tech. Soc. Proc. **4**, 1 (1957).
107. BOOTH, H. S., DUTTON, F. B.: J. Am. Chem. Soc. **61**, 2937 (1939).
108. BOŘKOVEC, A. B.: Science **137**, 1034 (1962).
109. BOŘKOVEC, A. B.: Residue Rev. **6**, 87 (1964).
110. BOŘKOVEC, A. B.: Advan. Pest Control Res. **7** (1966).
111. BOŘKOVEC, A. B., CHANG, S. C., LIMBURG, A. M.: J. Econ. Entomol. **57**, 815 (1964).
112. BOŘKOVEC, A. B., WOODS, C. W.: Advan. Chem. Ser. **41**, 47 (1963).
113. BOŘKOVEC, A. B., WOODS, C. W., BROWN, R. T.: J. Med. Chem. **9**, 522 (1966).
114. VAN DEN BOS, B. G.: Rec. Trav. Chim. **79**, 1129 (1960).
115. VAN DEN BOS, B. G., KOOPMANS, M. J., HUISMAN, H. O.: Rec. Trav. Chim. **79**, 807 (1960).
116. VAN DEN BOS, B. G., MELTZER, J. (North American Philip Co.): A. P. 3.326.751 (1963/1967).
117. VAN DEN BOS, B. G., SCHIPPERHEYN, A., VAN DEURSEN, F. W.: Rec. Trav. Chim. **85**, 429 (1966).
118. VAN DEN BOS, B. G., SCHOOT, C. J., KOOPMANS, M. J., MELTZER, J.: Rec. Trav. Chim. **80**, No. 9–10, 1040 (1961).
119. BOTER, H. L., OOMS, A. J. J., VAN DEN BERG, G. R., VAN DIJK, C.: Rec. Trav. Chim. **85**, 147 (1966).
120. BOUSH, G. M., MATSUMURA, F.: J. Econ. Entomol. **60**, 918 (1967).
121. BOYD, D. B.: J. Am. Chem. Soc. **91**, 1200 (1969).
122. BOYD, N. R., JR., ARTHUR, B. W.: J. Econ. Entomol. **53**, 848 (1960).
123. BOYER, W. P. (Virginia Carolina Chemical Corp.): A. P. 2.761.806 (1954/1956), E. P. 791.818 (1955/1958).
124. BRACHA, P., O'BRIEN, R. D.: Biochemistry **7**, 1545 (1968).
125. BRACHA, P., O'BRIEN, R. D.: Biochemistry **7**, 1555 (1968).
126. BRADY, U. E., JR., ARTHUR, B. W.: J. Econ. Entomol. **56**, 477 (1963).
127. BRADY, U. E., STERNBURG, J.: J. Insect Physiol. **12**, 1171 (1966).
128. BRADY, U. E., STERNBURG, J.: J. Insect Physiol. **13**, 369 (1967).
129. BRADY, J. G., SZABO, K. (Stauffer Chemical Co.): A. P. 3.128.224 (1962/1964).
130. BRÄHLER, B., ZIMMERMANN, R., REESE, J. (Chem. Werke Albert): D. A. S. 1.125.929 (1958/1962).
131. BRAID, P. E., NIX, M.: Can. J. Biochem. **47**, 1 (1969).
132. BRAY, R. S., GARNHAM, P. C. C.: Bull. World Health Organ. **31**, (1), 143 (1964).
133. BRODIE, B. B., GILLETTE, J. R., LA DU, B. N.: Ann. Rev. Biochem. **27**, 427 (1958).
134. BROWN, N. C., HOLLINSHEAD, D. T. (Cooper, Mc Dougall & Robertson Ltd.): A. P. 3.257.416 (1963/1966), Belg. P. 610.896 (1961/1962).
135. BROWN, N. C., HOLLINSHEAD, D. T., KINGSBURY, P. A., MALONE, I. C.: Nature **194**, 379 (1962).
136. BROWN, B. B., LUTZ, L. J., SMITH, C. J., JR., WEIHMÜNSTER, E. A. (Olin Mathieson Corp.): A. P. 3.167.574 (1954/1965).
137. BRYNER, F. (Dow Chemical Co.): A. P. 2.726.270 (1951/1955).
138. BUCHANAN, G. H. (American Cyanamid Co.): A. P. 1.868.192 (1930/1932).
139. BUCHNER, B., LOCKHART, L. B., JR.: J. Am. Chem. Soc. **73**, 755 (1951); Org. Syn. **31**, 88 (1951).
140. BUDDE, P. B., TOLKMITH, H. (Dow Chemical Co.): A. P. 3.323.990 (1965/1967).
141. BULL, D. L.: J. Econ. Entomol. **58**, No. 2, 249 (1965).
142. BULL, D. L., LINDQUIST, D. A.: J. Agr. Food Chem. **12**, No. 4, 311 (1964).
143. BULL, D. L., LINDQUIST, D. A.: J. Agr. Food Chem. **14**, No. 2, 105 (1966).
144. BULL, D. L., LINDQUIST, D. A., GRABBE, R. R.: J. Econ. Entomol. **60**, No. 2, 332 (1967).
145. BULL, D. L., LINDQUIST, D. A., HACSKAYLO, J.: J. Econ. Entomol. **56**, No. 2, 129 (1963).
146. BUNTON, C. A.: J. Chem. Educ. **45**, 21 (1968).
147. BUNTON, C. A., ROBINSON, L.: J. Am. Chem. Soc. **91**, 6072 (1969).

148. Burt, P. E.: Pestic. Sci. **1**, 88 (1970).
149. Burt, P. E., Gregory, G. E., Molloy, F. M.: Ann. Appl. Biol. **59**, 1 (1967).
150. Busbey, R. L.: J. Econ. Entomol. **54**, 254 (1961).
151. Bushland, R. C.: Advan. Vet. Sci. **6**, 1 (1960).
152. Bushland, R. C., Hopkins, D. E.: J. Econ. Entomol. **44**, 725 (1951); **46**, 648 (1953).
153. Bushland, R. C., Lindquist, A. W., Knipling, E. F.: Science **122**, 287 (1955).
154. Busvine, J. R.: Bull. World Health Organ. **29**, Suppl. 31 – 40 (1963).
155. Buttram, J. R., Arthur, B. W.: J. Econ. Entomol. **54**, No. 3, 446 (1961).
156. Calderbank, A., Edgar, E. C., Silk, J. A.: Chem. Ind. (London) **1961**, 630.
157. Calderbank, A., Ghosh, R.: J. Chem. Soc. **1960**, 637.
158. Campion, D. G.: DANS-Pest Articles and News Summaries Sect. A-Insect Control **11** (4), 467 (1965).
159. Caseley, J.: Pestic. Sci. **1**, 114 (1970).
160. Casida, J. E.: Biochem. Pharmacol. **5**, 332 (1961).
161. Casida, J. E., Arthur, B. W. (Wisconsin Alumni Research Foundation): A. P. 2.911.435 (1957/1959), A. P. 2.927.881 (1956/1960).
162. Casida, J. E., Baron, R. L., Eto, M., Engel, J. L.: Biochem. Pharmacol. **12**, 73 (1963).
163. Casida, J. E., McBride, L., Niedermeier, R. P.: J. Agr. Food Chem. **10**, No. 5, 370 (1962).
164. Casida, J. E., Gatterdam, P. E., Knaak, J. B., Lance, R. D., Niedermeier, R. P.: J. Agr. Food Chem. **6**, No. 9, 658 (1958).
165. Casida, J. E., Sanderson, D. M.: Nature **189**, 507 (1961).
166. Casida, J. E., Stahmann, M. A.: J. Agr. Food Chem. **1**, 883 (1953).
167. Cassaday, J. T. (American Cyanamid Co.): A. P. 2.578.652 (1950/1951), D. B. P. 847.897 (1951/1952).
168. Cassaday, J. T., Hoegberg, E. I., Gleissner, B. D. (American Cyanamid Co.): A. P. 2.494.283 (1948/1950), D. B. P. 819.998 (1949/1951).
169. Cassidy, J. E., Ryskiewich, D. P., Murphy, R. T.: J. Agr. Food Chem. **17**, No. 3, 558 (1969).
170. Castle, R. E., Ristich, S. S.: J. Agr. Food Chem. **14**, 301 (1966).
171. Cavanagh, J. B.: J. Neurol. Neurosurg. Psychiat. **17**, 163 (1954).
172. Chaboussou, F.: Z. Angew. Zool. **53**, 257 (1966).
173. Chamberlain, W. F.: J. Econ. Entomol. **55**, 240 (1962).
174. Chamberlain, W. F., Gatterdam, P. E., Hopkins, D. E.: J. Econ. Entomol. **54**, No. 4, 733 (1961).
175. Chance, L. H. (Secretary of Agriculture): A. P. 3.146.228 (1952/1964).
176. Chang, S. C., Bořkovec, A. B.: J. Econ. Entomol. **57**, 488 (1964).
177. Chang, S. C., Bořkovec, A. B.: J. Econ. Entomol. **59**, 1359 (1966).
178. Chang, S. C., Terry, P. H., Bořkovec, A. B.: Science **144**, 57 (1964).
179. Cheymol, J., Chabrier, P., Selim, M., Leduc, P.: Compt. Rend. **247**, 1014 (1958).
180. Chiu, Y. C., Hassan, A., Guthrie, F. E., Dautermann, W. C.: Toxicol. Appl. Pharmacol. **12**, 219 (1968).
181. Chopard, P. A., Clark, V. M., Hudson, R. F., Kirby, A. I.: Tetrahedron **21**, 1961 (1965).
182. Christen, P. J., van den Muysenberg, J. A. C. M.: Biochim. Biophys. Acta **110**, 217 (1965).
183. Christensen, B. G. (Merck & Co. Inc.): Offenlegungsschrift 1.805.685 (1968/1970).
184. Christensen, B. G., Leanza, W. J., Beattie, T. R., Patchett, A. A., Arison, B. H., Ormond, R. E., Kuehl, F. A., Jr., Albers-Schonberg, G., Jardetzky, O.: Science **166**, 123 (1969).
185. Christmann, L. J. (American Cyanamid Co.): A. P. 1.893.018 (1928/1933).
186. Chupp, J. P., Newallis, P. E. (Monsanto Chem. Co.): A. P. 3.099.597 (1961/1963), E. P. 995.359 (1962/1965).
187. Chupp, J. P., Newallis, P. E. (Monsanto Chemical Co.): Belg. P. 615.988 (1962/1962), A. P. 3.145.140 (1962/1964).
188. Chupp, J. P., Speziale, A. J. (Monsanto Chemical Co.): A. P. 3.102.018 (1960/1963).
189. Chupp, J. P., Speziale, A. J. (Monsanto Chemical Co.): A. P. 3.102.022 (1960/1963).
190. Clark, V. M., Hutchinson, D. W., Kirby, A. I., Warren, S. G.: Angew. Chem. **76**, 704 (1964).
191. Clay, J. P.: J. Org. Chem. **16**, 892 (1951).
192. Cleland, W. W.: Biochim. Biophys. Acta **67**, 104 (1963).
193. Cleland, W. W.: Biochim. Biophys. Acta **67**, 173 (1963).
194. Cleland, W. W.: Biochim Biophys. Acta **67**, 188 (1963).
195. de Clermont, Ph.: Compt. Rend. **39**, 338 (1854).

196. CLITHEROW, J. W., MITCHARD, M., HARPER, N. J.: Nature **199**, H. 4897, 1000 (1963).
197. CLOEZ, M.: Compt. Rend. **24**, 388 (1847).
198. COATES, H., CROFTS, P. C. (Albright & Wilson Ltd.): E. P. 713.669 (1951/1954).
199. COE, D. G., HURTIG, H., PERRY, B. J., SHERLOCK, E. S.: J. Agr. Food Chem. **7**, 251 (1959).
200. COE, D. G., PERRY, B. J., BROWN, R. K.: J. Chem. Soc. **1957**, 3604.
201. COLEMAN, I. W., PATTON, G. E., BANNARD, R. A. B.: Can. J. Physiol. Pharmacol. **46**, 109 (1967).
202. COLHOUN, E. H.: Advan. Insect Physiol., vol. I, p. 1. London-New York: Academic Press 1964.
203. COLLIN, R. L.: J. Am. Chem. Soc. **88**, 3281 (1966).
204. CÖLLN, R. (Farbenfabriken Bayer AG): internal report 1963.
205. CÖLLN, R. (Farbenfabriken Bayer AG): D. A. S. 1.216.278 (1963/1966).
206. CÖLLN, R. (Farbenfabriken Bayer AG): Belg. P. 700.436 (1967/1967).
207. CÖLLN, R., SCHRADER, G. (Farbenfabriken Bayer AG): D. A. S. 1.059.425 (1957/1959).
208. CÖLLN, R., SCHRADER, G. (Farbenfabriken Bayer AG): D. A. S. 1.099.535 (1959/1961).
209. CÖLLN, R., SCHRADER, G. (Farbenfabriken Bayer AG): D. A. S. 1.132.132 (1959/1962).
210. CÖLLN, R., SCHRADER, G. (Farbenfabriken Bayer AG): D. A. S. 1.150.979 (1959/1963).
211. COLLOMP: Les Trilons, Bull. Inf. Scient. Min. Guerre (Sect. Techn. de l'Armée) Paris 1949, No. 23 G, 1.
212. COOK, J. W., BLAKE, J. R., YIP, G., WILLIAMS, M.: J. Assoc. Offic. Agr. Chemists **41**, Nr. 2, 400 (1958).
213. COREY, R. A., J. Econ. Entomol. **58**, 112 (1965).
214. COREY, R. A., DORMAN, S. C., HALL, W. E., GLOVER, L. C., WHETSTONE, R. R., Science **118**, 28 (1953).
215. COWEN, F. M. (American Cyanamid Co.): D. B. P. 938.548 (1953/1956).
216. COX, J. R., JR., RAMSAY, O. B.: Chem. Rev. **64**, 317 (1964).
217. CRAMER, F.: Angew. Chem. **72**, 236 (1960).
218. CRAMER, H. H.: Pflanzenschutz-Nachr. Bayer **20** (1), 1 (1967).
219. CRAMER, H. H.: 6. Int. Welt-Getreide-und-Brot-Kongreß, Dresden 24. Mai 1970 (in press).
220. CRAMER, F., HETTLER, H.: Chem. Ber. **91**, 1181 (1958).
221. CRAMER, F., MACKENSEN, G.: Angew. Chem. **78**, 641 (1966).
222. CRAMER, F., WINTER, M.: Chem. Ber. **94**, 989 (1961).
223. CREMER, J.: Chem. Ingr.-Tech. **37**, 705 (1965).
224. CRYSTAL, M. M.: J. Econ. Entomol. **56**, 468 (1963).
225. CRYSTAL, M. M.: J. Econ. Entomol. **57**, 726 (1964).
226. CUNNINGHAM, L. W.: Science **125**, 1145 (1957).
227. DAUTERMANN, W. C., MAIN, A. R.: Toxicol. Appl. Pharmacol. **9**, 408 (1966).
228. DAVEY, K. G.: Advan. Insect Physiol., vol. II, p. 219. London-New York: Academic Press 1964.
229. DAVICH, T. B., APPLE, J. W.: J. Econ. Entomol. **44**, 528 (1951).
230. DAVICH, T. B., KELLER, J. C., MITCHELL, E. B., HUDDLESTON, P., HILL, R., LINDQUIST, D. A.: J. Econ. Entomol. **58**, 127 (1965).
231. DAVID, W. A. L., KILBY, B. A.: Nature **164**, 522 (1949).
232. DAVIES, N., DICK, J. H.: J. Chem. Soc. **1932**, 483.
233. DAVIES, D. R., HOLLAND, P., RUMENS, M. J.: Brit. J. Pharmacol. **15**, 271 (1960).
234. DAVISON, A. N.: Brit. J. Pharmacol. **8**, 212 (1953).
235. DAWSON, TH. P., ARMSTRONG, J. W. (Secretary of the Army): A. P. 2.847.469 (1955/(1958).
236. DEDEK, W., SCHWARZ, H.: Atompraxis **12**, H. 12, 603 (1966).
237. DELFS, D., WEDEMEYER, K. (Farbenfabriken Bayer AG): D. A. S. 1.051.863 (1957/1959).
238. DELFS, D., WEDEMEYER, K. (Farbenfabriken Bayer AG): D. A. S. 1.063.177 (1957/1959).
239. DEWALD, C. L., FANCHER, L. W. (Stauffer Chemical Co.): F. P. 1.327.963 (1962/1963).
240. DITTRICH, V.: J. Econ. Entomol. **56**, 182 (1963).
241. DITTRICH, V.: J. Econ. Entomol. **62**, 44 (1969).
242. DIVELEY, W. R. (Hercules Powder Co.): A. P. 2.864.826 (1956/1958).
243. DIVELEY, W. R., BUNTIN, G. A., HAUBEIN, A. H.,MCCLELLAN SPECK, RH. (Hercules Powder Co.): D. A. S. 1.052.161 (1956/1959).
244. DIVELEY, W. R., HAUBEIN, A. H., LOHR, A. D., MOSELEY, P. B.: J. Am. Chem. Soc. **81**, 139 (1959).
245. DIVELEY, W. R., LOHR, A. D. (Hercules Powder Co.): A. P. 2.725.328 (1954/1955).

246. Dixon, J. K., DuBreuil, Sh., Boardway, N. L., Gordon, F. M. (American Cyanamid Co.): D. A. S. 1.156.274 (1958/1963), Belg. P. 570.965 (1958/1959).
247. Dorough, H. W., Randolph, N. M.: J. Econ. Entomol. **60**, 285 (1967).
248. Dortmann, H. A., Delfs, D., Schrader, G. (Farbenfabriken Bayer AG): Belg. P. 558.542 (1957/1957).
249. Drábek, J., Pelikán, J.: Chem. Průmysl **6**/31, 293 (1956), C. 11841 (1957).
250. Drago, R. S., Mode, V. A., Kay, J. G., Lydy, D. L.: J. Am. Chem. Soc. **87**, 5010 (1965).
251. Drummond, R. O.: J. Econ. Entomol. **53**, 1125 (1960).
252. DuBois, K. P., Kinoshita, F. K., Frawley, J. P.: Toxicol. Appl. Pharmacol. **12**, 273 (1968).
253. DuBreuil, Sh., Young, R. W., Ailman, D. E., Berkelhammer, G. (American Cyanamid Co.): D. A. S. 1.164.408 (1961/1964).
254. Dudek, G. O., Westheimer, F. H.: J. Am. Chem. Soc. **81**, 2641 (1959).
255. Duffey, G. H.: J. Chem. Phys. **17**, 196 (1949).
256. Ebeling, W., Wagner, R. E., Reierson, D. A.: J. Econ. Entomol. **59**, 1374 (1966).
257. Eccles, J. C.: Angew. Chem. **76**, 674 (1964).
258. Edwards, J. S., Gomez, D.: J. Insect Physiol. **12**, 1061 (1966).
259. Edwards, J. O., Pearson, R. G.: J. Am. Chem. Soc. **84**, 16 (1962).
260. Engelhard, N., Prchal, K., Nenner, M.: Angew. Chem. **79**, 604 (1967).
261. Ephraim, F.: Ber. **44**, 631 (1911).
262. Erdmann, W. D.: Arzneimittel-Forsch. **15**, 135 (1965).
263. Erdmann, W. D., v. Clarmann, M.: Deut. Med. Woch.schr. **88**, 2201 (1963).
264. Erdmann, W. D., Engelhard, H.: Arzneimittel-Forsch. **14**, 5 (1964).
265. Erdmann, W. D., et al., Oettel, H.: Dtsch. Ärzteblatt **31**, 1644 (1967).
266. Erdmann, W. D., Zech, R., Franke, P., Bosse, I.: Arzneimittel-Forsch. **16**, 492 (1966).
267. Erlenmeyer-Kimling, L., Hirsch, J., Weiss, J. M.: J. Comp. Physiol. Psychol. **55**, 722 (1962).
268. Eto, M., Casida, J. E., Eto, T.: Biochem. Pharmacol. **11**, 337 (1962).
269. Eto, M., Kinoshita, Y., Kato, T., Oshima, Y.: Agr. Biol. Chem. (Tokyo) **27**, 789 (1963).
270. Eto, M., Kishimoto, K., Matsumura, K., Ohshita, N., Oshima, Y.: Agr. Biol. Chem. (Tokyo) **30**, 181 (1966).
271. Eto, M., Ohkawa, H., in: O'Brien, Biochemical toxicology of insecticides. New York: Academic Press Inc. 1970.
272. Eto, M., Oshima, Y.: Agr. Biol. Chem. (Tokyo) **26**, 452 (1962).
273. Fancher, L. W. (Stauffer Chemical Co.): A. P. 2.767.194 (1955/1956).
274. Fancher, L. W. (Stauffer Chemical Co.): A. P. 2.793.224 (1954/1957).
275. Fancher, L. W. (Stauffer Chemical Co.): D. A. S. 1.112.735 (1958/1961).
276. Farooq, M., Hairston, N. G., Samaan, S. A.: Pflanzenschutz-Nachr. Bayer **19**, (1) 43 (1966).
277. Fay, R. W., Kilpatrick, J. W., Morris, G. C.: J. Econ. Entomol. **51**, 452 (1958).
278. Fenwick, M. L.: Biochem. J. **70**, 373 (1958).
279. Fest, C., Schrader, G. (Farbenfabriken Bayer AG): E. P. 979.568 (1963/1965), A. P. 3.284.455 (1963/1966).
280. Finger, H.: J. Prakt. Chem. **37**, 431 (1888).
281. Fink, H., Hund, G.: Arzneimittel-Forsch. **15**, 624 (1965).
282. Fink, H., Hund, G., Meysing, D.: Method. Inform. Med. **5**, 19 (1966).
283. Finney, D. J.: Probit analysis, p. 88. London: Cambridge Univ. Press 1952.
284. Fiszer, B., Michalski, J.: Roczniki Chem. **25**, 514 (1951).
285. Fiszer, B., Michalski, J.: Roczniki Chem. **26**, 688 (1952).
286. Fiszer, B., Michalski, J., Wieczorkowski, J.: Roczniki Chem. **27**, 482 (1953).
287. Fletcher, J. H. (American Cyanamid Co.): A. P. 2.664.437 (1950/1953).
288. Fletcher, J. H., Hamilton, J. C., Hechenbleikner, I., Hoegberg, E. I., Sertl, B. J., Cassaday, J. T.: J. Am. Chem. Soc. **70**, 3943 (1948).
289. Fletcher, J. H., Hamilton, J. C., Hechenbleikner, I., Hoegberg, E. I., Sertl, B. J., Cassaday, J. T.: J. Am. Chem. Soc. **72**, 2461 (1950).
290. Florkin, M., Stolz, E. H. (eds.): Comprehensive biochemistry, vol. 13. Report of the Commission on Enzymes of the International Union of Biochemistry. Amsterdam-London-New York: Elsevier Publishing Comp. 1964.
291. Fluck, E., Binder, H.: Angew. Chem. **79**, 243 (1967).
292. Flynn, A. D., Eden, W. G.: J. Econ. Entomol. **53**, 692 (1960).

293. Folckemer, F. B., Hanson, R. E., Miller, A. (Shell Oil Co.): A. P. 3.318.769 (1963/1967).
294. Forgash, A. J.: J. Econ. Entomol. **60**, 1596 (1967).
295. Foss, O.: Acta Chem. Scand. **1**, 310 (1947).
296. Fox, D. L., Turner, E. E.: J. Chem. Soc. **1930**, 1858.
297. Francis, J. I., Barnes, J. M.: Bull. World Health Organ. **29**, 205 (1963).
298. Frear, D. E. H.: Pesticide handbook-entoma. 20th edition. College Science Publisher, State College Pennsylvania 1968.
299. Fricke, G., Georgi, W.: J. Prakt. Chem. **20**, 4th Ser. 250 (1963).
300. Frohberger, P. E.: Hoefchen Briefe **2** (2), 10 (1949).
301. Fucks, W.: Formeln zur Macht, p. 143, 144. Stuttgart: Deutsche Verlagsanstalt 1966.
302. Fujita, T.: J. Med. Chem. **9**, 797 (1966).
303. Fukami, J., Shishido, T.: Botyu-Kagaku **28**, 63 (1963).
304. Fukami, J., Shishido, T.: Botyu-Kagaku **28**, 77 (1963).
305. Fukami, J., Shishido, T.: J. Econ. Entomol. **59**, 1338 (1966).
306. Fukuda, K., Hayashi, Sh., Owada, T., Munekata, T. (Kureha Kagaku Kogyo Kabushiki Kaisha): D. A. S. 1.250.425 (1965/1967).
307. Fukuto, T. R.: Advances in pest control research. vol. I, p. 165 (ed. R. L. Metcalf). New York-London: Interscience Publishers 1957.
308. Fukuto, T. R.: Advances in pest control research. vol. I, p. 172 (ed. R. L. Metcalf). New York-London: Interscience Publishers 1957.
309. Fukuto, T. R.: Ann. Rev. Entomol. **6**, 313 (1961).
310. Fukuto, T. R., Metcalf, R. L.: J. Am. Chem. Soc. **76**, 5103 (1954).
311. Fukuto, T. R., Metcalf, R. L.: J. Agr. Food Chem. **4**, 930 (1956).
312. Fukuto, T. R., Metcalf, R. L.: J. Am. Chem. Soc. **81**, 372 (1959).
313. Fukuto, T. R., Metcalf, R. L. Frederickson, M., Winton, M. Y.: J. Agr. Food Chem. **12**, 228 (1964).
314. Fukuto, T. R., Metcalf, R. L., Jones, R. L., Myers, R. O.: J. Agr. Food Chem. **17**, 923 (1969).
315. Fukuto, T. R., Metcalf, R. L., March, R. B., Maxon, M. G.: J. Econ. Entomol. **48**, 347 (1955); **49**, 147 (1956).
316. Fukuto, T. R., Metcalf, R. L., Winton, M. Y.: J. Econ. Entomol. **54**, 955 (1961).
317. Fukuto, T. R., Metcalf, R. L., Winton, M. Y., March, R. B.: J. Econ. Entomol. **56**, 808 (1963).
318. Fukuto, T. R., Stafford, E. M.: J. Am. Chem. Soc. **79**, 6083 (1957).
319. Fukuto, T. R., Wolf, I. P., III, Metcalf, R. L., March, R. B.: J. Econ. Entomol. **49**, 147 (1956).
320. Fusco, R., Losco, G., Perini, M. (Montecatini): F. P. 1.153.596 (1956/1958).
321. Gaines, J. C., Ivy, E. E., King, C. E.: J. Econ. Entomol. **44**, 750 (1951).
322. Gaines, T. B., Kimbrough, R. D.: Bull. World Health Organ. **34**, (2), 317 (1966).
323. Gann, P. W., Heider, R. L. (Monsanto Chemical Co.): A. P. 2.692.890 (1949/1954).
324. Gardiner, J. E., Kilby, B. A.: J. Chem. Soc. **1950**, 1769.
325. Garrison, A. W., Boozer, C. E.: J. Am. Chem. Soc. **90**, 3486 (1968).
326. Gatterdam, P. E.: Entomologica Society of America, St. Louis, December 1963.
327. Gatterdam, P. E., Wozniak, L. A., Bullock, M. W., Parks, G. L., Boyd, J. E.: J. Agr. Food Chem. **15**, 845 (1967).
328. Gätzi, K., Müller, P. (Geigy AG): D. B. P. 957.213 (1955/1957).
329. Geoghegan, J. T., McPherson, J. B. (American Cyanamid Co.): A. P. 2.784.207 (1954/1957).
330. Gerolt, P.: J. Insect Physiol. **15**, 563 (1969).
331. Gerrard, W.: J. Chem. Soc. **1940**, 1464.
332. Gersdorff, W. A., Mitlin, N.: J. Econ. Entomol. **44**, 474 (1951).
333. Getzin, L. W.: J. Econ. Entomol. **60**, 505 (1967).
334. Ghosh, R. (Imperial Chemical Industries): E. P. 738.839 (1953/1955).
335. Ghosh, R. (Imperial Chemical Industries): E. P. 763.516 (1955/1956).
336. Ghosh, R., Newman, J. F.: Chem. Ind. (London) **1955**, 118.
337. Gilbert, E. E. (Allied Chemical Co.): A. P. 2.891.887 (1957/1959).
338. Gilbert, E. E., Otto, J. A. (Allied Chemical Co.): A. P. 3.151.022 (1962/1964).
339. Gilmore, J. E., Munger, F.: J. Econ. Entomol. **60**, 52 (1967).
340. Glees, P.: Deut. Med. Wochschr. **86**, 1175 (1961).
341. Glees, P., Janzik, H.: Progr. Brain Res. **14**, 97 (1965).
342. Glees, P., Nauta, W. J. H.: Monatsschr. Psychiat. Neurol. **129**, 74 (1955).

343. Godovikov, N. N., Kabachnik, M. I.: Zh. Obshch. Khim. **31**, 1628 (1961).
344. Gönnert, R.,Webbe, G.: Pflanzenschutz-Nachr. Bayer **19**, (1) 36 (1965).
345. Gordon: "Information on poison gas manufacture in Germany". Report 12. U.S. Department of Commerce Washington D.C. (1945).
346. Gottlieb, H. B.: J. Am. Chem. Soc. **54**, 748 (1932).
347. Goubeau, J.: Angew. Chem. **81**, 343 (1969).
348. Goulding, R. L.: J. Econ. Entomol. **55**, 577 (1962).
349. Graf, R. (Farbwerke Hoechst AG): Chem. Ber. **85**, 9, (1952); D. B. P. 910.649 (1943/1954).
350. Graham, O. H.: J. Econ. Entomol. **54**, 1046 (1961).
351. Graham, H. M., Martin, D. F., Ouye, M. T., Hardman, R. M.: J. Econ. Entomol. **59**, 950 (1966).
352. Grayson, J. M.,Cochran, D. G.: World Rev. Pest Contr. **7**, 172 (1968).
353. Green, M., Hudson, R. F.: J. Chem. Soc. **1963**, 540, 3883.
354. Gretillat, S.: Agr. Vet. Chem. **7** (3), 61 (1966).
355. Groenweghe, L. C. D., Payne, J. H.: J. Am. Chem. Soc. **83**, 1811 (1961).
356. Gross, E.: see Monograph G. Schrader, 3rd edition 1963, p. 4.
357. Groves, K.: J. Agr. Food Chem. 6, 30 (1958).
358. Gruner, L.: Rev. Zool. Agr. Appl. **65**, (1 – 3), 1 (1966).
359. Gunther, F. A., Blinn, R. C.: Analysis of insecticides and acaricides, Vol. 6, p. 476. New York-London: Interscience Publishers 1955.
360. Gysin, H., Margot, A. (Geigy AG): D. B. P. 910.652 (1952/1954), A. P. 2.754.243 (1953/1956).
361. Haake, P. C., Westheimer, F. H.: J. Am. Chem. Soc. **83**, 1102 (1961).
362. Hagedorn, I., Gündel, W. H., Schoene, K.: Arzneimittel-Forsch. **19**, 603 (1969).
363. Hall, S. A., Jacobson, M.: Ind. Eng. Chem. **40**, I, 694 (1948).
364. Hall, W. E., Sun, Y. P.: J. Econ. Entomol. **58**, 845 (1965).
365. Hammann, I.: Pflanzenschutz-Nachr. Bayer **23** (2), 140 (1970).
366. Hammond, P. R.: J. Chem. Soc. **1962**, 2521.
367. Hamon, J.: Bull. World Health Organ. **29** Suppl., 115 (1963).
368. Hansch, C., Deutsch, E. W.: Biochim. Biophys. Acta **112**, 381 (1966); **126**, 117 (1966).
369. Hansen, R.: Acta Chem. Scand. **16**, 1593 (1962).
370. Hardy, E. E., Kosolapoff, G. M. (Monsanto Chemical Co.): A. P. 2.409.039 (1944/1946).
371. Harman, D., Stiles, A. R. (Shell Development Co.): A. P. 2.630.450 (1949/1953).
372. Harris, C. R.: J. Econ. Entomol. **62**, 1437 (1969).
373. Harris, E. E. (Merck & Co. Incorporated): Offenlegungsschrift 1.924.259 (1969/1970).
374. Harris, W. D., Feldman, A. W. (U.S. Rubber Co., now Uniroyal Incorp.): A. P. 2.828.198 (1955/1958).
375. Harris, C. R., Hitchon, J. L.: J. Econ. Entomol. **63**, 2, (1970).
376. Hartley, G. S. (Pest Control Ltd.): D. B. P. 895.228 (1950/1953).
377. Hartley, G. S., Pound, D. W. (Pest Control Ltd.): E. P. 652.981 (1948/1951), A. P. 2.671.109 (1949/1954).
378. Hasan, S., Vago, C.: Plant Disease Reptr. **50** (10), 764 (1966).
379. Hassan, A., Dautermann, W. C.: Biochem. Pharmacol. **17**, 1431 (1968).
380. Hata, T., Mukaiyama, T.: Bull. Chem. Soc. Japan **35**, II, 1106 (1962).
381. Haubein, A. H., Buntin, G. A., Diveley, W. R., Lohr, A. D. (Hercules Powder Co.): D. A. S. 1.017.848 (1955/1957).
382. Hayes, W. J., Jr.: Bull. World Health Organ. **31** (5), 721 (1964).
383. Haynes, J. W., Hedin, P. A., Davich, T. B.: J. Econ. Entomol. **59**, 1014 (1966).
384. Heap, R., Saunders, B. C.: J. Chem. Soc. **1948**, 1313.
385. Heath, D. F.: J. Chem. Soc. **1956**, 3796.
386. Heath, D. F.: J. Chem. Soc. **1958**, 1643.
387. Heath, D. F.: Organophosphorus poisons. Oxford-London-New York-Paris: Pergamon Press 1961.
388. Heath, D. F., Vandekar, M.: Biochem. J. **67**, 187 (1957).
389. Hechenbleikner, J. (American Cyanamid Co.): A. P. 2.482.063 (1947/1949).
390. Hechenbleikner, J. (American Cyanamid Co.): D. B. P. 819.999 (1949/1951), A. P. 2.692.893 (1952/1954).
391. Hedin, P. A., Cody, C. P., Thompson, A. C.,Jr.: J. Econ. Entomol. **57**, 270 (1964).

392. Hedin, P. A., Wiygul, G., Mitlin, N.: J. Econ. Entomol. **60**, 215 (1967).
393. Hedin, P. A., Wiygul, G., Vickers, D. A., Bartlett, A. C., Mitlin, N.: J. Econ. Entomol. **60**, 209 (1967).
394. Heilbronn, E.: Acta Chem. Scand. **13**, 1547 (1959).
395. Heilbronn, E.: Biochem. Pharmacol. **12**, 25 (1963).
396. Heilbronn-Wikström, E.: Svensk. Kem. Tidskr. **77**, 598 (1965).
397. Helfenberger, H., Lutz, K. (Sandoz AG): E. P. 1.085.340 (1966/1967).
398. Heller, G., Siller, A.: J. Prakt. Chem. **116**, 1 (1927).
399. Hendlin, D., Stapley, E. O., Jackson, M., Wallick, H., Miller, A. K., Wolf, F. J., Miller, T. W., Chaiet, L., Kahan, F. M., Foltz, E. L., Woodruff, H. B., Mata, J. M., Hernandez, S., Mochales, S.: Science **166**, 122 (1969).
400. Henglein, A., Schrader, G.: Z. Naturforsch. **10b** (1), 12 (1955).
401. Henneberry, T. J., Smith, W. R.: J. Econ. Entomol. **58**, 312 (1965).
402. Herne, D. H. C., Brown, A. W. A.: J. Econ. Entomol. **62**, 205 (1969).
403. van den Heuvel, M. J., Cochran, D. G.: J. Econ. Entomol. **58**, 872 (1965).
404. Hewitt, R. J., Waletzky, E. (American Cyanamid Co.): A. P. 3.007.845 (1958/1961).
405. Hewlett, P. S.: Chem. Ind. (London) **1968**, Nr. 22, 701.
406. Hilgetag, G., Lehmann, G., Martini, A., Schramm, G., Teichmann, H.: J. Prakt. Chem. **8**, 207 (1959).
407. Hilgetag, G., Teichmann, H.: Chem. Ber. **96**, 1446 (1963).
408. Hilgetag, G., Teichmann, H.: Angew. Chem. **77**, 1001 (1965).
409. Hinz, B.: Psychopathol. Z. **56**, (1), 54 (1966).
410. Hodgson, E., Casida, J. E.: J. Agr. Food Chem. **10**, 208 (1962).
411. Hodkin, A. L.: Angew. Chem. **76**, 661 (1964).
412. Hoffmann, F., Moore, T., Kagan, B.: J. Am. Chem. Soc. **78**, 6413 (1956).
413. Hofmann, A. W.: Ber. **5**, 104 (1872).
414. Hollingworth, R. M., Fukuto, T. R., Metcalf, R. L.: J. Agr. Food Chem. **15**, 235 (1967).
415. Hollingworth, R. M., Metcalf, R. L., Fukuto, T. R.: J. Agr. Food Chem. **15**, 1250 (1967).
416. Holmstedt, B.: Acta Physiol. Scand. **25**, Suppl. 90, 31 (1951).
417. Holmstedt, B.: Acta Physiol. Scand. **25**, Suppl. 90, 32 (1951).
418. Holmstedt, B.: Acta Physiol. Scand. **25**, Suppl. 90, 33 (1951).
419. Homeyer, B.: Mededelingen Rijksfakulteit Landbouwwetenschappen Gent **34**, (3) 1969.
420. Hook, E. O., Moss, Ph. H. (American Cyanamid Co.): A. P. 2.586.655 (1948/1952), D. B. P. 833.807 (1949/1951).
421. Houben-Weyl, Vol. 12/1, p. 1.
422. Houben-Weyl, Vol. 12/1, p. 313.
423. Houben-Weyl, Vol. 12/1, p. 340f.
424. Houben-Weyl, Vol. 12/2, p. 881.
425. Houben-Weyl, Vol. 12/2, p. 905.
426. Houben-Weyl, Vol. 12/2, p. 910
427. Houben-Weyl, Vol. 12/2, p. 943.
428. Houben-Weyl, Vol. 12/2, p. 945.
429. Hudson, R. F.: Chimia **16**, 173 (1962).
430. Hudson, R. F.: Organo-phosphorus compounds, p. 371 (IUPAC at Heidelberg). London: Butterworths 1964.
431. Hudson, R. F.: Structure and mechanism in organo-phosphorus chemistry. London-New York: Academic Press 1965.
432. Hudson, R. F., Green, M.: Angew. Chem. **75**, 47 (1963).
433. Huffman, K. R. (American Cyanamid Co.): A. P. 3.203.959 (1963/1965).
434. Huxley, A. F.: Angew. Chem. **76**, 668 (1964).
435. Jacobson, H. J., Harvey, R. G., Jensen, E. V.: J. Am. Chem. Soc. **77**, 6064 (1955).
436. Jaffé, H. H., Freedman, L. D., Doak, G. O.: J. Am. Chem. Soc. **75**, 2209 (1953).
437. Jamieson, D.: Biochem. Pharmacol. **12**, 693 (1963).
438. Janning, unpublished (I.G. Farben, Leverkusen) internal report Nr. 2 (1946).
439. Jarczyk, H. J.: unpublished (1960).
440. Jarczyk, H. J.: unpublished (1961).
441. Jarczyk, H. J.: unpublished (1962).

442. JARCZYK, H. J.: Z. Naturforsch. **20b**, 257 (1965).
443. JARCZYK, H. J.: Pflanzenschutz-Nachr. Bayer **19**, (1) 1 (1966).
444. JARCZYK, H. J.: 6. Int. Pflanzenschutz-Kongreß Wien 1967, Abstracta p. 294. Wien: Verlag Wiener Med. Akad. 1967.
445. JARCZYK, H. J.: Int. Pflanzenschutz-Kongreß Wien 1967, Lecture.
446. JELINEK, A. G. (DuPont de Nemours & Co.): A. P. 2.503.390 (1948/1950), A. P. 2.667.727 (1954/).
447. JENKINS, R. L. (Monsanto Chemical Co.): A. P. 2.426.691 (1944/1947).
448. JENSEN, W. L. (Continental Oil Co.): A. P. 2.662.917 (1951/1953).
449. JOESTEN, M. D., HILL, R. A.: J. Agr. Food Chem. **14**, 512 (1966).
450. JOHNSON, G. A. (American Cyanamid Co.): A. P. 2.713.018 (1953/1955), D. B. P. 1.011.660 (1954/1957).
451. JOHNSON, M. K.: Biochem. J. **98**, 44 (1966).
452. JOHNSON, M. K.: Biochem. J. **111**, 487 (1969).
453. JONAS, H. (Farbenfabriken Bayer AG): D. B. P. 835.145 (1949/1954).
454. JONAS, H., THRAUM, W. (Farbenfabriken Bayer AG): D. B. P. 834.243 (1948/1952).
455. JONAS, H., THRAUM, W. (Farbenfabriken Bayer AG): D. B. P. 835.297 (1950/1952).
456. JONES, W. E., KAUFFMAN, G. W. (Shell Intern. Research): D. A. S. 1.234.198 (1964/1967).
457. JOSEPHS, M. J. (Dow Chemical Co.): A. P. 3.156.552 (1962/1964).
458. JUCKER, O.: Mitt. Gebiete Lebensm. Hyg. **49**, 299 (1958).
459. JUNG, H. F.: Bull. World Health Organ. **21**, 215 (1959).
460. JUNG, H. F., KÜKENTHAL, F., TECHNAU, G.: Hoefchen Briefe **13**, 13 (1960).
461. KABACHNIK, M. I., *et al.*: Zh. Obshch. Khim. **29**, 2182 (1959).
462. KABACHNIK, M. I.: Z. Chem. **1**, 289 (1961).
463. KABACHNIK, M. I., ABDUVAKHABOV, A. A., AGABEKOVA, I. I., BRESTKIN, A. P., VOLKOVA, R. I., GODOVIKOV, N. N., GODYNA, E. I., MIKHAILOV, S. S., MIKHEL'SON, M. YA., ROZENGART, V. I., ROZENGART, E. V., SITKEVICH, R. V.: Russ. Chem. Rev. **39** (6), 485 (1970).
464. KABACHNIK, M. I., GODOVIKOV, N. N.: Ber. Akad. Wiss. UdSSR **110**, 217 (1956).
465. KABACHNIK, M. I., MASTRUKOVA, T. A., SHIPOV, A. E., MELENTYEVA, T. A.: Tetrahedron **9**, 10 (1960).
466. KADO, M.: Experimental approaches of pesticide metabolism, degradation and mode of action, U.S.-Japan Seminar 16–19. August 1967, Nikko, Japan, p. 121–125, Pflanzenschutzber. Nr. 12 (1968), Ref. 3900.
467. KADO, M., TSUBOI, T., MAEDA, T., YOSHINAGA, SH. and E. (Ihara Chemicals Co. Ltd.): A. P. 3.274.051 (1964/1966), D. A. S. 1.252.961 (1965/1967).
468. KAISER, E. T., KUDO, K.: J. Am. Chem. Soc. **89**, 6725 (1967).
469. KAISER, D. W., PETERS, G. A.: J. Org. Chem. **18**, 196 (1953).
470. KARAVANOV, K. V., IVIN, S. Z.: Zh. Obshch. Khim. **35**, 78 (1965).
471. KAUER, K. C., BRITTON, E. C. (Dow Chemical Co.): A. P. 2.836.612 (1956/1958), D. A. S. 1.075.609 (1957/1960).
472. KAYSER, H., SCHRADER, G. (Farbenfabriken Bayer AG): D. A. S. 1.121.882 (1958/1962), A. P. 2.978.479 (1959/1961).
473. KEIDING, J.: Bull. World Health Organ. **29**, Suppl., 51 (1963).
474. KEIDING, J.: World Rev. Pest Contr. **6**, 115 (1967).
475. KENAGA, E. E. (Dow Chemical Co.): A. P. 2.954.317 (1959/1960).
476. KENAGA, E. E., WHITNEY, W. K., HARDY, I. L., DOTY, A. E.: J. Econ. Entomol. **58**, 1043 (1965).
477. KEPLINGER, M. L., DEICHMANN, W. B.: Toxicol. Appl. Pharmacol. **10**, 586 (1967).
478. KERR, S. H.: J. Econ. Entomol. **61**, 523 (1968).
479. KHARASCH, M. S., BENGELSDORF, J. S.: J. Org. Chem. **20**, 1356 (1955).
480. KHARASCH, M. S., JENSEN, E. V., WEINHOUSE, S.: J. Org. Chem. **14**, 432 (1949).
481. KIERMEIER, F., KERN, R., WILDBRETT, G.: Naturwissenschaften **48**, 99 (1961).
482. KIERMEIER, F., KERN, R., WILDBRETT, G.: Z. Lebensm. Untersuch.-Forsch. **118** (3), 14 (1962).
483. KIERMEIER, F., WILDBRETT, G., LETTENMAYER, L.: Z. Lebensm. Unters.-Forsch. **133**, 22 (1966).
484. KILBY, B. A.: Advan. Insect Physiol. Vol. I, p. 112. London-New York: Academic Press 1964.
485. KINNEAR, A. M., PERREN, E. A. (Nation. Research Develop. Corp.): E. P. 707.961 (1948/1954).
486. KINNEAR, A. M., PERREN, E. A.: J. Chem. Soc. **1952**, 3437.
487. KIRBY, A. J., VARVOGLIS, A. G.: J. Am. Chem. Soc. **89**, 415 (1967).

488. Kirby, A. J., Warren, S. G.: The organic chemistry of phosphorus, p. 52. Amsterdam-London-New York: Elsevier Publishing Comp. 1967.
489. Kittredge, J. S., Isbell, A. F., Hughes, R. R.: Biochemistry **6**, 289 (1967).
490. Klement, R., Wild, A.: Chem. Ber. **96**, 1916 (1963).
491. Klimmer, O. R.: Pflanzenschutz- u. Schädlingsbekämpfungsmittel p. 44ff. Hattingen: Hundt-Verlag 1964.
492. Klöckner, H.: Chem. Ind. **18**, 682 (1966).
493. Kluger, R., Covitz, F., Dennis, E., Williams, L. D., Westheimer, F. H.: J. Am. Chem. Soc. **91**, 6066 (1969).
494. Knaak, J. B., O'Brien, R. D.: J. Agr. Food Chem. **8**, 198 (1960).
495. Knaak, J. B., Stahmann, M. A., Casida, J. E.: J. Agr. Food Chem. **10**, 154 (1962).
496. Knipling, E. F.: Science **139** (3380), 902 (1959).
497. Knipling, E. F.: Sci Am. **203**, 54 (1960).
498. Knipling, E. F.: J. Econ. Entomol. **53**, 415 (1960).
499. Knipling, E. F.: J. Econ. Entomol. **55**, 782 (1962).
500. Knipling, E. F.: Umschau Wiss. Tech. **63**, 632 (1963).
501. Knipling, E. F.: Agr. Res. **14** (9), 3 (1966).
502. Knowles, C. O., Casida, J. E.: J. Agr. Food Chem. **14**, 566 (1966).
503. Kochmann, W., Seelmann, H. G., Frotscher, F., Wetzke, G., Müller, I., Springer, R., Sass, D., Schmidt, P. (VEB Elektrochem. Kombinat Bitterfeld): D. P. (DDR) 49.605 (1965/1966).
504. Komkov, I. P., Ivin, S. Z., Karavanov, K. V.: Zh. Obshch. Khim. **28**, 2960 (1958).
505. Komkov, I. P., Karavanov, K. V., Ivin, S. Z.: Zh. Obshch. Khim. **28**, 2963 (1958).
506. Koopmans, M. J., Meltzer, J., Huisman, H. O., van den Bos, B. G., Wellinga, K. (North American Philips Co.): A. P. 3.121.090 (1958/1964).
507. Koshland, D. E.: Science **142**, 1533 (1963).
508. Kosolapoff, G. M. (Monsanto Chemical Co.): A. P. 2.486.658 (1947/1949).
509. Kosolapoff, G. M. (Monsanto Chemical Co.): A. P. 2.502.966 (1948/1950).
510. Kosolapoff, G. M. (Monsanto Chemical Co.): A. P. 2.503.204 (1947/1950).
511. Kosolapoff, G. M. (Monsanto Chemical Co.): A. P. 2.567.154 (1948/1951).
512. Kosolapoff, G. M. (Monsanto Chemical Co.): A. P. 2.582.204 (1948/1952).
513. Kosolapoff, G. M.: Organophosphorus Compounds, p. 3. New York-London: John Wiley & Sons 1950.
514. Kosolapoff, G. M.: J. Am. Chem. Soc. **73**, 4989 (1951).
515. Krawiecki, Cz., Michalski, J.: J. Chem. Soc. **1960**, 881.
516. Kreutzkamp, N., Kayser, H.: Chem. Ber. **89**, 1614 (1956).
517. Krueger, H. R., O'Brien, R. D., Dautermann, W. C.: J. Econ. Entomol. **53**, 25 (1960).
518. Krupka, R. M.: Arch. Biochem. Biophys. **97**, 277 (1962); Biochemistry **2**, 76 (1963); Biochemistry **3**, 1749 (1964); Can. J. Biochem. **42**, 677 (1964); Biochemistry **4**, 429 (1965); Biochemistry **5**, 1983 (1966); Biochemistry **5**, 1988 (1966).
519. Krupka, R. M., Laidler, K. J.: J. Am. Chem. Soc. **83**, 1445 (1961).
520. Krysan, J. L., Chadwick, L. E.: J. Insect Physiol. **12**, 781 (1966).
521. Kucera, J. J., Carpenter, D. C.: J. Am. Chem. Soc. **57**, 2346 (1935).
522. Kugel, L., Halmann, M.: J. Am. Chem. Soc. **89**, 4125 (1967).
523. Kühn, G., Fischer, G. W., Lohs, K. H.: Arch. Pharm. **300**, 363 (1967).
524. Kumber, W. D., Eiler, J. J.: J. Am. Chem. Soc. **65**, 2355 (1943).
525. Kuramoto, S., Nishizawa, Y., Sakamoto, H., Mizutani, T. (Sumitomo Chemical Co. Ltd.): A. P. 3.150.040 (1961/1964), F. P. 1.296.482 (1961/1962).
526. Kuramoto, S., Nishizawa, Y., Sakamoto, H., Mizutani, T. (Sumitomo Chemical Co. Ltd.): F. P. 1.302.593 (1961/1962), D. A. S. 1.214.040 (1961/1966).
527. LaBrecque, G. C.: J. Econ. Entomol. **54**, 684 (1961).
528. LaBrecque, G. C., Gouck, H. K.: J. Econ. Entomol. **56**, 476 (1963).
529. Ladd, E. C., Harvey, M. P. (U.S. Rubber Co., now Uniroyal Incorp.): A. P. 2.631.162 (1949/1953).
530. Lamb, J. C., Steinberg, G. M.: Biochim. Biophys. Acta **89**, 171 (1964).
531. Lamb, J. C., Steinberg, G. M., Hackley, B. E.: Biochim. Biophys. Acta **89**, 174 (1964).
532. Landram, J. F., Shaver, R. J.: J. Parasitol. **45**, 55 (1959).

533. LANE, D. W. J., HEATH, D. F. (Pest Control Ltd.): D. B. P. 961.670 (1953/1956), A. P. 2.791.599 (1953/1957).
534. LANGE, W., v. KRUEGER, G.: Ber. **65**, 1598 (1932).
535. LARSSON, L.: Acta Chem. Scand. **11**, 1131 (1957).
536. LARSSON, L., HOLMSTEDT, B., TJUS, E.: Acta Chem. Scand. **8**, 1563 (1954).
537. LASSAIGNE, I.-L.: Ann. Chimie Physique **13** (2. Série) 294 (1820).
538. LEASURE, J. K. (Dow Chemical Co.): A. P. 3.074.790 (1960/1963).
539. LEBER, J. L., LUTZ, K. (Sandoz AG): D. A. S. 1.116.217 (1958/1961).
540. LEE, R. M., BATHAM, P.: Entomol. Exptl. Appl. **9**, 13 (1966).
541. LEE, R. M., HODSDEN, M. R.: Biochem. Pharmacol. **12**, 1241 (1963).
542. LEGATOR, M., KREUTZER, W. A., SOLOWAY, S. B. (Shell Oil Co.): A. P. 3.080.274 (1960/1963), D. A. S. 1.073.237 (1959/1960).
543. Lendle, L.: Med. Welt **1962**, 1719.
544. LEUZINGER, W., BAKER, A. L.: Proc. Nat. Acad. Sci. Wash. **57**, 446 (1967).
545. LEUZINGER, W., BAKER, A. L., CAUVIN, E.: Proc. Nat. Acad. Sci. Wash. **59**, 620 (1968).
546. LEUZINGER, W., GOLDBERG, M., CAUVIN, E.: J. Mol. Biol. **40**, 217 (1969).
547. LEYDEN WEBB, J.: Enzyme and metabolic inhibitors, Vol. 1, p. 51–53. New York-London: Academic Press 1963.
548. LICHTENTHALER, F. W.: Chem. Rev. **61**, 607 (1961).
549. LICHTENTHALER, F. W.: Chem. Rev. **61**, 624 (1961).
550. LICHTENTHALER, F. W.: Chem. Rev. **61**, 626 (1961).
551. LINDQUIST, D. A., BULL, D. L.: J. Agr. Food Chem. **15**, 267 (1967).
552. LINDQUIST, D. A., GORZYCKI, L. J., MAYER, M. S., SCALES, A. L., DAVICH, T. B.: J. Econ. Entomol. **57**, 745 (1964).
553. LÖFROTH, G.: Naturwissenschaften **57**, 393 (1970).
554. LOHS, K. H.: Synthetische Gifte. 3rd edition, p. 187. Berlin: Deutscher Militärverlag 1967.
555. LOHS, K. H.: Synthetische Gifte. 3rd edition, p. 190, 204, 248. Berlin: Deutscher Militärverlag 1967.
556. LORENZ, W. (Farbenfabriken Bayer AG): A. P. 2.701.225 (1953/1955).
557. LORENZ, W. (Farbenfabriken Bayer AG): D. B. P. 927.270 (1953/1955), A. P. 2.758.115 (1955/1956).
558. LORENZ, W. (Farbenfabriken Bayer AG): D. B. P. 930.446 (1953/1955).
559. LORENZ, W. (Farbenfabriken Bayer AG): D. B. P. 1.003.720 (1954/1957), A. P. 2.865.943 (1954/1958).
560. LORENZ, W. (Farbenfabriken Bayer AG): D. A. S. 1.064.072 (1958/1959).
561. LORENZ, W. (Farbenfabriken Bayer AG): Belg. P. 596.091 (1960/1960).
562. LORENZ, W. (Farbenfabriken Bayer AG): Techn. Inform. of Chevron Chem. Co. (California), May 1970.
563. LORENZ, W., FEST, CH., HAMMANN, I., FEDERMANN, M., FLUCKE, W., STENDEL, W. (Farbenfabriken Bayer AG): D. A. S. 1.238.902 (1965/1967).
564. LORENZ, W., HENGLEIN, A., SCHRADER, G.: J. Am. Chem. Soc. **77**, 2554 (1955).
565. LORENZ, W., MÜHLMANN, R., SCHRADER, G. (Farbenfabriken Bayer AG): D. B. P. 964.045 (1955/1957).
566. LORENZ, W., MÜHLMANN, R., SCHRADER, G., TETTWEILER, K. (Farbenfabriken Bayer AG): D. B. P. 949.229 (1955/1956).
567. LORENZ, W., SCHRADER, G. (Farbenfabriken Bayer AG): D. B. P. 820.001 (1949/1951).
568. LORENZ, W., SCHRADER, G. (Farbenfabriken Bayer AG): D. B. P. 917.668 (1952/1954), A. P. 2.759.010 (1953/1956).
569. LORENZ, W., SCHRADER, G. (Farbenfabriken Bayer AG): A. P. 2.902.493 (1956/1959).
570. LORENZ, W., SCHRADER, G. (Farbenfabriken Bayer AG): D. B. P. 1.035.958 (1955/1958), A. P. 2.952.700 (1957/1960).
571. LORENZ, W., SCHRADER, G. (Farbenfabriken Bayer AG): D. A. S. 1.067.017 (1958/1959).
572. LORENZ, W., SCHRADER, G. (Farbenfabriken Bayer AG): F. P. 1.198.263 (1958/1959).
573. LORENZ, W., SCHRADER, G. (Farbenfabriken Bayer AG): D. A. S. 1.124.034 (1960/1962).
574. LORENZ, W., SCHRADER, G. (Farbenfabriken Bayer AG): D. A. S. 1.159.934 (1961/1963).
575. LORENZ, W., SCHRADER, G. (Farbenfabriken Bayer AG): A. P. 3.082.240 (1961/1963).
576. LORENZ, W., WEGLER, R. (Farbenfabriken Bayer AG): D. B. P. 962.608 (1954/1954).

577. LOSCO, G., ROSSI, G., MICHIELI, G. (Montecatini): I. P. 625.074 (1959/1961).
578. LOSHADKIN, N. A., SMIRNOV, V. V.: Review of modern literature on the chemistry and toxicology of organophosphorus inhibitors of cholinesterases. Translated by G. M. KOSOLAPOFF. Glen Ridge, New Jersey: Assoc. Techn. Services, Inc. 1962.
579. LOVELL, J. B. (American Cyanamid Co.): A. P. 3.197.365 (1963/1965).
580. LOVELL, J. B., BAER, R. W. (American Cyanamid Co.): Belg. P. 648.531 (1963/1963).
581. LUCKEY, T. D.: J. Econ. Entomol. **61**, 7 (1968).
582. LÜTTRINGHAUS, A., HAGEDORN, I.: Arzneimittel-Forsch. **14**, 1 (1964).
583. LUTZ, K., SCHULER, M. (Sandoz AG): D. A. S. 1.148.806 (1960/1963).
584. LUTZ, K., SCHULER, M., JUCKER, O. (Sandoz AG): Swiss P. 319.579 (1953/1957).
585. MACAFEE, M. W. (Dow Chemical Co.): A. P. 2.198.915 (1935/1940).
586. MAGEE, PH. S. (Chevron Res. Co.): A. P. 3.309.266 (1965/1967).
587. MAIN, A. R.: Science **144**, 992 (1964).
588. MAIN, A. R., BRAID, P. E.: Biochem. J. **84**, 255 (1962).
589. MAKLYAEV, F. L., SMIRNOV, I. V., MARKOV, S. M., LOSHADKIN, N. A., ANIKIENKO, K. A.: Zh. Obshch. Khim. **33**, 3833 (1963). (Engl. Transl. p. 3773).
590. MALATESTA, L.: G. **81**, 596 (1951).
591. MALATESTA, L., PIZZOTTI, R.: Chim. Ind. **27**, 6 (1945).
592. MALOWAN, I. E.: Inorg. Syn. **4**, 58 (1953).
593. MALZ, H., HERMANN, R. G. (Farbenfabriken Bayer AG): D. A. S. 1.196.897 (1962/1965), A. P. 3.281.231 (1963/1966).
594. MANGHAM, J. R. (Virginia-Carolina Chemical Corp.): A. P. 2.864.847 (1953/1958).
595. MARAMOROSCH, K.: Ann. Rev. Entomol. **8**, 369 (1963).
596. MARCH, R. B.: Misc. Publ. Entomol. Soc. Am. **1**, 13 (1959).
597. MARCO, G. J., JAWORSKI, E. G.: J. Agr. Food Chem. **12**, 305 (1964).
598. MARSH, J. E., BALLIOL, M. A., GARDNER, J. A.: J. Chem. Soc. **59**, 648 (1891).
599. MARSH, J. E., GARDNER, J. A.: J. Chem. Soc. **65**, 35 (1894).
600. MARTIN, H.: Guide to the chemicals used in crop protection, p. 167. 4th edition. London-Ontario: Canada Department of Agriculture 1961).
601. MARTIN, H.: The scientific principles of crop protection, p. 76. 5th edition. London: Edward Arnold (Publishers) Ltd. 1964.
602. MARTIN, H.: The scientific principles of crop protection, p. 109f. 5th edition. London: Edward Arnold (Publishers) Ltd. 1964.
603. MATSUMURA, F., BROWN, A. W. A.: J. Econ. Entomol. **54**, 1176 (1961).
604. MATSUMURA, F., BROWN, A. W. A.: J. Econ. Entomol. **56**, 381 (1963).
605. MATSUMURA, F., HOGENDIJK, C. J.: J. Agr. Food Chem. **12**, 447 (1964).
606. MATSUMURA, F., SAKAI, K.: J. Econ. Entomol. **61**, 598 (1968).
607. MATSUMURA, F., VOSS, G.: J. Econ. Entomol. **57**, 911 (1964).
608. MATTSON, A. M., FAY, R. W., GAINES, T. B., PEARCE, G. W.: J. Agr. Food Chem. **8**, 196 (1960).
609. MATTSON, A. M., SPILLANE, J. T., PEARCE, G. W.: J. Agr. Food Chem. **3**, 319 (1955).
610. MAZUR, A.: J. Biol. Chem. **164**, 271 (1946).
611. MCCOMBIE, H., SAUNDERS, B. C.: Nature **157**, 287 (1946).
612. MCCOMBIE, H., SAUNDERS, B. C.: Nature **157**, 776 (1946).
613. MCCOMBIE, H., SAUNDERS, B. C., CHAPMAN, N. B., HEAP, R., PRATT, D. (Ministry of Supply): E. P. 602.446 (1944/1948).
614. MCCOMBIE, H., SAUNDERS, B. C., STACEY, G. J.: J. Chem. Soc. **1945**, 380.
615. MCCOMBIE, H., SAUNDERS, B. C., STACEY, G. J.: J. Chem. Soc. **1945**, 921.
616. MCDANIEL, D. H., BROWN, H. C.: J. Org. Chem. **23**, 420 (1958).
617. MCHATTIE, G. V. (Imperial Chemical Industries Ltd.): NE. P. 6.704.665 (1966/1967), E. P. 1.019.227 (1963/1966).
618 MCIVOR, R. A., MCCARTHY, G. D., GRANT, G. A.: Can. J. Chem. **34**, 1819 (1956).
619. MCKINSTRY, D. N., KOENIG, E., KOELLE, W. A., KOELLE, G. B.: Can. J. Biochem. Physiol. **41**, 2599 (1963).
620. MCPHERSON, J. B., JOHNSON, G. A.: J. Agr. Food Chem. **4**, 42 (1956).
621. MELNIKOV, N. N., MANDELBAUM, J. A., ZAKS, P. G.: Zh. Obshch. Khim. **29**, 522 (1959); USSR P. 107.007 (1959/1958).

622. MELNIKOV, N. N., SHVETSOVA-SHILOVSKAYA, K. D., KAGAN, M. YA.: Zh. Obshch. Khim. **30**, 200 (1960).
623. MENN, J. J., HOSKINS, W. H.: J. Econ. Entomol. **55**, 90 (1962).
624. MENN, J. J., MCBAIN, J. B.: J. Agr. Food Chem. **12**, 163 (1964).
625. MENN, J. J., MCBAIN, J. B., ADELSON, B. J., PATCHETT, G. G.: J. Econ. Entomol. **58**, 875 (1965).
626. MENN, J. J., SZABO, K.: J. Econ. Entomol. **58**, 734 (1965).
627. MENZER, R. E., CASIDA, J. E.: J. Agr. Food Chem. **13**, 102 (1965).
628. MERSCH, R., DELFS, D. (Farbenfabriken Bayer AG): D. B. P. 1.024.978 (1953/1958).
629. METCALF, R. L., FREDERICKSON, M.: J. Econ. Entomol. **58**, 143 (1965).
630. METCALF, R. L., FUKUTO, T. R.: J. Econ. Entomol. **53**, 127 (1960).
631. METCALF, R. L., FUKUTO, T. R.: J. Econ. Entomol. **55**, 340 (1962).
632. METCALF, R. L., FUKUTO, T. R., MARCH, R. B.: J. Econ. Entomol. **50**, 338 (1957).
633. METCALF, R. L., MARCH, R. B.: J. Econ. Entomol. **42**, 721 (1949).
634. METCALF, R. L., MARCH, R. B., FUKUTO, T. R., MAXON, M. G.: J. Econ. Entomol. **48**, 364 (1955).
635. METCALF, R. L., REYNOLDS, H. T., FUKUTO, T. R., COLLINS, C.: J. Econ. Entomol. **57**, 531 (1964).
636. MÉTIVIER, J. (Rhône-Poulenc): D. B. P. 1.004.422 (1955/1957), F. P. 1.125.943 (1955/1956).
637. MÉTIVIER, J. (Rhône-Poulenc): D. A. S. 1.123.863 (1959/1962), D. A. S. 1.153.568 (1961/1963).
638. MÉTIVIER, J. (Rhône-Poulenc): F. P. 1.482.025 (1966/1967).
639. MICHAELIS, A.: Liebigs Ann. Chem. **326**, 129 (1903).
640. MICHAELIS, A., Liebigs Ann. Chem. **326**, 172 (1903).
641. MICHAELIS, A., BECKER, TH.: Ber. **30**, 1003 (1897).
642. MICHAELIS, A., KAEHNE, R.: Ber. **31**, 1048 (1898).
643. MICHAELIS, A., LUXEMBOURG, K.: Ber. **29**, 710 (1896).
644. MICHAELIS, A., SCHULZE, G.: Ber. **27**, 2572 (1894).
645. MICHAELIS, A., SILBERSTEIN, E.: Ber. **29**, 720 (1896).
646. MICHAELIS, L., MENTEN, M. L.: Biochem. Z. **49**, 333 (1913)
647. MICHALSKI, J.: Roczniki Chem. **29**, 960 (1955).
648. MICHALSKI, J., KRAWIECKI, CZ.: Chem. Ind. (London) **1957**, 1323.
649. MICHALSKI, J., WIECZORKOWSKI, J.: Bull. Acad. Polon. Sci. Classe III, 273 (1956).
650. MICHALSKI, J., ZWIERZAK, A.: Roczniki Chem. **36**, 97 (1962).
651. MILLER, B. (American Cyanamid Co.): D. A. S. 1.150.069 (1961/1963).
652. MILLER, B.: Proc. Chem. Soc. **1962**, 303.
653. MILLER, B.: J. Am. Chem. Soc. **84**, 403 (1962).
654. MILLER, B.: J. Org. Chem. **28**, 345 (1963).
655. MILOBEDZKI, T., SACHNOWSKI, A.: Chem. Polski **15**, 34 (1917), C. 1918 I, 911.
656. MILSTIEN, J. B., FIFE, T. H.: J. Am. Chem. Soc. **90**, 2164 (1968).
657. MONRO, J.: Science **151** (3717), 1536 (1966).
658. MORELLO, A., SPENCER, E. Y., VARDANIS, A.: Biochem. Pharmacol. **16**, 1703 (1967).
659. MORELLO, A., VARDANIS, A., SPENCER, E. Y.: Biochem. Biophys. Res. Commun. **29**, 241 (1967).
660. MORELLO, A., VARDANIS, A., SPENCER, E. Y.: Can. J. Biochem. **46**, 885 (1968).
661. MORGAN, P. B., LABRECQUE, G. C.: J. Econ. Entomol. **57**, 896 (1964).
662. MORIKAWA, O., SAITO, T.: Botyu-Kagaku **31**, 130 (1966).
663. MORRIS, R. C., VAN WINKLE, J. L. (Shell Development Co.): A. P. 2.744.128 (29. 2. 1952/ 1. 5. 1956).
664. MORRISON, D.: J. Am. Chem. Soc. **77**, 181 (1955).
665. MOSCHNIN: Liebigs Ann. Chem. **91**, 375 (1854).
666. MÜHLMANN, R., LORENZ, W., SCHRADER, G. (Farbenfabriken Bayer AG): D. B. P. 947.368 (1954/1956).
667. MÜHLMANN, R., SCHRADER, G. (Farbenfabriken Bayer AG): D. B. P. 947.366 (1954/1956).
668. MÜHLMANN, R., TIETZ, H.: Hoefchen Briefe **9**, 116 (1956).
669. MUKAIYAMA, T., FUJISAWA, T.: Bull. Chem. Soc. Japan **34**, 812 (1961).
670. MURDOCK, L. L., HOPKINS, T. L.: J. Agr. Food Chem. **16**, 954 (1968).
671. MUSHER, J. I.: Angew. Chem. **81**, 68 (1969).
672. MYERS, D. K., TOE, J. W., DEJONGE, M. H. T.: Biochem. J. **65**, 223 (1957).
673. NAKATSUGAWA, T., DAHM, P. A.: J. Econ. Entomol. **55**, 594 (1962).
674. NAKATSUGAWA, T., DAHM, P. A.: J. Econ. Entomol. **58**, 500 (1965).

675. NARAHASHI, T.: Advan. Insect Physiol. Vol. I, p. 175. London-New York: Academic Press 1964.
676. NENNER, M.: Thesis Univ. Göttingen 1969.
677. NEWALLIS, P. E., LOMBARDO, P., GILBERT, E. E., SPENCER, E. Y., MORELLO, A.: J. Agr. Food Chem. **15**, 940 (1967).
678. NIERMANN, H., CREMER, J., HARNISCH, H. (Knapsack-Griesheim AG): D. A. S. 1.191.369 (1963/1965).
679. NIESSEN, H., TIETZ, H., FREHSE, H.: Pflanzenschutz-Nachr. Bayer **15** (3), 148 (1962).
680. Nihon Kayaku Kogyo, Techn. Inform.
681. NOBORU SHINDO et al. (Nissan Chem. Ind. Ltd.); Jap. P. 18.543 (1965/1968).
682. NOELLING, E., KOPP, E.: Ber. **38**, 3510 (1905).
683. NORRIS, M. G., WADE, L. L. (Dow Chem. Co.): A. P. 2.811.480 (1956/1957).
684. NYLÉN, P.: Ber. **57**, 1023 (1924).
685. NYLÉN, P.: Studien über organische Phosphorverbindungen. (Inaugural Dissertation, Uppsala 1930).
686. O'BRIEN, R. D.: Toxic phosphorus esters. New York-London: Academic Press 1960.
687. O'BRIEN, R. D.: Biochem. J. **79**, 229 (1961).
688. O'BRIEN, R. D.: J. Agr. Food Chem. **11**, 163 (1963).
689. O'BRIEN, R. D., HILTON, R. D.: J. Agr. Food Chem. **12**, 53 (1964).
690. O'BRIEN, R. D., KIMMEL, E. C., SFERRA, P. R.: J. Agr. Food Chem. **13**, 366 (1965).
691. O'BRIEN, R. D., SMITH, E. H.: J. Econ. Entomol. **54**, 187 (1961).
692. OERTEL, G., MALZ, H. (Farbenfabriken Bayer AG): D. A. S. 1.163.310 (1962/1964).
693. OERTEL, G., MALZ, H., DÖRKEN, A. (Farbenfabriken Bayer AG): D. A. S. 1.251.304 (1962/1967).
694. OLAH, G. A., KUHN, S. J. (Dow Chem. Co.): A. P. 3.096.371 (1960/1963).
695. OPPENOORTH, F. J., VAN ASPEREN, K.: Science **132**, 298 (1960).
696. OPPENOORTH, F. J., VAN ASPEREN, K.: Entomol. Exptl. Appl. **4**, 311 (1961).
697. OSBORN, CH. W., GOODHUE, L. D. (Philips Petroleum Comp.): A. P. 2.841.486 (1953/1958).
698. OSBORNE, D. W., SENKBEIL, H. O., WASCO, J. L.: J. Org. Chem. **31**, 192 (1966).
699. OSBORNE, M. P.: J. Insect Physiol. **12**, 1503 (1966).
700. OSBORNE, M. P.: J. Insect Physiol. **13**, 827 (1967).
701. OSHIMA, Y., ETO, M. (Sumitomo Chem. Co.): E. P. 987.378 (1963/1965), F. P. 1.360.130 (1963/1964).
702. OSPENSON, J. N., KOHN, G. K. (California Res. Corp.): D. A. S. 1.190.246 (1958/1965).
703. OSSIANNILSSON, F.: Ann. Rev. Entomol. **11**, 213 (1966).
704. OTTMANN, G. F., AGAHIGIAN, H., HOOKS, H., VICKERS, G. D., KOBER, E. H., RÄTZ, R. F. W.: Inorg. Chem. **3**, 753 (1964).
705. OTTMANN, G. F., HOOKS, H., JR., KOBER, E. H., RÄTZ, R. F. W., RISTICH, S. S. (Olin Mathieson Corp.): A. P. 3.136.754 (1962/1964).
706. OTTMANN, G. F., HOOKS, H., JR., KOBER, E. H., RÄTZ, R. F. W., RISTICH, S. S. (Olin Mathieson Corp.): A. P. 3.197.464 (1962/1965).
707. OUYE, M. T., GRAHAM, H. M.: J. Econ. Entomol. **60**, 244 (1967).
708. PAETZOLD, R.: Z. Chem. **4**, 321 (1964).
709. PARISH, J. C., ARTHUR, B. W.: J. Econ. Entomol. **58**, 699 (1965).
710. PARISH, J. C., ARTHUR, B. W.: J. Econ. Entomol. **58**, 976 (1965).
711. PATHAK, M. D., VEA, E., JOHN, V. T.: J. Econ. Entomol. **60**, 218 (1967).
712. PAULING, L.: The nature of the chemical bond. 3rd edition. Ithaca, New York: Cornell University Press 1960.
713. PEARSON, R. G.: J. Am. Chem. Soc. **85**, 3533 (1963).
714. PEARSON, R. G.: Science **151**, 172 (1966).
715. PEARSON, R. G.: J. Chem. Educ. **45**, 581 (1968).
716. PEARSON, R. G.: J. Chem. Educ. **45**, 643 (1968).
717. V. PECHMANN, H., DUISBERG, C.: Ber. **16**, 2119 (1883).
718. PEISSKER, H., JÄGER, A., STEINHAUSEN, W., BOROSCHEWSKI, G. (Schering AG): D. A. S. 1.169.194 (1962/1964).
719. PERINI, M., SPERONI, G. (Montecatini): A. P. 3.047.459 (1955/1962).
720. PERINI, M., SPERONI, G. (Montecatini): D. A. S. 1.060.659 (1957/1959).
721. PERKOW, W.: Ber. **87**, 755 (1954).

722. Perkow, W., Koddebusch, H. (Nordd. Affinerie – Spieß u. Sohn): D. B. P. 944.430 (16. 1. 1951/24. 5. 1956).
723. Perkow, W., Koddebusch, H. (Nordd. Affinerie – Spieß u. Sohn): E. P. 739.726 (5.2.1953/ 2.11.1955).
724. Perkow, W., Ullerich, K., Meyer, F.: Naturwissenschaften **39**, 353 (1952) (received 14.6.1952).
725. Perry, A. S.: J. Agr. Food Chem. **8**, 266 (1960).
726. Pest Control Ltd., Cambridge (1950).
727. Peters, R. A.: Proc. Roy. Soc. (London) B **139**, 143 (1952).
728. Pfaff, W.: Thesis Univ. Bonn 1952.
729. Philip, C. B., Burgdorfer, W.: Ann. Rev. Entomol. **6**, 391 (1961).
730. Phillips, D. D., Ward, L. F. (Shell Oil Comp.): A. P. 3.102.842 (1962/1963).
731. Pianka, M.: Chem. Ind. (London) **1961**, 324.
732. Pianka, M., Polton, D. J. (Murphy Chem. Co.): F. P. 1.206.931 (1958/1960), E. P. 867.780 (1958/1961), D. A. S. 1.143.052 (1958/1963).
733. de Pietri Tonelli, P., Losco, G., Rossi, G., Santi, R. (Montecatini): F. P. 1.220.962 (1959/1960).
734. Pinner, E.: Naturw. Rundschau **19**, 469 (1966).
735. Pistschimuka, P.: Ber. **41**, 3854 (1908), J. pr. Chem. [2] **84**, 746 (1911).
736. Plapp, F. W., Jr., Bigley, W. S., Chapman, G. A., Eddy, G. W.: J. Econ. Entomol. **55**, 607 (1962).
737. Plapp, F. W., Jr., Bigley, W. S., Chapman, G. A., Eddy, G. W.: J. Econ. Entomol. **56**, 643 (1963).
738. Plapp, F. W., Jr., Bigley, W. S., Darrow, D. J.: J. Econ. Entomol. **53**, 60 (1960).
739. Plapp, F. W., Jr., Bigley, W. S., Darrow, D. J., Eddy, G. W.: J. Econ. Entomol. **54**, 389 (1961).
740. Plapp, F. W., Casida, J. E.: J. Agr. Food Chem. **6**, 662 (1958).
741. Plapp, F. W., Jr., Eddy, G. W.: Science **134**, 2043 (1961).
742. Plapp, F. W., Jr., Tong, H. H. C.: J. Econ. Entomol. **59**, 11 (1966).
743. Plapp, F. W., Jr., Valega, T. M.: J. Econ. Entomol. **60**, 1094 (1967).
744. Polles, S. G., Vinson, S. B.: J. Econ. Entomol. **62**, 89 (1969).
745. Porter, G. R., Rydon, H. N., Schofield, J. A.: Nature **1958** (4640), 927.
746. Poulsen, E., Aldridge, W. N.: Biochem. J. **90**, 182 (1964); **93**, 619 (1964).
747. Pound, D. W., Hartley, G. S., Heath, D. F., Hulme, J. M. (Pest Control Ltd.): E. P. 688.787 (1949/1953), A. P. 2.678.334 (1951/1954).
748. Pound, D. W., Saunders, B. C. (Pest Control Ltd.): E. P. 631.549 (1947/1948).
749. Preussmann, R., Schneider, H., Epple, F.: Arzneimittel-Forsch. **19** (7), 1057 (1970).
750. Price, Ch. C., Wakefield, L. B.: J. Org. Chem. **12**, 232 (1947).
751. Price, G. R. (Stauffer Chem. Comp.): D. A. S. 1.240.523 (1965/1967).
752. Price, G. R., Walsh, E. N., Dewald, C. L., Soong, S. Y. C. (Stauffer Chem. Comp.): A.P. 3.161.498 (1962/1964).
753. Proeseler, G.: Nachr. bl. Deut. Pflanzenschutzdienst (Berlin) **18** (2), 31 (1964).
754. Pudovik, A. N., Biktimirova, L. G.: J. Gen. Chem. USSR **27**, 2161 (1957).
755. Pudovik, A. N., Cebotareva, E. G.: J. Gen. Chem. USSR **28**, 2527 (1958).
756. Pudovik, A. N.: J. Gen. Chem. USSR **26**, 2503 (1956).
757. Quin, L. D.: Biochemistry **4**, 324 (1965).
758. Quistad, G. B., Fukuto, T. R., Metcalf, R. L.: J. Agr. Food Chem. **18**, 189 (1970).
759. Railton, R.: Liebigs Ann. Chem. **92**, 348 (1854).
760. Ralls, J. W., Cortes, A.: J. Econ. Entomol. **59**, 1296 (1966).
761. Ralls, J. W., Gilmore, D. R., Cortes, A.: J. Agr. Food. Chem. **14**, 387 (1966).
762. Ramirez, F.: IUPAC-Symposium on Organophosphorus Compounds, p. 337 (Heidelberg 1964).
763. Ramirez, F., Bigler, A. J., Smith, C. P.: J. Am. Chem. Soc. **90**, 3507 (1968).
764. Rätz, R. F. W. (Olin Mathieson Chem. Corp.): A. P. 3.180.793 (1963/1965).
765. Rätz, R. F. W. (Olin Mathieson Chem. Corp.): A. P. 3.210.398 (1963/1965).
766. Rätz, R. F. W., Gruber, M. J. (Olin Mathieson Chem. Corp.): A. P. 3.163.640 (1963/1964).
767. Rätz, R. F. W., Gruber, M. J. (Olin Mathieson Chem. Corp.): A. P. 3.314.848 (1964/1967).
768. Rätz, R. F. W., Grundmann, C. J. (Olin Mathieson Chem. Corp.): A. P. 2.858.306 (1957/1958).
769. Rätz, R. F. W., Kober, E. H., Grundmann, C. J., Ottmann, G. F.: Inorg. Chem. **3**, 757 (1964).
770. Richter, S. B. (Velsicol Chem. Corp.): A. P. 3.459.836 (1965/1969), A. P. 3.551.563 (1968/1970).
771. Rigterink, R. H. (Dow Chem. Co.): A. P. 3.228.960 (1962/1966).
772. Rigterink, R. H. (Dow Chem. Co.): F. P. 1.360.901 (1963/1964).

773. RIGTERINK, R. H., KENAGA, E. E.: J. Agr. Food Chem. **14**, 304 (1966).
774. RILEY, G., TURNBULL, J. H., WILSON, W.: J. Chem. Soc. **1957**, 1373.
775. RISER, A.: Protar **16** (11/12), 132 (1950).
776. RISTICH, S. S., RATCLIFFE, R. H., PERLMAN, D.: J. Econ. Entomol. **58**, 929 (1965).
777. ROGERS, T. H., LOANE, CL. M., SHOEMAKER, B. H. (Standard Oil Co.): A. P. 2.175.509 (1936/1939).
778. DEROOS, A. M., TOET, H. J.: Rec. Trav. Chim. **78**, 59 (1959).
779. ROOT, R. B., SKELSEY, J. J.: J. Econ. Entomol. **62**, 223 (1969).
780. ROSIN, J., HAUS, J. (Montrose Chem. Comp.): A. P. 2.899.456 (1956/1959).
781. ROULSTON, W. J., SCHUNTNER, C. A., SCHNITZERLING, H. J.: Australian J. Biol. Sci. **19**, 619 (1966).
782. RUF, E. (Th. Goldschmidt AG, Essen): D. A. S. 1.173.088 (1961/1964).
783. RÜFENACHT, K.: Helv. Chim. Acta **51**, 518 (1968).
784. RUSSELL, P. F.: Ann. Rev. Entomol. **4**, 415 (1959).
785. RYDON, H. N.: Nature **182** (4640), 928 (1958).
786. SAKIMURA, K., In: Biol. transmission of disease agents (Symp. of the Ent. Soc. of Am., Atlantic City, 1960), p. 33–40. New York-London: Academic Press 1962.
787. SAKSENA, K. N., SILL, W. H., JR., KAINSKI, J. M.: Plant Disease Reptr. **48**, 756 (1964).
788. SALLMANN, R. (Ciba AG): D. B. P. 974.569 (1952/1961).
789. SALLMANN, R. (Ciba AG): D. B. P. 975.092 (1952/1961).
790. SALLMANN, R. (Ciba AG): D. A. S. 1.036.251 (1956/1958).
791. SALLMANN, R., KOHLER, P. (Ciba AG): D. A. S. 1.050.330 (1955/1959).
792. SANTI, R., GIACOMELLI, R.: J. Agr. Food Chem. **10**, 257 (1962).
793. SAUL, G. A., GODFREY, K. L. (Monsanto Chem. Co.): E. P. 744.484 (1953/1956).
794. SAUNDERS, B. C.: Some aspects of the chemistry and toxic action of organic compounds containing phosphorus and fluorine, p. 91. Cambridge (Great Britain): Cambridge University Press 1957.
795. SAUNDERS, B. C., STACEY, G. J.: Report to the Ministry of Supply No. 19 (1945).
796. SAWICKI, R. M.: Pestic. Sci. **1**, 84 (1970).
797. SCHEGK, E., SCHRADER, G. (Farbenfabriken Bayer AG): D. A. S. 1.089.376 (1958/1960).
798. SCHEGK, E., SCHRADER, G. (Farbenfabriken Bayer AG): A. P. 3.042.703 (1960/1962).
799. SCHEGK, E., SCHRADER, G. (Farbenfabriken Bayer AG): D. A. S. 1.101.406 (1956/1961).
800. SCHEGK, E., SCHRADER, G. (Farbenfabriken Bayer AG): D. A. S. 1.116.656 (1958/1961).
801. SCHEINPFLUG, H., JUNG, H. F.: Pflanzenschutz-Nachr. Bayer **21**, 79 (1968).
802. SCHEINPFLUG, H., JUNG, H. F., SCHRADER, G. (Farbenfabriken Bayer AG): D. A. S. 1.213.664 (1963/1966).
803. SCHEINPFLUG, H., JUNG, H. F., SCHRADER, G. (Farbenfabriken Bayer AG): Belg. P. 655.301 (1964/1965).
804. SCHERER, O., HAHN, H. (Farbwerke Hoechst AG): D. A. S. 1.025.200 (1956/1958).
805. SCHICKE, H. G., BERGER, A. (Farbenfabriken Bayer AG): D. A. S. 1.197.878 (1963/1965).
806. SCHLIEBS, R. (Farbenfabriken Bayer AG): D. A. S. 1.079.022 (1958/1960).
807. SCHLIEBS, R. (Farbenfabriken Bayer AG): D. A. S. 1.119.861 (1959/1961).
808. SCHLIEBS, R. (Farbenfabriken Bayer AG): D. A. S. 1.119.862 (1959/1961).
809. SCHLIEBS, R. (Farbenfabriken Bayer AG): D. A. S. 1.131.670 (1960/1962).
810. SCHLIEBS, R. (Farbenfabriken Bayer AG): D. A. S. 1.165.596 (1962/1964).
811. SCHLIEBS, R., KAISER, H. (Farbenfabriken Bayer AG): D. A. S. 1.119.860 (1959/1961).
812. SCHLÖR, H., SCHEGK, E., SCHRADER, G. (Farbenfabriken Bayer AG): D. A. S. 1.099.530 (1959/1961).
813. SCHMERLING, L. (Universal Oil Prod. Comp.): A. P. 2.986.579 (1954/1961).
814. SCHMIDT, K. J.: Rep. 2nd IUPAC Congress Pest. Chem. Israel (Tel Aviv, 1971 in press).
815. SCHMIDT, K. J., HAMMANN, I. (Farbenfabriken Bayer AG): Belg. P. 681.443 (1966/1966), E. P. 1.081.249 (1966/1967).
816. SCHMIDT, K. J., HAMMANN, I. (Farbenfabriken Bayer AG): Belg. P. 711.819 (1968/1968).
817. SCHMIDT, K. J., HAMMANN, I.: Pflanzenschutz Ber. (Wien) **40**, 165 (1969).
818. SCHMIDT, K. J., HAMMANN, I., UNTERSTENHÖFER, G. (Farbenfabriken Bayer AG): Ne. P. 6.810.870 (1968/1969).
819. SCHMIR, G. L., ZIOUDROU, C.: Biochemistry **2**, 1305 (1963).
820. SCHMITT, J. B.: Ann. Rev. Entomol. **7**, 137 (1962).

821. SCHOOT, C. J., KOOPMANS, M. J., VAN DEN BOS, B. G. (Philips Gloeilampen Fabrieken): D. A. S. 1.201.116 (1959/1965).
822. SCHRADER, G.: cited in Houben-Weyl, Vol. XII/1, p. 415 (1963).
823. SCHRADER, G.: cited in Houben-Weyl, Vol. XII/2, p. 288 (1964).
824. SCHRADER, G.: cited in Houben-Weyl, Vol. XII/2, p. 385 (1964).
825. SCHRADER, G.: BIOS Final Rep. **714** (1947), **1095** (1947).
826. SCHRADER, G. (Farbenfabriken Bayer AG): D. R. P. 675.303 (1937/1939).
827. SCHRADER, G. (I.G. Farbenindustrie AG): A. P. 2.336.302 (1939/1943).
828. SCHRADER, G. (Farbenfabriken Bayer AG): D. B. P. 767.511 (1937/1952).
829. SCHRADER, G. (Farbenfabriken Bayer AG): BIOS Final Rep. **1095** (1946); D. B. P. 814.152 (1948/1951).
830. SCHRADER, G. (Farbenfabriken Bayer AG): D. B. P. 814.152 (1948/1951).
831. SCHRADER, G. (Farbenfabriken Bayer AG): D. B. P. 814.297 (1948/1951).
832. SCHRADER, G. (Farbenfabriken Bayer AG): D. B. P. 818.352 (1949/1951).
833. SCHRADER, G. (Farbenfabriken Bayer AG): D. B. P. 814.152 (1948/1951), D. B. P. 830.261 (1949/1952).
834. SCHRADER, G. (Farbenfabriken Bayer AG): D. B. P. 830.508 (1950/1952).
835. SCHRADER, G. (Farbenfabriken Bayer AG): D. B. P. 830.509 (1950/1952).
836. SCHRADER, G. (Farbenfabriken Bayer AG): D. B. P. 836.349 (1950/1952), A. P. 2.571.989 (1950/1951).
837. SCHRADER, G.: Angew. Chem. **62**, 471 (1950).
838. SCHRADER, G. (Farbenfabriken Bayer AG): D. B. P. 881.194 (1951/1953), A. P. 2.748.146 (1952/1956).
839. SCHRADER, G. (Farbenfabriken Bayer AG): D. B. P. 896.644 (1944/1953), BIOS Final Rep. **714** (1947).
840. SCHRADER, G. (Farbenfabriken Bayer AG): D. B. P. 900.814 (1942/1954).
841. SCHRADER, G.: BIOS Final Rep. **1095** (1946).
842. SCHRADER, G.: BIOS Final Rep. **949** (1946), p. 16.
843. SCHRADER, G.: BIOS Final Rep. **714** (1947).
844. SCHRADER, G.: BIOS Final Rep. **714** (1947), p. 33.
845. SCHRADER, G.: BIOS Final Rep. **714** (1947), p. 41.
846. SCHRADER, G. (Farbenfabriken Bayer AG): A. P. 2.597.534 (1950/1952), A. P. 2.640.847 (1949/1953).
847. SCHRADER, G. (Farbenfabriken Bayer AG): D. B. P. 921.870 (1952/1954), A. P. 2.701.259 (1953/1955), Angew. Chem. **66**, 265 (1954).
848. SCHRADER, G. (Farbenfabriken Bayer AG): D. B. P. 935.432 (1953/1955).
849. SCHRADER, G. (Farbenfabriken Bayer AG): D. B. P. 942.988 (1953/1956).
850. SCHRADER, G. (Farbenfabriken Bayer AG): D. B. P. 954.415 (1953/1956).
851. SCHRADER, G. (Farbenfabriken Bayer AG): D. B. P. 961.083 (1954/1957).
852. SCHRADER, G.: Angew. Chem. **66**, 265 (1954).
853. SCHRADER, G. (Farbenfabriken Bayer AG): D. A. S. 1.011.416 (1955/1957).
854. SCHRADER, G. (Farbenfabriken Bayer AG): D. A. S. 1.050.768 (1957/1959).
855. SCHRADER, G.: Angew. Chem. **70**, 690 (1958).
856. SCHRADER, G. (Farbenfabriken Bayer AG): D. A. S. 1.072.245 (1957/1959).
857. SCHRADER, G. (Farbenfabriken Bayer AG): D. B. P. 1.104.506 (1959/1961).
858. SCHRADER, G. (Farbenfabriken Bayer AG): D. A. S. 1.145.615 (1959/1963).
859. SCHRADER, G.: Hoefchen Briefe **1**, 1 (1960).
860. SCHRADER, G. (Farbenfabriken Bayer AG): Belg. P. 620.758 (1962/1963).
861. SCHRADER, G. (Farbenfabriken Bayer AG): D. A. S. 1.175.659 (1962/1964).
862. SCHRADER, G. (Farbenfabriken Bayer AG): D. A. S. 1.212.554 (1961/1966).
863. SCHRADER, G.: Die Entwicklung neuer insektizider Phosphorsäureester. 3rd edition. Weinheim: Verlag Chemie 1963.
863a. SCHRADER, G.: Die Entwicklung neuer Insektizide auf Grundlage organischer Fluor- und Phosphorverbindungen. 2nd edition, p. 50. Weinheim: Verlag Chemie 1952.
864. SCHRADER, G.: Z. Naturforsch. **18b**, 965 (1963).
865. SCHRADER, G. (Farbenfabriken Bayer AG): A. P. 3.201.444 (1963/1965).
866. SCHRADER, G.: World Rev. Pest Contr. **4**, 140 (1965).

867. SCHRADER, G. (Farbenfabriken Bayer AG): D. A. S. 1.240.850 (1966/1967).
868. SCHRADER, G.: Pflanzenschutz Ber. (Wien) **36**, 1 (1967) (Sonderheft).
869. SCHRADER, G.: Pflanzenschutz Ber. (Wien) **36**, 14 (1967) (Sonderheft).
870. SCHRADER, G., BAYER, O., KÜKENTHAL, H. (Farbenfabriken Bayer AG): D. R. P. 664.062 (1935/1938).
871. SCHRADER, G., CÖLLN, R. (Farbenfabriken Bayer AG): D. A. S. 1.101.417 (1959/1961).
872. SCHRADER, G., DÖRKEN, A. (Farbenfabriken Bayer AG): D. B. P. 936.037 (1953/1955).
873. SCHRADER, G., EUE, L., HACK, H. (Farbenfabriken Bayer AG): D. A. S. 1.245.952 (1965/1967).
874. SCHRADER, G., FROHBERGER, P. E. (Farbenfabriken Bayer AG): Belg. P. 648.813 (1964/1964), E. P. 1.018.314 (1964/1966).
875. SCHRADER, G., GEBHARDT, H. (Farbenfabriken Bayer AG): D. B. P. 767.830 (1939/1953).
876. SCHRADER, G., HAMMANN, I., STENDEL, W. (Farbenfabriken Bayer AG): Belg. P. 724.681 (1968/1969), F. P. 1.600.932 (1968/1970).
877. SCHRADER, G., KÜKENTHAL, H. (Farbenfabriken Bayer AG): D. B. P. 767.153 (1938/1952).
878. SCHRADER, G., KÜKENTHAL, H. (Farbenfabriken Bayer AG): D. B. P. 767.723 (1937/1952).
879. SCHRADER, G., KÜKENTHAL, H. (Farbenfabriken Bayer AG): D. B. P. 811.514 (1948/1951).
880. SCHRADER, G., KÜKENTHAL, H. (Farbenfabriken Bayer AG): D. B. P. 833.270 (1948/1952), A. P. 2.583.744 (1949/1952).
881. SCHRADER, G., KÜKENTHAL, H. (Farbenfabriken Bayer AG): D. B. P. 918.603 (1941/1954); BIOS Final Rep. **1095** (1947).
882. SCHRADER, G., LORENZ, W. (Farbenfabriken Bayer AG): D. B. P. 926.488 (1953/1955).
883. SCHRADER, G., LORENZ, W. (Farbenfabriken Bayer AG): D. B. P. 947.369 (1955/1956).
884. SCHRADER, G., LORENZ, W. (Farbenfabriken Bayer AG): D. B. P. 949.231 (1955/1956), A. P. 2.862.017 (1956/1958).
885. SCHRADER, G., LORENZ, W. (Farbenfabriken Bayer AG): D. A. S. 1.064.064 (1956/1959).
886. SCHRADER, G., LORENZ, W., CÖLLN, R., SCHLÖR, H. (Farbenfabriken Bayer AG): Belg. P. 576.811 (1959/1959).
887. SCHRADER, G., LORENZ, W., MÜHLMANN, R.: Angew. Chem. **70**, 690 (1958).
888. SCHRADER, G., LORENZ, W., UNTERSTENHÖFER, G., HAMMANN, I. (Farbenfabriken Bayer AG): D. A. S. 1.210.835 (1964/1966).
889. SCHRADER, G., MANNES, K., SCHEINPFLUG, H. (Farbenfabriken Bayer AG): Belg. P. 686.048 (1966/1967), E. P. 1.083.377 (1966/1967).
890. SCHRADER, G., MÜHLMANN, R. (Farbenfabriken Bayer AG): D. B. P. 848.812 (1950/1952).
891. SCHRADER, G., SCHEINPFLUG, H. (Farbenfabriken Bayer AG): Belg. P. 710.882 (1968/1968).
892. SCHRADER, G., SCHLIEBS, R. (Farbenfabriken Bayer AG): D. A. S. 1.233.390 (1966/1967).
893. SCOTT, C. B., YALE, J. W., HASHIMOTO, S.: J. Agr. Food Chem. **8**, 303 (1960).
894. SEHRING, R., ZEILE, K. (C. H. Boehringer, Sohn): D. A. S. 1.076.662 (1958/1960).
895. SEHRING, R., ZEILE, K. (C. H. Boehringer, Sohn): D. A. S. 1.078.558 (1958/1960).
896. SEHRING, R., ZEILE, K. (C. H. Boehringer, Sohn): D. A. S. 1.174.104 (1961/1964), Belg. P. 625.198 (1961/1962).
897. SELIM, M., THANK, T. N.: Compt. Rend. **250**, 2377 (1960).
898. SEUME, F. W., CASIDA, J. E., O'BRIEN, R. D.: J. Agr. Food Chem. **8**, 43 (1960).
899. SEUME, F. W., O'BRIEN, R. D.: J. Agr. Food Chem. **8**, 36 (1960).
900. SHALITIN, Y., BERNHARD, S. A.: J. Am. Chem. Soc. **88**, 4711 (1966).
901. SHISHIDO, T., FUKAMI, J.: Botyu-Kagaku **28**, 69 (1963).
902. SHUSTER, L.: Ann. Rev. Biochem. **33**, 571 (1964).
903. SIEBERT, H.: Anwendung der Schwingungs-Spektroskopie in der anorganischen Chemie p. 35. Berlin-Heidelberg-New York: Springer 1966.
904. SIRRENBERG, W., LORENZ, W. (Farbenfabriken Bayer AG): D. A. S. 1.170.401 (1961/1964).
905. SMISSAERT, H. R.: Nature **143** (3602), 129 (1964).
906. SMITH, C. L.: J. Econ. Entomol. **53**, 1110 (1960).
907. SMITH, C. N.: Advan. Chem. Ser. **41**, 36 (1963).
908. SMITH, C. N., LABRECQUE, G. C., BOŘKOVEC, A. B.: Ann. Rev. Entomol. **9**, 269 (1964).
909. SMITH, D. S., TREHERNE, J. E.: Advan. Insect Physiol. Vol. I. London-New York: Academic Press 1964.
910. SMITH, G. U., WATSON, B. S., FISCHER, F. S.: J. Agr. Food Chem. **15**, 127 (1967); J. Econ. Entomol. **59**, 1464 (1966).

911. SMITH, H. V., SPALDING, J. M. K.: Lancet **277**, 1019 (1959).
912. SMITH, J. N.: Ann. Rev. Entomol. **7**, 465 (1962).
913. SMITH, M. I., ELVOVE, E., FRAZIER, W. H.: U.S. Public Health Rept. XIV, 2509.
914. SPEZIALE, A. J., CHUPP, J. P. (Monsanto Chem. Co.): A. P. 3.102.019 (1961/1963).
915. SPEZIALE, A. J., CHUPP, J. P. (Monsanto Chem. Co.): A. P. 3.102.020 (1960/1963).
916. SPEZIALE, A. J., CHUPP, J. P. (Monsanto Chem. Co.): A. P. 3.102.021 (1960/1963).
917. SPEZIALE, A. J., CHUPP, J. P. (Monsanto Chem. Co.): A. P. 3.102.023 (1960/1963).
918. SPEZIALE, A. J., CHUPP, J. P. (Monsanto Chem. Co.): Can. P. 710.340 (1960/1965).
919. STAAB, H. A.: Einführung in die theoretische Organische Chemie. Weinheim: Verlag Chemie 1959.
920. STEEGER, O., KÖTZ, G., SEEGER, P. (Farbenfabrik Wolfen): D. P. (DDR) 37.845 (1963/1965), D. A. S. 1.242.602 (1964/1967).
921. STEINBERG, G. M.: J. Org. Chem. **15**, 637 (1950).
922. STEINHAUSEN, W. R.: Abstr. 6th Int. Pflanzenschutz-Kongr. (Wien), p. 303. Wien: Verlag d. Wiener Medizin. Akad. 1967; Z. Angew. Zool. **55**, 107 (1968).
923. STIASNI, M., REHBINDER, D., DECKERS, W.: J. Agr. Food Chem. **15**, 474 (1967).
924. STIEMERLING, K.: Pflanzenschutz-Nachr. Bayer **19**, 53 (1966).
925. STIEMERLING, K.: Pflanzenschutz-Nachr. Bayer **19**, 73 (1966).
926. STILES, A. R. (Shell Devel. Corp.): A. P. 2.685.552 (1952/1954).
927. STOKES, H. N.: Ber. **28**, 437 (1895).
928. STONE, B. F.: J. Econ. Entomol. **62**, 977 (1969).
929. STRECKER, W., GROSSMANN, CH.: Ber. **49**, 63 (1916).
930. STREET, L., WAY, O. (California Spray Chem. Co.): Belg. P. 585.182 (1957/1959).
931. STURTEVANT, J. M.: J. Chem. Educ. **44**, 184 (1967).
932. SUN, Y. P.: J. Econ. Entomol. **61**, 949 (1968).
933. SUN, Y. P., JOHNSON, E. R.: J. Agr. Food Chem. **8**, 261 (1960).
934. SUN, Y. P., JOHNSON, E. R.: J. Econ. Entomol. **53**, 887 (1960).
935. SUN, Y. P., JOHNSON, E. R., WARD, L. F., JR.: J. Econ. Entomol. **60**, 828 (1967).
936. SUZUKI, SH., FUJII, K., NISHIZAWA, Y., KADOTA, T. (Sumitomo Chem. Corp.): A. P. 3.091.565 (1960/1963), A. P. 3.135.780 (1960/1964).
937. SYLVESTER, E. S., in: Biol. transmission of disease agents, p. 11–31. Symp. Ent. Soc. Am., Atlantic City 1960. London-New York: Academic Press 1962.
938. SYNNATSCHKE, G.: Die BASF **16** (4), 182 (1966).
939. SZABO, K., BRADY, J. G., WILLIAMSON, T. B. (Stauffer Chem. Comp.): A. P. 2.988.474 (1960/1961).
940. TABOR, E. G. (Dow Chemical Co.): A. P. 2.928.864 (1958/1960).
941. TAMMELIN, L. E.: Acta Chem. Scand. **11**, 859 (1957).
942. TAMMELIN, L. E.: Acta Chem. Scand. **11**, 859, 1738 (1957).
943. TAMMELIN, L. E.: Acta Chem. Scand. **11**, 1340 (1957).
944. TAMMELIN, L. E.: Acta Chem. Scand. **11**, 1738 (1957).
945. TAMMELIN, L. E.: Svensk Kem. Tidskr. **70**, 157 (1958).
946. Techn. Release, Nat. Pest. Contr. Assoc. No. 11, 67 (1967).
947. TEICHMANN, H., HILGETAG, G.: Angew. Chem. **79**, 1077 (1967).
948. TÉLIZ, D., GROGAN, R. G., LOWNSBERRY, B. F.: Phytopathology **56**, 658 (1966).
949. TERRIERE, L., BOOSE, R. B., ROUBAL, W. T.: Biochem. J. **79**, 620 (1961).
950. THÉNARD, P.: Compt. Rend. **25**, 892 (1847).
951. THOMAS, J. G., BRAZZEL, J. R.: J. Econ. Entomol. **54**, 417 (1961).
952. THURSTON, J. T.: U.S. Dept. Commerce OTS, PB 60890 (1964).
953. TIEMAN, CH. H., STILES, A. R. (Shell Oil Co.): A. P. 3.068.268 (1961/1962).
954. TOLKMITH, H.: J. Org. Chem. **23**, 1682 (1958).
955. TOLKMITH, H.: J. Org. Chem. **23**, 1685 (1958).
956. TOLKMITH, H.: Nature **211** (5048), 522 (1966).
957. TOLKMITH, H., MUSSELL, D. R.: World Rev. Pest Contr. **6**, 74 (1967).
958. TOLKMITH, H., ORVILLE, H. (Dow Chemical Co.): Belg. P. 661.891 (1964/1964), E. P. 1.034.493 (1964/1966).
959. TOLKMITH, H., SENKBEIL, H. O., MUSSELL, D. R.: Science **155**, 85 (1967).
960. TOLKMITH, H., SENKBEIL, H. O., MUSSELL, D. R.: Science **155**, 3758 (1967).
961. TOPLEY, B.: Chem. Ind. (London) **1950**, 859.

962. Toppozada, A., Abdallah, S., Eldefrawi, M. E.: J. Econ. Entomol. **59**, 1125 (1966).
963. Toy, A. D. F. (Victor Chem. Works): A. P. 2.504.165 (1947/1950), J. Am. Chem. Soc. **70**, 3882 (1948).
964. Toy, A. D. F.: J. Am. Chem. Soc. **73**, 4670 (1951).
965. Toy, A. D. F., McDonald, G. A. (Victor Chem. Works): A. P. 2.715.136 (1949/1955).
966. Trademan, L., Yagelowich, F. R. (Chemagro Corp.): A. P. 3.089.807 (1961/1963).
967. Treherne, J. E., Beament, J. W. L., (eds.): The physiology of the insect central nervous system. New York: Academic Press 1965.
968. Triggle, D. J.: Chemical aspects of the autonomic nervous system. In: Theoretical and experimental biology, Vol. 4. London-New York: Academic Press 1965.
969. Triolo, A. J., Coon, J. M.: J. Agr. Food Chem. **14**, 549 (1966).
970. Uchida, T., Dautermann, W. C., O'Brien, R. D.: J. Agr. Food Chem. **12**, 49 (1964).
971. Uchida, T., Rahmati, H. S., O'Brien, R. D.: J. Econ. Entomol. **58**, 831 (1965).
972. Unterstenhöfer, G.: Hoefchen Briefe **1** (1), 20 (1948).
973. Unterstenhöfer, G.: Hoefchen Briefe **1** (2), 20 (1948).
974. Unterstenhöfer, G.: Z. Pflanzenkrankh. Pflanzenschutz **57**, 272 (1950).
975. Unterstenhöfer, G.: Z. Pflanzenkrankh. Pflanzenschutz **58**, 268 (1951).
976. Unterstenhöfer, G.: Hoefchen Briefe **6**, 121 (1953).
977. Unterstenhöfer, G.: Anz. Schaedlingskunde **27**, 134 (1954).
978. Unterstenhöfer, G.: Anz. Schaedlingskunde **30**, 7 (1957).
979. Unterstenhöfer, G.: Z. Pflanzenkrankh. Pflanzenschutz **64**, 619 (1957).
980. Unterstenhöfer, G.: Mededel. Landgeb. Gent **23**, 770 (1958).
981. Unterstenhöfer, G.: Verh. IVth Int. Pflanzenschutz-Kongr. Hamburg, Vol. II, p. 1243. Braunschweig: Selbstverlag d. IV. Kongresses 1960.
982. Unterstenhöfer, G.: Hoefchen Briefe **13**, 141 (1960).
983. Unterstenhöfer, G.: Hoefchen Briefe **13**, 44 (1960).
984. Unterstenhöfer, G.: Proc. Ist Irish Crop Prot. Conf. Dublin, p. 126 (1962).
985. Unterstenhöfer, G.: Pflanzenschutz-Nachr. Bayer **16**, (3) 81 (1963).
986. Unterstenhöfer, G.: Mededel. Landgeb. Gent **26**, 1033 (1961).
987. Unterstenhöfer, G., Frehse, H.: Pflanzenschutz-Nachr. Bayer **16**, (4) 171, 172 (1963).
988. Unterstenhöfer, G., Frehse, H.: Pflanzenschutz-Nachr. Bayer **16**, (4) 174 (1963).
989. Usherwood, P. N. R., Machili, P.: J. Exptl. Biol. **49**, 341 (1968).
990. Vardanis, A., Crawford, L. G.: J. Econ. Entomol. **57**, 136 (1964).
991. Vives, J. P., Mathis, F.: Compt. Rend. **246**, 1879 (1958).
992. Voss, G.: Naturwissenschaften **46**, 652 (1959).
993. Voss, G., Dautermann, W. C., Matsumura, F.: J. Econ. Entomol. **57**, 808 (1964).
994. Wadley, F. M.: U.S. Dep. Agr. Bur. Ent. and Plant Quar. ET 223 (June 1945)
995. Wadley, F. M.: U.S. Dep. Agr. Bur. Ent. and Plant Quar. ET 275 (August 1949).
996. Wagner, F. A., Baer, R. W., Berkelhammer, G.: J. Med. Chem. **8**, 377 (1965).
997. Walsh, E. N. (Stauffer Chem. Co.): D. A. S. 1.131.652 (1960/1962).
998. Walsh, E. N., Toy, A. D. F.: Biochem. Prep. **7**, 71 (1960).
999. Ward, L. F., Phillips, D. D. (Shell Oil Comp.): A. P. 3.174.990 (1963/1965), A. P. 3.242.043 (1964/1966), E. P. 990.327 (1964/1965).
1000. Warren, S. G.: Angew. Chem. **80**, 649 (1968).
1001. Watanabe, H., Kobara, R.: Jap. J. Appl. Entomol. Zool. **11**, 9 (1967).
1002. van Wazer, J. R.: Phosphorus and its compounds, Vol. I, p. 88. New York-London: Interscience Publishers, Inc. 1958.
1003. van Wazer, J. R.: Phosphorus and its compounds, Vol. I, p. 348. New York-London: Interscience Publishers, Inc. 1958.
1004. van Wazer, J. R.: Phosphorus and its compounds, Vol. II, p. 989, 1171. New York-London: Interscience Publishers, Inc. 1961.
1005. Weddige, A., Finger, H.: J. Prakt. Chem. **35**, 262 (1887).
1006. Wedemeyer, K., Delfs, D. (Farbenfabriken Bayer AG): D. A. S. 1.088.980 (1957/1960).
1007. Wegler, R., Grewe, F., Mehlhose, K. (Farbenfabriken Bayer AG): D. B. P. 943.050 (1953/1956).
1008. Weischer, B.: Mitt. Biol. Bundesanstalt Land- Forstwirtsch. Berlin-Dahlem 1964 (111), p. 98.
1009. Wenzl, H., Pflanzenarzt **17**, 20 (1964).

1010. WESTHEIMER, F. H.: Chem. Soc. (London), Spec. Publ. **8**, 180 (1957).
1011. WESTHEIMER, F. H.: Acc. Chem. Res. **1**, 70 (1968).
1012. WHEATLEY, P. J.: J. Chem. Soc. **1962**, 300
1013. WHETSTONE, R. R., HARMAN, D. (Shell Devel. Co.): A. P. 2.765.331 (1952/1956).
1014. WHETSTONE, R. R., HARMAN, D. (Shell Devel. Co.): A. P. 2.956.073 (1960/1960), A. P. 3.116.201 (1952/1963).
1015. WHETSTONE, R. R., HARMAN, D. (Shell Devel. Co.): A. P. 3.116.201 (1952/1963).
1016. WHETSTONE, R. R., PHILLIPS, D. D., SUN, Y. P., WARD, L. F., SCHWELLENBERGER, T. (Shell Oil Comp.): Abstr. Pap. 149th Meeting 1965; J. Agr. Food Chem. **14**, 352 (1966).
1017. World Health Organ. – Exp. Comm. Insect., 7th Rep. World Health Organ. – Tech. Rept. Ser. Genf 1957, p. 125.
1018. WICHELHAUS, H.: Liebigs Ann. Chem. (Suppl.) **6**, 257, 263 (1868).
1019. WILDBRETT, G., KIERMEIER, F., LETTENMAYER, L.: Z. Naturforsch. **22b**, 307 (1967).
1020. WILKINSON, C. F.: World Rev. Pest Contr. **7**, 155 (1968).
1021. WILLARD, J. R., ALLEN, J. F., HOLDEN, K. R. (Food Mach. Chem. Corp.): A. P. 2.983.644 (1955/1961).
1022. WILLARD, J. R., HENAHAN, J. F. (Food Mach. Chem. Corp.): A. P. 2.873.228 (1956/1959).
1023. WILLIAMS, O. F., ALLEN, J. F., RACINE, F. R. (Food Mach. Chem. Corp.): D. A. S. 1.022.412 (1955/1958), A. P. 2.798.880 (1953/1957).
1024. WILLIS, N. E. (Monsanto Chem. Co.): A. P. 2.523.243 (1948/1950).
1025. WILMS, H., BAYER, O., WEGLER, R. (Farbenfabriken Bayer AG): D. B. P. 949.948 (1952/1956).
1026. WILSON, I. B.: J. Biol. Chem. **190**, 111 (1951).
1027. WILSON, I. B., GINSBURG, S.: Biochem. Biophys. Acta **18**, 168 (1955).
1028. WINTERINGHAM, F. P. W., DISNEY, R. W.: Biochem. J. **91**, 506 (1964).
1029. WINTERINGHAM, F. P. W., HEWLETT, P. S.: Chem. Ind. (London) **1964**, 1512.
1030. WIRTH, W.: Arch. Exptl. Pathol. Pharmakol. **207**, 545 (1949); **217**, 144 (1953).
1031. WITTER, R. F., GAINES, T. B.: Biochem. Pharmacol. **12**, 1377 (1963).
1032. WITTER, R. F., GAINES, T. B.: Biochem. Pharmacol. **12**, 1421 (1963).
1033. WOODSTOCK, W. H. (Victor Chemical Works): A. P. 2.402.703 (1942/1946).
1034. WORK, T. H.: Bull. World Health Organ. **29**, 59 (1963).
1035. WU, CH. (Union Carbide Corp.): A. P. 3.341.632 (1964/1967).
1036. WURTZ, A.: Liebigs Ann. Chem. **58**, 49 (1846).
1037. WYBOU, A., HAMMANN, I.: Mededel. Rijksf. Landbouw. Gent. **33** (3), 817 (1968).
1038. YAMASAKI, T.: Sci. Rept. Tohoku Univ. **11**, 73 (1959).
1039. YOUNG, R. W., BERKELHAMMER, G. (American Cyanamid Co.): A. P. 2.959.610 (1959/1960).
1040. YOUNG, S. Y., BERGER, R. S.: J. Econ. Entomol. **62**, 727 (1969).
1041. YOUNG, S. Y., BERGER, R. S.: J. Econ. Entomol. **62**, 929 (1969).
1042. YOUNGSON, CH. R. (Dow Chemical Co.): A. P. 3.005.749 (1959/1961), D. A. S. 1.158.745 (1960/1963).
1043. ZACHARIAS, E.: J. Prakt. Chem. **43**, 446 (1891).

5.4. Author Index

The page numbers in *italics* refer to the section 5.4. References

5.5. Subject Index

Common names and trade names see pp. 275–293